T0331279

Entrepreneurship Development in Food Processing

About the Editors

Prof. (Dr.) K.P. Sudheer, obtained his B Tech Agricultural Engineering degree from Kerala Agricultural University (KAU), M Tech Agricultural Processing from Tamil Nadu Agricultural University(TNAU), PhD Agricultural Engineering from IARI New Delhi and Post doctorate from KU Leuven Belgium. Prof. Sudheer started his career as scientist in Agro Processing Division, CIAE (ICAR), Bhopal. He is presently working as ICAR National Fellow and Project Coordinator of Centre of Excellence in Post-harvest Technology, College of Horticulture, Vellanikkara, KAU.

Professor Sudheer is actively involved in teaching various subjects for B Tech Food Engineering, B Tech Agricultural Engineering, M. Tech Agricultural Processing and Food Engineering, Ph D Agricultural Processing and Food Engineering, and in research programmes under AICRP on Post-harvest Engineering & Technology. He is also guiding Doctoral and Masters research scholars and has many external aided research projects funded by Ministry of Food Processing Industries, Kerala State Council for Science, Technology & Environment, Ministry of Rural Development, NABARD, Food Corporation of India, Government of Kerala. Prof. Sudheer is also acting as the Project Coordinator of Food and Agricultural Process Engineering research group in KAU. He has 150 research papers in reputed national & international journals, and proceedings of seminar/workshops. He authored five text books and many bulletins in the field of Post-harvest Technology. Prof. Sudheer is the recipient of prestigious Normal E Borlaug Fellowship by USDA. He has also received many international and national fellowships for his research programmes including NUFFIC Fellowship from Netherlands, VLIR-UDC fellowship from Belgium, CINADCO fellowship from Israel, ERASMUS MUNDUS Fellowship from Sweden. Prof. Sudheer has received the Krishi Vigyan Award - 2015 for the best Agricultural Scientist, from the Government of Kerala.

Prof. (Dr.) V. Indira, took her Master's degree in Foods and Nutrition from Sri. Avinashilingam Home Science College for Women, Coimbatore in 1978. She joined KAU as Junior Assistant Professor of Nutrition in the Department of Processing Technology, College of Horticulture in 1979. In 1993, she took her Ph.D. in Foods and Nutrition from KAU. She was former Professor & Head, Department of Home Science, College of Horticulture, Vellanikkara.

Dr. Indira, had extensive teaching, research and extension experience in KAU. She handled UG courses in food and nutrition for Agriculture, Dairy Science and Technology, and Veterinary and Animal Science students, and M. Sc. and Ph.D. courses for the students of Department of Home Science, College of Horticulture, KAU, Vellanikkara. She implemented two externally aided projects and guided four Ph.D and 24 M. Sc. students in the Department of Home Science as major advisor. She organized various training programmes in fruit and vegetable preservation, mushroom cultivation, food and nutrition. She has in her credit about 50 research publications and many bulletins and books.

Entrepreneurship Development in Food Processing

K.P. Sudheer *(ICAR National Fellow)*
Professor & Head, Department of Agricultural Engineering
Project Coordinator, Centre of Excellence in Post-harvest Technology
College of Horticulture, Vellanikkara
Kerala Agricultural University
Thrissur, Kerala- 680 656

V. Indira
Former Professor & Head
Department of Home Science
College of Horticulture, Vellanikkara
Kerala Agricultural University
Thrissur, Kerala- 680 656

CRC Press
Taylor & Francis Group
Boca Raton London New York

CRC Press is an imprint of the
Taylor & Francis Group, an **informa** business

NEW INDIA PUBLISHING AGENCY
New Delhi – 110 034

First published 2022
by CRC Press
2 Park Square, Milton Park, Abingdon, Oxon, OX14 4RN

and by CRC Press
6000 Broken Sound Parkway NW, Suite 300, Boca Raton, FL 33487-2742

British Library Cataloguing-in-Publication Data
A catalogue record for this book is available from the British Library

Library of Congress Cataloging-in-Publication Data
A catalog record has been requested

ISBN: 978-1-032-15866-2 (hbk)
ISBN: 978-1-003-24602-2 (ebk)

DOI: 10.1201/9781003246022

Dedicated to
our loving family

Dr. K. Alagusundaram
DDG (Agril. Engineering)

Foreword

India is the second largest producer of foods in the world and also one of the greatest looser of foods in the post-production system. Nearly 6 to 18% foods are lost due to inadequate, poor and improper post-harvest handling and processing facilities. The value of these losses in 2014 price levels translate to a whopping Rs.920 billion per annum.

The Indian Council of Agricultural Research through its All India Coordinated Research Project on Post-Harvest Engineering and Technology have developed numerous technologies, gadgets, and machinery for processing, preservation and value addition of agricultural and horticultural produce. ICAR technologies and the technologies developed by State Agricultural Universities and other research organizations, if properly taken to the people, will help in reducing the post-harvest losses phenomenally, will support in increasing the farmers' income levels, and will make more food available for our consumption and export. We need also to create awareness among the public on venturing in food processing businesses, the supports extended by the Government for Start-ups and the benefits of becoming an entrepreneur.

Text books and publications on subjects of food processing and value addition and entrepreneurships are important for creating such awareness. I am glad to see Dr. K.P. Sudheer, Professor and Project Coordinator, Centre of Excellence in Post-harvest Technology, Kerala Agricultural University, and Dr.V. Indira, former Professor & Head, Department of Home Science, College of Horticulture, Vellanikkara, are bringing out a text book '**Entrepreneurship Development in Food Processing**'. I have had the opportunity to glance through the book. The topics chosen are timely and relevant and the presentation is lucid and extremely nice. I am sure this book will be a good reading material for students, teachers and researchers in food processing area and the new entrepreneurs on the subject.

I congratulate the authors of the chapters for their excellent contribution and the Editors for excellently presenting them together.

New Delhi

Dr. K. Alagusundaram
DDG (Agril. Engineering)

Preface

Food processing is now regarded as the sunrise sector of the Indian economy in view of its large potential for growth and likely socio-economic impact specifically on employment and income generation. In total per cent of country GDP, fourteen per cent of our GDP is contributed by agro based sector. During and after the green revolution, India has witnessed a tremendous growth in its agricultural production. On one hand, India moved from the status of insufficiency to surplus in its food availability due to adoption of latest technologies in production systems and on the other huge quantities of our produce is lost due to poor post-harvest handling and management. It is imperative to process and preserve the available food in seasons of plenty not only to avoid wastage but also to confirm that it is available during off seasons and at reasonable price. Poverty and unemployment in rural areas continue to be the two most important challenges. Secondary agriculture and product diversification has been considered as a panacea for the present state of affairs in agriculture sector. New business ventures for processing sector must be established to avert this situation. The promotion of value addition technologies and agripreneurship offers the key for ensuring better income for the farming community and for the sustainability of farming to produce more food. This ensures nutritional security as well.

Food processing is considered to be the most important sector which adds value to the products. However, it requires professional skills and resources in order to bring benefits to the smaller scale stakeholders in developing countries. Keeping this in mind, government has initiated Technology Business Incubation units to promote entrepreneurship and agro- industry which will open the vistas of incubation landscape to the micro segment of the vast rural economy. Agripreneurship based on latest developments in food processing sector has the potential to contribute to a range of social and economic development such as employment generation, income generation, poverty reduction and improvements in nutrition, health and overall food security in the national economy. The growth of this industry will bring immense benefits to the economy, raising agricultural yields, enhancing productivity, reducing post-harvest losses, creating employment and raising life-standards of a large number of people across the country, especially those in rural areas or production catchments. It is therefore imperative that post-harvest interventions of food crops are essential, without which the food processing industry cannot thrive. It is against this background that the present book is conceived.

This book mainly comprises of new process protocols and equipment details for installation of concerned food industry. The book provides the scope and opportunities of entrepreneurship in the major food crops like rice, pulses, millets, soybean, groundnut, milk, meat, fish, and jaggery. The book also enlightens the readers about the marketing strategies, business plan preparation, etc which are essential for starting a new business venture. Its special feature is the treatment given to the food processing and entrepreneurship in an integrated manner. The authorship of various chapters comes from renowned professors, scientists and research scholars who have compiled the scattered information from different sources at one place. Presenting vital information in an accessible and objective way has been our prime intention.

We acknowledge from the core of our heart to all the contributors of this book who have put-in a lot of scholarly efforts in bringing out a state-of-art technology. Their sincere effort made towards collection of information from literature consulted during the preparation of the manuscript is greatly appreciated. With deep sense of pride and dignity, editors express heartfelt sense of gratitude and regards to Dr. K. Alagusundaram, Deputy Director General (Engg), ICAR, New Delhi, Prof. (Dr.) P. Rajendran, Vice-Chancellor, Kerala Agricultural University, Prof. P Indira Devi, Director of Research, KAU, Prof. Jiju P Alex, Director of Extension, KAU, Prof. George Thomas C, Associate Dean, College of Horticulture, and Prof. K.P. Visalakshi, Former Head, Department of Agricultural Engineering, College of Horticulture, with whose guidance, scientific knowledge, constructive criticism and constant encouragement, we have been able to publish this manuscript.

It is expected that the book would assist to streamline the entrepreneurial perspectives of food processing in the most befitting manner to the concerned reader.

Prof. (Dr.). K. P. Sudheer
Prof. (Dr.). V. Indira

Contents

4. Trends in Food Processing– An Opportunity to Entrepreneur ... 47

Sudheer K P, Saranya S, Ranasalva N & Seema B R

5. Rice Milling Sector– A Prospective Avenue for Entrepreneurs . 69

Sudheer K P and Ravindra Naik

6. Entrepreneurial Opportunities in Rice Processing 97

Durgadevi M, Sinija V R, Hema V and Anandharamakrishnan C

**7. An Insight to Entrepreneurial Opportunities in Millet
   Processing 119**

*Udaykumar Nidoni, Mouneshwari K, Ambrish G,
Mathad P F, Shruthi V H and Anupamac C*

List of Contributors

1. **Ajesh Kumar V,** Scientist, Centre of Excellence in Soyabean Processing and utilization Central Institute of Agricultural Engineering, Bhopal, Madhya Pradesh-462 038

2. **Ambarish Ganachari,** Assistant Professor, Department of Processing and Food Engineering, College of Agricultural Engineering, UAS, Raichur-584 104

3. **Anandharamakrishnan C.,** Director, Indian Institute of Food Processing Technology Pudukkottai Road, Thanjavur, Tamil Nadu - 613 005

4. **Anupama C,** Scientist, Krishi Vigyan Kendra, Gangavathi-583 227

5. **Aparna Sudhakaran V,** Assistant Professor, Department of Dairy Microbiology College of Diary Science & Technology, Kerala Veterinary and Animal Sciences University Mannuthy, Thrissur, Kerala-680 651

6. **Aswin S Warrier,** Assistant Professor, Department of Dairy Engineering, KVASU Dairy Plant Kerala Veterinary and Animal Sciences University, Mannuthy, Thrissur Kerala -680 651

7. **Borkar, P A,** Head & Research Engineer, AICRP on PHET, Department of Agricultural Process Engineering, Dr. Panjabrao Deshmukh Krishi Vidyapeeth, Akola

8. **Durgadevi, M,** Assistant Professor, Food Science & Nutrition, Indian Institute of Food Processing Technology, Pudukkottai Road, Thanjavur, Tamil Nadu - 613 005

9. **George Ninan,** Principal Scientist, Fish Processing, Central Institute of Fisheries Technology, Cochin, Kerala- 682 029

10. **Hema, V,** Assistant Professor, Food Science & Nutrition, Indian Institute of Food Processing Technology, Pudukkottai Road, Thanjavur, Tamil Nadu - 613 005

11. **Indira V,** Former Professor & Head, Department of Home Science, College of Horticulture, Vellanikkara, Kerala Agricultural University, Thrissur, Kerala- 680 656

12. **Jagannadha Rao, P V K,** Research Engineer, AICRP on PHET, Regional Agricultural Research Station, Anakapalle, Acharya N.G. Ranga Agricultural University, Andhra Pradesh.

13. **Manoj P. Samuel,** Head, Engineering Division, Central Institute of Fisheries Technology Cochin, Kerala- 682 029

14. **Mathad P F,** Assistant Professor, Department of Processing and Food Engineering College of Agricultural Engineering, UAS, Raichur-584 104

15. **Mouneshwari Kammar,** Scientist, Krishi Vigyan Kendra, University of Agricultural Sciences, Raichur-584 104

16. **Ranasalva N,** Ph D Scholar, Kelappaji College of Agricultural Engineering & Technology Kerala Agricultural University, Tavanur, Kerala- 679 573

17. **Ravindra Naik,** Principal Scientist, Central Institute of Agricultural Engineering, Regional Centre, Coimbatore, Tamilnadu-641 007

18. **Ravishankar, C N,** Director, Central Institute of Fisheries Technology, Cochin Kerala- 682 029

19. **Renuka Nayar,** Assistant Professor, Department of Livestock Products Technology College of Veterinary & Animal Sciences, Pookode, Wayanad, Kerala- 673576

20. **Sangeetha K Prathap,** Assistant Professor, School of Management Studies, Cochin University of Science & Technology, Cochin, Kerala -682 022

21. **Saranya S,** Research Associate, Centre of Excellence in Post Harvest Technology, College of Horticulture, Vellanikkara, Kerala Agricultural University, Thrissur- 680 656

22. **Seema B R,** Teaching Assistant, Regional Agricultural Research Station, Pilicode, Kerala Agricultural University, Kerala- 671 310

23. **Shruthi V.H,** Ph D Scholar, Department of Processing and Food Engineering, College of Agricultural Engineering, UAS, Raichur-584 104

24. **Sinija V R,** Associate Professor & Head, Post Harvest Engineering, Indian Institute of Food Processing Technology, Pudukkottai Road, Thanjavur, Tamil Nadu - 613 005

25. **Sudheer, K P,** Professor & Head, ICAR National Fellow, Project Coordinator, Centre of Excellence in Post-harvest Technology, Department of Agricultural Engineering, College of Horticulture, Vellanikkara, Kerala Agricultural University, Thrissur, Kerala- 680 656

26. **Sumeda S Deshpande,** Principal Scientist, Centre of Excellence in Soyabean Processing and Utilization, Central Institute of Agricultural Engineering, Bhopal, Madhya Pradesh- 462 038

27. **Udaykumar Nidoni,** Professor & Head, Department of Processing and Food Engineering University of Agricultural Science, Raichur, Karnataka- 584 104

28. **Vasudevan V N,** Assistant Professor, Department of Livestock Products Technology College of Veterinary & Animal Sciences, Mannuthy, Thrissur, Kerala- 680 561

29. **Vennila P,** Professor and Head, Department of Home Science Extension, Home Science College and Research Institute, Tamil Nadu Agricultural University, Madurai -625 104 Tamil Nadu

30. **Vishnu Vardhan S,** Assistant Professor, Post Harvest Technology, ANGRAU, Bapatla Guntur (A.P.)

31. **Visvananthan R,** Professor (Agricultural Processing), Anbil Dharmalingam Agricultural College and Research Institute, Thiruchirappalli - 620 009 Tamil Nadu

1

Introduction to Entrepreneurship Development in Food Processing

Sudheer K P and Sangeetha K Prathap

Introduction

World's population is continuously on the rise, triggering the demand for food. The Food and Agriculture Organization of the United Nations (FAO, 2011) observed that 70% increase in food production will be needed to sustain the globe's population by 2050. Though, increase in agricultural production and productivity is said to be the panacea to solve food security issues, the greatest challenge that need to be addressed is the inefficiencies in the supply chain; instigated by insufficient storage capacities and perishability of food products. Food preservation and processing is highlighted in this context, as it serves to conserve/improve the quality and reduce food loss due to wastage as well as add nutritional security.

Post green-revolution India has witnessed a tremendous growth in its agricultural production. With more than five-fold increase in food grain production from 50 million tons in 1950-51 to about 273 million tons in 2016-17, India has considerably improved its position from 'food aid dependent' to 'net food exporter'(United Nations in India, 2017). However, almost one third of India's population falls below the poverty line and half of the children are malnourished poses the threat of nutritional insecurity. Food processing technologies and promotion of agripreneurship can resolve inefficiencies in storage; promote value chain in agriculture, where ultimate consumers are benefitted of getting better quality products. This can also serve as a key to ensure food and nutritional security and better income for farming community.

A farmer is exposed to callous competition in an agri-produce market, wherein unpredictable demand and supply conditions can destabilize his earnings. In 2016, the government launched a number of programmes to double farmers' income by 2022 envisaging removal of bottlenecks for greater agricultural productivity, especially in rain-fed areas. Processing technologies for location specific agri products may be promoted in a region, adding to the value in terms of quality, shelf life and nutrition, thus enabling the farmer to gain better margins by adopting diversification in his entrepreneurial options along with his farm business.

Entrepreneurship development in agriculture

When we speak of agriculture (including farming, manufacturing, processing, distribution *etc.*) in business terms; as to what revenues it might earn, at what costs, what effects can environment factors bring about in business, it can be termed as "agribusiness."

An "agricultural entrepreneur" is an individual or group with the right to use or exploit the land or other related elements required to carry out agricultural, forestry or mixed activities (Suarez, 1972). Wortman, 1990 defines a related term, much more encompassing term "rural enterprise" as "…..creation of new organizations that introduce new products, create new markets or use new technologies from rural areas."

In clear terms, agri-business comprises organizations and enterprises that contribute to value chain in agriculture such as production, processing, handling, marketing, packaging, transportation, and wholesale and retail trade. A typical agri value chain includes the entire spectrum from farm gate to plate. In a sense, agri-business includes all operations involved in the manufacture and distribution of farm supplies.

Amongst the looming concerns to address food and nutritional security, increase in the supplies of farm outputs, reversing trend of reducing farm activities and shifting of population from agriculture has been observed. As per the census of 2011, 263 million people are engaged in the agriculture sector and over half of them are agricultural labourers indicating that farmers are forced to quit agricultural activities due to declining profits in the sector. Diversifying agriculture with emphasis on value addition is a typical solution to solve economic interests of the farmers. In this context, farmers and rural people may be encouraged to take up value addition of locally grown agri-based products adopting commercially viable technologies available with research and development agencies. Establishment of agribusiness enterprises not only generates income for the rural entrepreneur, but also assumes the role of job provider in rural transects.

Agribusiness has the potential to contribute to creation of employment opportunities, income generation, poverty reduction, nutritional security, health and overall food security. The growth of this industry can also contribute to reduce post-harvest losses and raise life-standards of large number of people across the country, especially those in rural areas or production catchments.

Potential areas of entrepreneurship in agriculture sector

On-farm activities: the possible areas of entrepreneurship in agriculture are given below:

i.   *Agro-processing centre's (APC's)*: The APC's focus on primary processing, rather than secondary and tertiary processing. They merely process the agriculture produce. e.g., spice powdering units, mini mills, dal mills, decorticating mills etc.

ii. *Agro produce manufacturing centre*: These centres produce entirely novel products based on the agricultural produce as the main raw material. e.g., sugar factories, bakery, confectionaries, etc.

iii. *Agro-inputs manufacturing units*: These units produce goods either for mechanization of agriculture or for increasing productivity. e.g., fertiliser manufacturing plants, insecticide production units, food processing units, agricultural implements etc.

iv. *Agro-service centres:* These include the workshops and service centres for repairing and servicing the implements used in agriculture.

The possible areas of entrepreneurship in allied sector: This includes the activities like, dairying, sericulture, goat rearing, rabbit rearing, floriculture, fisheries, poultry farming, olericulture, grafting/budding etc...

Off-farm activities: Entrepreneurship development is also profitable in different off-farm activities like rope making, basket making, bamboo work, and other rural handicrafts.

Food processing industry in India

India has the potential to be the leading global food supplier if it can remove the inefficiencies in the supply chain. The country is one of the largest producers of milk, wheat & rice, second largest producer of groundnut, fruits & vegetables and tops in production of mangoes and bananas.

With a huge population of 1.08 billion, India has a large growing market with 350 million strong urban middle class. A large part of shift in food consumption due to changing food habits, is to processed food market, which accounts to 32% of the total food market. According to the Confederation of Indian Industry, food processing sector has the potential of attracting US $33 billion investment in ten years and with an estimated employment potential of nine million-man days (Negi, 2013).

The food processing industry in India ranks fifth in terms of production, consumption and export. Food processing sector is highly fragmented comprising fruits & vegetables, milk & milk products, meat & poultry, marine products, grain processing, beer & alcoholic beverages, and convenience food & drink. Majority of the entrepreneurs in food processing sector (42%) belongs to small and unorganized section. Though, organized sector is small (25%), it is growing at a much faster rate. Small scale Industries in food processing sector constitutes 33%. Fruit & vegetable processing industry is also fragmented with units of low production capacity of up to 250 tons per annum. Primary milling of grains is an important activity of grain processing industry. Oil seed processing is largely concentrated as a cottage industry. Branded grains as well as processed products are gaining popularity due to hygienic packaging. Products of bakery & bread manufacturing are reserved for small-scale sector. India's dairy industry is considered as one of the successful development industries in post-Independent era with 35% of the total milk being processed, of which organized sector accounts for 13% and the remaining in unorganized sector. It is notable that dairy cooperatives account for major share in organized sector.

Changing life styles, food habits, post liberalization trends and penetration of organized food retail have given a boost to processing sector. In India, processing sector is characterized by poor infrastructure, inadequate quality control, inefficient supply chain, high transportation, high taxation & packaging cost. Availability of raw materials, priority sector status to agro processing and vast domestic markets are major strength of processing industries. Setting up of special economic zones/agri export zones, food parks & mega food parks and promotional schemes and opening of global markets provide lot of opportunity for entrepreneurs in this sector.

Pre requisites for food processing industries

The food processing industry in the country is in a nascent and primitive stage. Number of establishments in the organized sector is far too few compared with several developed and developing nations. The technologies adopted for processing, preservation and value addition in medium, small and tiny industries are primitive and outdated. A pre requisite for adapting to newer technologies is lack of awareness about recent developments in the areas of post-harvest technology and food engineering. Irrespective of the technology adopted, nature of raw materials or finished products, the food processing industry should have the basic features.

- Food processing industry should be set up and run in a hygienic environment
- Availability of raw materials and infrastructural facilities must be ensured
- Ensure the quality control all through supply chain (farm gate to plate)
- Enhance shelf-life of products without any contamination/deterioration
- The food product should be hygienic, tasty, and acceptable to consumers
- The produce should have good market demand and generate employment
- The venture should be economically feasible and socially pleasing

Policy initiatives for food processing industry

Government policies are essential for strengthening sectors and successive Governments have designed and implemented several initiatives for encouraging entrepreneurship along with special schemes targeted towards entrepreneurship. The policy initiatives under the 'Make in India' regime has improved all business processes and procedures which have opened up new areas of opportunities and created confidence among entrepreneurs. It is interesting to note that India has moved up the ladder in the World Bank's 'Ease of Doing Business' ranking from 134 in 2013 to 131 in 2017. The processes of doing business has made significant improvement as in the case of

- Incorporation of a company reduced to 1 day instead of 10 days
- Power connection provided within a mandated time frame of 15 days instead of 180 days
- No. of documents for exports and imports reduced from 11 to 3
- Validity of industrial license extended to 7 years from 3 years
- Bankruptcy Code 2015 – New bankruptcy law, providing for simple and time-bound insolvency process to be operational by 2017
- Goods and Services Tax – Single tax frame work in July, 2017
- Permanent Residency Status for foreign investors for 10 years

The initiatives under the Make in India programme include Digital India, Start-up India and Skill India programmes. Digital India was launched on 1st July 2015 to ensure that government services are made available to citizens electronically by improving online infrastructure and by increasing internet connectivity. The Ministry of Electronics and IT (MeitY) expects a threefold increase in the use of electronic point of sale (PoS) machines by the end of year 2017. Start-up India launched in the same year (15th August 2015) aims at start-up ventures to boost entrepreneurship with support for bank financing for encouraging start-ups. A start-up (only for the purpose of Government schemes) means an entity incorporated or registered in India: i)not prior to five years ii)with an annual turnover not exceeding INR 25 crore in any preceding financial year and iii) working towards innovation, development, deployment or commercialisation of new products, processes or services driven by technology or intellectual property. Skill India is a campaign with an aim to train over 400 million people in India in different skills by 2022.

While the above said initiatives are contributing to the businesses in general, there are specific programmes targeted at the food processing sector that are summarised in Table.1.

In order to promote investment in the food processing sector, several policy initiatives have been taken. These include

- Food processing industry has been declared a priority area. So, it qualifies for a number of fiscal relief and incentives to encourage commercialization and value addition to agricultural products
- Full repatriation of profits and capital is allowed
- Almost the entire sector is delicensed, freeing it from bureaucratic hassles
- Automatic approvals for foreign investment up to 100 per cent, except in few cases, and also technology transfer
- Zero-duty import of capital goods and raw materials for 100 per cent export-oriented units
- Government grants given for setting up common facilities in Agro Food Park

The vision 2015 adopted by the Ministry of Food Processing Industries envisages the following achievable targets to enhance this sector.

- Trebling the size of the processed food sector
- Increasing level of processing of perishables from 6 per cent to 20 per cent
- Value addition to increase from 20 per cent to 35 per cent
- Share in global food trade to increase from 1.5 per cent to 3 per cent

Government schemes and incentives for promotion of MSMEs

While there is no dearth of schemes for promotion of small and medium enterprises, many a times people are not able to get adequate information on schemes offered by various plan and other programmes. An attempt is made to compile various schemes that offer assistance to entrepreneurs, along with necessary informative content regarding the schemes and required contact

Table 1: Policy initiatives for food processing sector (2014-18)

Year of Implementation	Policy Initiative	Description
2014-15	Creation of special fund of Rs. 2000 crore in NABARD	Availability of affordable credit to mega food parks and food processing units set up therein
	Launched investors portal	Information on potential and opportunities for investment in the food processing sector and incentives provided by the Central and State Govts were made available to the prospective investors at a single point.
	Strengthened grievance redressal system	A committee of three independent monitors was constituted to address the grievances of the applicants whose proposal for mega food parks and cold chains that could not be selected
	Food map of India	It maps the potential food processing in surplus production areas
	Streamlining of monitoring process	Monitoring to ensure utilisation of funds
	Excise duty of machinery reduced from 10% to 6%	Reduction in cost of investment in food processing projects and enhancing viability of projects
2015-16	Inclusion of food and agro-based processing unit and cold chain infrastructure in priority sector lending	Availability of additional credit facility
	Service tax on pre conditioning, pre coding, ripening, waxing, retail packaging and labelling of fruits and vegetables exempted in cold chain projects	Encourages more investment in cold chain
2016-17	100% FDI in retail trade including E-Commerce of food products manufactured and or produced in India	For encouraging foreign investment in India
	Online for filing claims for infrastructure development projects	Quicker disposal of claims
	Setting up of investment tracking and facilitation desk of invest India in the Ministry	Desk will identify potential investors and approach them in focused and structured manner for investment and follow-up the investment cases by providing hand holding services

Contd.

Year of Implementation	Policy Initiative	Description
	SAMPADA scheme launched restructuring existing schemes and designing new schemes	SAMPADA scheme to target creation of infrastructure and increasing capacities of processing and preservation in the entire supply chain of food processing sector right from farm gate to retail outlets.
	Excise duty reduced from 12.5% to 6% on refrigerated containers	Reduction of cost of investment in food processing projects and enhancing viability.
2017-18	e-NAM to be expanded from 250 to 585 APMCs, to have primary processing facilities and assistance for primary processing to cleaning, grading and packaging	Encourage direct procuring by processing units and retail traders
	Model law on contract farming to be prepared to integrate farmers	Will attract investment in post harvest management activities
	Dairy processing and development fund of Rs. 8000 crore to be setup in NABARD	Modernise old and obsolete milk processing units in co-operative sector
	National policy on food processing	Policy for holistic development for food processing sector, create conducive environment for the growth of the supply chain and promote food processing in the country
	World Food India, 2017	Ministry is organising World Food India Expo 2017 from 3-5 Nov. 2017, New Delhi.

Source: Press Information Bureau, GoI, MoFPI

information for assistance. Details of schemes available from the respective websites along with contact information are appended as Appendix-1.

Entrepreneurship development programmes

Entrepreneurship development is recognised as one of the most promising initiatives in the present day context, earmarking its relevance in the economic scenario of the country. Today, there are concerted efforts from the part of the Government to promote entrepreneurship through its flagship initiative of "Make in India". Several other initiatives like Start up India, Stand up India, Digital India etc supplement the Make in India programme. It is worthwhile to look into the history of such initiatives in the country.

Small Scale Industries have dominant space in the industrial map of India, due to its contribution to gross value added (almost 40% of the gross value added) and employment generation. After agriculture, SSI remains one of the largest employers of middle class and mediocre population. While organised sector could add only 53.66 lakh jobs during the span of 1980 to 1991, SSI sector employed almost 80 lakh additional people on their rolls.

The Entrepreneurship Development Programmes pioneered by the Gujarat Industrial Investment Corporation was the first of its kind to be offered in India, for the purpose of promoting entrepreneurial culture by imparting skill training. Today, institutes like SISI, SIDO, TCOs, EDI, NIESBUD etc take up promotion of entrepreneurial skills by conducting entrepreneurship motivation campaigns and trainings. Entrepreneurship Development Programmes have the objective of accelerating industrial development by enlarging supply of entrepreneurs; essentially focusing on motivating and developing entrepreneurial qualities with a view to provide self-employment avenues to educated youth.

Phases of EDPs

EDPs in general have three phases; (i) Pre-training phase (ii) Training phase and (iii) Post training/Follow up phase.

Pre training phase: As can be presaged by the taxonomy, pre training phase connotes the stage before training which can be essentially termed as a preparatory phase. Basically, it involves activities which range from arrangement of facilities for training, call for participants and selection of participants for the programme.

Following steps can be assimilated as preparatory steps in launching an EDP. These steps follow the presumption that infrastructure for imparting training is available with the host institution or can be acquired on rental basis fitting into the budgeted expenditure.

- Selection of course director/Co-ordinator who will be responsible for the conduct of the programme
- Approval of infrastructural facilities/arrangement for requisite physical infrastructure for conduct of trainings
- Preparation and approval of syllabus of the programmes, coverage required, arranging for faculty for handling sessions (in-house/outsourced), budget of the programme, administrative approval etc
- Establishing contacts with business personalities and support agencies in the field of entrepreneurship who can contribute to programme content/ exposure visit/implementation support
- Planning for EDP
    - Printing of brochure and publicity through formal announcement in websites and media
    - Inviting applications from potential beneficiaries through open/closed notifications
    - Selection of trainees after screening of applications
    - Confirming trainees participation and preparation of final list of participants

Training phase: Training phase involves the actual conduct of the programme, from receiving trainees at the venue to closing the training programme with a formal valedictory. EDP may be organised for any duration, depending upon the scope of the programme. Motivation training may be organised for a weeks' time, while full-fledged EDP can be run for a period of three months. EDP may be organised with the following objectives, inputs and outcome either separately or in aggregate (Table.2).

Post training phase: It is a phase of follow-up support. EDP becomes successful when the trainees of the programme turn into entrepreneurs. However, after training, potential entrepreneurs usually get stuck up at various stages of initiation of the enterprise. Unless hurdles faced are solved, this may even lead to lack of interest in business. Hence, it is essential to have post training follow up to help the potential entrepreneurs in establishing their business. Handholding at times of difficulties can contribute to reduction of attrition rates. Post training follow up can be done over telephones, personal contact by trainer, mailing feedback forms and conduct of group meetings at regular intervals.

Table 2: Entrepreneurship Development Programme Design

Objectives	Inputs	Outcome
Motivation for entrepreneurship Initiating business ventures	Behavioural inputs Identification of business opportunities, pre requisites for establishing business: form of organisation, mandatory licences, linkages- input, marketing and other functional areas, project management, exposure visits etc	Participant is motivated to take up entrepreneurship Participant gets an overall view to develop an idea into a workable business preposition
Imparting technical skills	Specific skill training for project management, technology training by livelihood incubator models	Participant gets exposure to available technology products for adoption

Success factors of EDPs

Following are the factors responsible for success of Entrepreneurial Development Programmes.

Comprehensive & integrated approach: EDP should take a comprehensive approach that it covers all elements for establishing an enterprise. Hasty approach of conceding contents will offer a destructive effect on attitude of potential entrepreneurs. Programme should be offered in an integrated manner, not on a piece meal approach. Integration implies ensuring the flow of activities while traversing down the flow from pre training phase to the post training phase.

Goal oriented approach: Entire process of EDP should be focussed on the goal of enterprise creation. The programme should be successful in orienting the entrepreneur pitch his idea into an enterprise, the ultimate objective of the programme.

Mentoring support: In addition to formal classroom training, special emphasis should be given for individual counselling/mentoring. This is because every potential entrepreneur has got his unique idea to be converted into a full-fledged enterprise which requires one to one mentoring.

Customising/need based offerings: A pre designed rigid curriculum is not advisable for EDPs. This is because target trainees may have special interests which need to be catered for successful culmination of the programme. Customised deviations from the envisaged programme might be offered from the feedback of the participants or need based curriculum may be developed for each training module.

Evaluation and Review: For success of EDP activity, programmes should be strategically modified, updated and developed to suit the needs of circumstances. Hence, programme contents, tools and techniques should be subject to review at regular intervals.

Outside the content and delivery, the most important element that influences effectiveness of Entrepreneurship Development Programmes are the supporting linkages, that have an active role in supporting the entrepreneur while pitching the idea into a business preposition. In many a case, entrepreneur fail to graduate through the preliminary phases due to lethargic attitude of these support agencies.

Problems of EDP

Even after concerted efforts to promote entrepreneurship through specialised training programmes meant for the purpose, it is disheartening that there are very low adoption rates. Many a time effectiveness of such programmes range between 2 to 10 per cent (connoting two/ten persons out of 100 persons trained start enterprises), which can be rated as very meager.

Some of the common hurdles faced by the EDP programmes are

- Shortage of committed and specialized organisations for conduct of EDP
- Insufficient skill set of trainers and they fail to motivate the trainees
- Identification and selection of wrong projects
- Lack of conducive entrepreneurial environment
- Apathetic attitude of supportive institutions/infrastructure
- Lack of linkages
- Inadequate follow up after training
- High attrition rate of potential entrepreneurs

Role of business incubators in promoting entrepreneurships

A shift from agriculture to agribusiness is an essential pathway to revitalize Indian agriculture and to make more attractive and profitable venture. Keeping this in mind, Government has initiated technology business incubation units to promote entrepreneurship and agro- industry which will open the vistas of incubation landscape to the micro segment of the vast rural economy.

Business incubators are powerful economic development tool to promote entrepreneurship development in food processing sector. They promote growth through innovation and application of technology, support economic development strategies for small business development, encourage growth from within local economies, and provide mechanism for technology transfer. The agri business incubators (ABI's) would primarily focus on those technologies which needs support for commercialisation and further proliferation. These can act as a growth driver in the low end spectrum of the incubation eco-system. The components under ABI include mentoring support in business and technology plans, entrepreneurship cum skill development, identification of appropriate technology, hands on experience on processing machineries, product development (process protocols for various value added products), project report preparation, marketing assistance and professional assistance to make the enterprise successful and achieve higher growth.

The ABI facility at Kerala Agricultural University also envisages designing agri-market-oriented development plan that seeks to improve farmers' livelihoods. Agri business incubator aims to set a benchmark in the field of food processing, to check post harvest losses of the state. The ABI at Kerala Agricultural University is a premier incubation centre for post harvest technology research with its effective and cost-attractive processing systems for agricultural commodities, particularly fruits and vegetables, spices, coconut and rice. The potential of rural food processing industry to tackle this challenge is yet to be fully exploited in the country. Employment is much higher in the food sector than any other sectors. Therefore, role of ABI becomes vital for a rapid transformation of the rural economy in a country like India.

Conclusion

The food industry's primary role is to produce safe and wholesome foods. Providing consumers with a wide variety of foods designed to meet their needs as well as their demands is also a high priority. This is not an easy task. In an era where attention to diet and health is at an all-time high, the food industry must deliver foods that not only meet consumer's basic nutritional requirements, but also their demands for products that are safe, nutritious, convenient, and of good taste. Consequently, the food industry consistently takes proactive steps in the identified three key areas namely technology, education and policy to provide consumers with the products they need and want, and the information necessary to make sound food.

Food processing research Institutes in India with their model research, extension and advocacy capacities is the right agencies to serve as platform to bring together and catalyse the various stakeholders in food supply chain. The ICAR, CSIR, DRDO, MoFPI and State Agricultural Universities are committed to undertake pioneering research and development activities in cutting edge, technologies of food processing, infrastructure development on agribusiness and food supply chain, technology promotion and refinement, field demonstration units showcasing advanced technologies, skill development, awareness campaigns etc required for promoting agri-food entrepreneurship development, thus mitigating post harvest losses in agriculture sector.

To withhold farmers in farming, revive farming and lure new generation into farming, the value added product in its final marketable form and its business/sales as a potential to increase income should be demonstrated first. Then, the production and adoption of production technologies will ensue naturally as a logical sequel as it creates more local demand for raw materials and attracts not only farmers but local entrepreneurs. It will fetch price to food crops, and farmers derive additional income and the nation get nutritive food items ensuring the nutritional security as well.

References

Awasthi, D. 2011. Approaches to Entrepreneurship Development: The Indian Experience. *J Global Entrepreneurship Res.*, 1 (1), 107-124.

Bairwa, S. L., S. K. Lakra., L. K. KushwahaMeena. & P. Kumar. 2014. Agripreneurship Development as a Tool to Upliftment of Agriculture. *Int. J. Sci. Res. Publ.* 4(3):1-4.

Bansal, S., D. Garg., & S. K. Saini. 2012. Agri-Business Practices and Rural Development in India - Issues and Challenges, *Int. J. Applied Engg. Res*, 7 (11).

Bnerjee, G. D. 2011. *Rural Entrepreneurship Development Programme in India – An Impact Assessment*, National Bank for Agriculture and Rural Development.

FAO. 2011. The State of the World's Land and Water Resources for Food and Agriculture. *Summary Report*, Page 9.

Gavane, S. S. 2012. Food Processing: The Biggest Agro-Based Industry in India, *Golden Research Thoughts*, 1 (x), 1-4.

Government of India. 2012. *Annual Report* 2011-'12, Ministry of Food Processing Industries, New Delhi.

Kachru, R. P. 2010. *Agro-Processing Industries in India-Growth, Status and Prospects*. Asstt. Director General (Process Engineering), ICAR, New Delhi.

Negi, S. 2013. Food Processing Entrepreneurship for Rural Development: Drivers and Challenges. In: *IIM, SUSCON III Third International Conference on Sustainability: Ecology, Economy & Ethics* (pp. 186-197). Tata McGraw Hill Education New Delhi.

Press Information Bureau, GoI, MoFPI.

Sherawat, P. S. 2006. Agro-Processing Industries-A Challenging Entrepreneurship For Rural Development. *The Int. Indigenous J. Entrepreneurship, Advancement, Strategy & Education*, 2006.

Suarez, 1972. Campesino Communitarian Enterprises in Latin America, In: *The Community Enterprise*, USA, IICA.

Sudheer, K. P. 2015. Emerging Trends in Grain Based Ready to Eat Foods- an Opportunity for Agripreneurship Development. In: *Proceedings of the National Seminar on Whole Grain – October16th, 2015*, held at IICPT Thanjavur.

Sudheer, K. P. 2016. Prospects of Value addition and Role of Agri Business Incubation Centre's. In: *Proceedings of VAIGA- 2016- International Workshop on Agro Processing & Value Addition*, held at Thruvananthapuram during 1st to 5th December 2016, pp:49-50.

Sudheer, K. P. & S. M. Mathew. 2016. *Recent Developments in Post Harvest Technology*, Published by Director of Extension, Kerala Agricultural University, Thrissur.

Tan, W. L. 2007. Entrepreneurship as a Wealth Creation and Value Adding Process. *J. Enterprising Culture*.15 (2):101-105.

United Nations in India, 2017. *Nutrition and Food Security*, http://in.one.un.org/.

Viswanathan, R. 2016. Entrepreneurship Through Value Addition of Major Millets, In: *Proceedings of VAIGA- 2016- International Workshop on Agro Processing & Value Addition*, held at Thruvananthapuram during 1st to 5th December 2016. Pp: 54-63.

Wortman, M. 1990. Rural Entrepreneurship Research: An Integration into Entrepreneurship Field, *Agribusiness* 6(4): 329-344.

2

Making of An Entrepreneur – Concepts and Functions

Sangeetha K Prathap

Introduction

It is common that after a rocket start, entrepreneur's lack the motivation to scale up business and eventually fails to succeed in the market. There are many factors that may lead to such failures or even success stories, the most important being the entrepreneurial skills and traits that enable business to sustain in the competitive world.

What makes an entrepreneur competent in running a business? According to Fisher and Koch (2008), 'Entrepreneurs are born; the advantage that they derive by birth being the natural advantage. However, we have changed our mindset now-a-days, that it is possible to nurture (make) an entrepreneur; the connotation points out that, right kind of education and training can inspire people with capacities to become motivated to be an entrepreneur.

In this chapter, our focus will be on identifying the entrepreneurial capacities in oneself, after reviewing the essential characteristics and traits that an entrepreneur should possess. Also we try to put forward a framework for entrepreneurship, fitting into the canvas of a food based entrepreneur.

> *"People are always blaming for circumstances for what they are. I don't believe in circumstances. The people who get on in this world are the people who get up and look for the circumstances they want and if they can't find them, they make them."*
>
> *George Bernad Shaw*

Most often it happens that people wait for the last piece of information to start their business; the wait runs for years together that renders the business preposition almost impossible. In this context the afore mentioned quote by Shaw is relevant, which states that circumstances may not be there, if not, creation of one adds to one's merit than remaining to complain about it.

Now, when it comes to the matter of taking decisions, entrepreneurs have to land upon several bewildering decisions; remember you have to take the right decisions. In fact, there is nothing called as prescribed decision in businesses. Each of your context and business may be different, demanding tailor made solutions to problems that may arise. Hence, textual knowledge of entrepreneurial contexts are certainly not going to help you in with readymade solutions, but can offer you capabilities that make you empowered in making decisions.

Do you have the right entrepreneurial spirit? This is a very pertinent question to be answered by each one of you as this could exactly phrase out whether an entrepreneurial activity promoted by you will be successful.

For the reader's understanding, ensuing section deals with entrepreneurial acumen that is depicted in terms of simple questions designed to bring out inherent capabilities of the pursuer.

Concept of entrepreneurship

Entrepreneurship is a word derived from the French word "Entreprendre" meaning 'to undertake'. Thus, conventionally, the entrepreneur is a person who undertakes to organize, manage, and assume necessary risks of a business. However, recent advances have broadened the definition of entrepreneurship; he is considered more as an innovator or developer who recognizes and seizes opportunities, converts such opportunities into ideas, furthering ideas into workable propositions by adding value through effort, skills, collaborations, funding, etc and finally enters the real world of markets with his 'product' to earn a reward on his effort.

As in case any other related aspects of business, entrepreneurship also derives its roots from economics; the 'economists'approach to entrepreneurship is that of "............consists of doing things that are not generally done in the ordinary course of business routine; it is essentially a phenomenon that comes under the wider aspect of leadership."

Joseph Schumpeter

"In ...entrepreneurship there is an agreement that we are talking about a kind of behaviour that includes (1) initiative taking (2) organising or re-organising of socio economic mechanisms to turn resources and situations to practical account and (3) acceptance of risk of failure."

Arthur Cole

From the pure economic approach speaking about input output relationships, entrepreneurship in modern times have advanced to encompass innovation/ creativeness, helping business grow, become profitable etc. among a few to list.

The modern definition relevant to a twenty first century context takes an integrated view "Entrepreneurship is a dynamic process of vision, change and creation. It requires an application of energy and passion towards the creation and implementation of new ideas and creative solutions. Essential ingredients include the willingness to take calculated risks- in terms of time, equity, or career; the ability to formulate an effective venture team; the creative skill to marshal needed resources; the fundamental skill of building solid business plan and finally the vision to recognise opportunity where others see chaos, contradiction and confusion".

Kurtako and Hodgetts, 2005

Reasons for starting business

People tell many reasons why they have chosen to start an enterprise; however it may not be possible to generalize as to why persons choose to be an entrepreneur. Research has proved that a decision to start an enterprise is influenced by the basic characteristics of an individual, his environment and support system. Some of the factors that motivate individuals to opt for starting business are

Need to be independent: Many people find themselves dejected with the present job when they understand that getting fired in the name of unwieldy targets and pressure may undermine longevity.

Live the life I want: Some of them exercise entrepreneurial option when they are set off for living a passionate life; the life they want to live, not suppressing the feelings for achieving somebody's (employer's) vision.

Chance to pursue my passion and use my creative energy: What promotes a few is the passion, to put in practice wild creativity that does not often take place in places that employ us. By and large, bosses under rate the employee's idea to be 'not requiring attention' or 'rubbish' which may harm the employee's sentimental attachment to his creativity.

My family is doing business: A lucky few who are born to business families continue the tradition by taking up business as the avocation. It goes without saying that Dhirubhai Ambani's son would opt for nothing other than business.

Spotted a potential opportunity: An opportunity may spur your idea of starting a business. For example, the new age entrepreneur, James Joseph, of 'Jack 365' was inspired by a jackfruit that caught his sight while sitting in his balcony for writing a book, taking a holiday from his job responsibilities at Microsoft, the topnotch IT company in the world.

Qualities of an entrepreneur

Whatever be the reasons why you opt for entrepreneurship, it goes without saying that certain innate qualities are also going to help you while doing business. It is pertinent to look upon the essential qualities that an entrepreneur need to have.

Need to achieve: Do you have that fire in you?

It's common that people have lovely dreams, but few make sincere efforts to achieve them. For example, we take New Year resolutions year after year, however few put in efforts to see that the dreams come true. An entrepreneurial spirit motivates one to pursue the goals with perseverance.

Perseverance: Are you able to run miles together till you reach the end point?

Once committed towards resorting to entrepreneurship, one should devote an extra effort of pursuing the dream opportunity. As regards entrepreneurship, it goes without saying that there can be failures, but a true entrepreneur will not be desperate about unwelcome happenings and should keep his pace until the ultimate goal is accomplished.

Risk taker: Do turn your back at opportunities that are currently submerged, but may arise in future

An entrepreneur needs to be risk savvy; however that does not mean gambling. An entrepreneur should be a moderate risk taker, taking calculated risks with considerable chance of success.

Ability to find and explore opportunity: Are you willing to ponder various innovative practice in and out of the arena in which you act?

The entrepreneur is very curious to find out opportunities that might lead to a business preposition. It is possible that a keen person will be able to find the business potential with every opportunity that he/she comes across. A person with a curious mind attending a training programme for canning of jackfruit may think how he can use this technology for other agricultural products available in plenty in his vicinity.

Independence: What do you value the most, freedom with risk, or security without freedom?

Entrepreneurs are characteristically independence loving, they are their own masters and are responsible for their own decisions.

Using feedback: Being a critique, an efficient appraiser, able to positively listen what others have say about us.

Taking feedback and analyzing them to identify the strengths and weakness is another characteristic feature of entrepreneurs, no matter the feedback is favourable or not.

Leadership, Interpersonal skills & influencing ability: Conquering stakeholder's attitude that he/she can drive the organisation towards its goals.

An entrepreneur has to have excellent interpersonal skills for working with people and motivating them to achieve goals. This will ultimately lead the organisation to success.

A host of other personal traits may also be observed in an entrepreneur like initiative, persistence, information gathering, concern for quality work, commitment, efficiency orientation, planning, problem solving, expertise, self critical, persuasion, assertiveness, monitoring, credibility, building of image, concern for employee welfare etc.

The characteristics and traits identified in entrepreneurs prepared at a workshop on entrepreneurship conducted at the East West Centre, Honolulu in 1977 is depicted in Table 1.

Table 1: Characteristics and traits of an entrepreneur

Characteristic	Trait
Self confidence	Confidence, independence, individuality, optimism
Task oriented	Need for achievement, profit oriented persistence, perseverance, determination, hard work, drive energy, initiative
Risk bearer	Risk taking ability, likes, challenges
Leadership	Leadership behaviour, inter personal relations, responsive to suggestions and criticisms
Originality	Innovative, creative, flexible, resourceful, versatile, knowledgeable
Future oriented	Foresight perspective

Source: Adapted from Mohanthy, 2007

Creating a framework for entrepreneurship

When a person thinks of starting the business, what is that comes to the mind in the first instance?

In the ensuing section, let us examine the framework for entrepreneurship. Entrepreneurial activity takes its cue from strategic management, where in business decisions have to be envisioned with a futuristic perspective to ensure that firms have a walk over on the ladder to success.

In his framework for Entrepreneur's guide to the Big Issues in Harvard Business Review, Amar Bhide envisages three step sequences

Step 1. Clarifies entrepreneur's current goals

Step 2. Do I have the right strategy?

Step 3. Assess the capacity to execute strategies

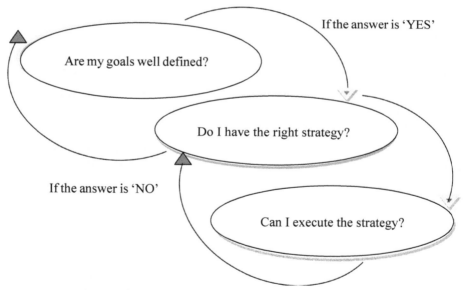

Adapted from: Bhide, 1999, Harvard Business Review on Entrepreneurship
Fig. 1: An entrepreneur's guide to big issues

An attempt is made here to adapt the framework when it applies to a typical agri business enterprise/food based enterprise.

Step1. Are my goals well defined?
What business?....... Type?...Size?...........

When you are thinking about business, you have to be very cautious as to what are you heading towards. This is important in the sense that your personal penchants are going to have serious implications on selection of the business enterprise. For example, if you are a person having affinity towards the natural environment, there is a chance that you will be attracted to jackfruit business which claims that it helps people utilize the most valued fruit which are available in plenty, but wasted. Another person who seeks a value based orientation in his life may end up setting up business which might contribute to the society by offering trustworthy products and employ a few. Likewise individual personality traits can be a deciding factor, while one chooses the suitable kind of a product for his business.

For example, the motivation behind starting a food processing plant may arise from the concern for wastage of agricultural produce during seasons. The jackfruit processing business has mushroomed from the realization that we are wasting tones of jackfruit, the one fruit available in plenty, free from unwanted pesticides and chemicals. Also there are philanthropists who think of feeding the society with available quality materials that would not harm people's health. In both these cases, perhaps profit takes the back bench, while community concern occupies highest priority. If such an objective is there in your mind, ideas can be executed by means of tracking location specific and product specific interventions enabled by appropriate Transfer of Technology (ToT). As earlier pointed out, the business volume you intend to achieve depends on the type of enterprise option you are exercising; if you are a life style entrepreneur, your concentration will be on sustaining the statuesque of the enterprise, rather than eventually expanding the business to soaring heights. On the other hand, an entrepreneur interested in capital gains having extended objectives heading for seamless business growth, may or may not entertain philanthropism to go hand in hand with businesses.

Risk identification....Risk taking

There is no business devoid of risk. Risk taking is a quality that has been pertinent in what makes an entrepreneur. Of course, future is always unpredictable. However, businesses render itself most risky, as we intend to invest the scarce capital available with us (foregoing the opportunity costs[1]) expecting fair returns

---

[1]Opportunity Cost is the cost of the next best alternative foregone. Opportunity costs need to be considered when we calculate economic profits, as foregoing the revenue adds to the cost.

in future. Also, risk varies with the type of project you are likely to embark upon. Hence, identification of risk and sensing your ability to contain risks is important in selection of business enterprises.

Clarence Danhof has classified entrepreneurs into four; Innovative, Adoptive, Fabian and Drone entrepreneurs. The most risk taking category of entrepreneurs are the innovative entrepreneurs, where they sense opportunities in the form of new ideas, innovative technology, new markets or new organisations. Whatever be the opportunity, when you are trying out something new, the risk factor assumes its top gear. Adoptive entrepreneurs are less risk savvy than the innovators; imitate the existing successful entrepreneurs, so that chances of going wrong are lesser. Majority of the entrepreneurs belong to this category. A Fabian entrepreneur is one who is very much skeptical to changes; for example, Fabians normally turn their face towards technological innovations. However Fabian's ultimately might give way for technological breakthroughs when the situation severely demands. The recent demonetization initiative by the Government of India has driven majority of the entrepreneurs, including the Fabians to opt for cashless (digital) transactions. Drone entrepreneurs; on the other hand, is strictly conservative, that they run only conventional businesses on traditional lines.

From the above information, you might certainly be able to examine what type of shoes might be matching for you. If you are an innovative entrepreneur, you might certainly be attracted towards new research and developments happening in R&D institutions and universities and will be certainly be prompted to adopt one of these proven commercially viable technologies.

Step 2. Do I have the right strategy?

Once you decide upon what business to do, the next step is thinking about making strategies. One of the common mistakes that an entrepreneur commits when starting a business is that he thinks and makes short term decisions based on present opportunities without thinking about future. Strategic decisions with future outlook are essential for building a strong enterprise.

Being an entrepreneur does not require any essential formal education; however, there are some specialized inputs that management education can contribute, adding to your unraveled expertise in managing the business. Here, we try explaining how systematic view of business will help you to launch and carry your business forward.

Starting from the common mistakes, short term thinking is most unwelcome when you start a business enterprise, as anybody who aspires to become an entrepreneur may not choose to stop it within a short span of operation. Hence, it is most important to have long term plan for setting up your business. For this purpose, you have to set vision, mission and create objectives for achieving such vision in a systematic manner. Let's have a look upon the vision, mission, core values and strategies of one of the biggest business houses in India, The TATA group.

TATA: Leadership with Trust

Vision

At the Tata group we are committed to improving the quality of life of the communities we serve. We do this by striving for leadership and global competitiveness in the business sectors in which we operate.

Our practice of returning to society what we earn evokes trust among consumers, employees, shareholders and the community. We are committed to protecting this heritage of leadership with trust through the manner in which we conduct our business.

Mission

To improve the quality of life of the communities we serve globally through longterm stakeholder value creation based on Leadership with Trust.

Core values

Tata has always been values driven. These values continue to direct the growth and business of Tata companies. The five core Tata values underpinning the way we do business are:

Pioneering: We will be bold and agile, courageously taking on challenges, using deep customer insight to develop innovative solutions.

Integrity: We will be fair, honest, transparent and ethical in our conduct~ Everything we do must stand the test of public scrutiny.

Excellence: We will be passionate about achieving the highest standards of quality, always promoting meritocracy.

Unity: We will invest in our people and partners, enable continuous learning, and build caring and collaborative relationships based on trust and mutual respect.

Responsibility: We will integrate environmental and social principles in our businesses, ensuring that what comes from the people goes back to the people many times over.

Source: As depicted on www.tata.com website

A vision tells us 'to what organisation is heading towards in future?' Mission, on the other hand, provides information about the unique identity of the firm as to 'who we are, what we do, and why we are here'. Your strategic role does not end with creating vision and mission; but you have to design ways and means for achieving it. The first step in such process is setting of objectives which will take you to the distant vision that you have created. Objectives provide you with timely, measurable performance targets. Once objectives are set, you have to think of alternatives that will help you achieve the vision. Next comes the part of creating strategies that would help you in achieving the stated objectives. You may think of alternative strategies that would lead to achievement of the said vision. A decision may be taken after evaluating the alternative strategies carefully. You may also have to look upon deciding what the core values are. Communication of the vision, mission and core values to all stakeholders including employees, customers, suppliers and others is also important, because all stakeholders need to work together for achievement of said perspectives.

Setting strategies

Deciding on what strategy to use is an important question to be pondered. For example, an entrepreneur may be interested in expanding his business. So what he can do is, search for right type of opportunity which may help you in extending the product line. An entrepreneur doing manufacturing and trading of raw banana chips can extend his product line to include raw jackfruit chips and the like. Also, he may think of introducing specialty premium item of vacuum fried chips (using innovative technology). The vacuum frying technology enables generating premium quality products with added advantages of maintaining the colour of the item and lesser oil absorption. Thus, an entrepreneur who is having a vision of serving the community may think of adopting this technology which may offer the community with premium quality product. While selecting strategies, important questions raised are whether the strategy selected can sustain the business, whether it will be able to generate expected profitability, whether I have the ability to pursue such winning strategies etc.

Step 3. What of executing the strategy?

Once you have decided what your business is and your strategies, the next step is execution of strategies. Firstly, we have to look into our strengths; in terms of resources, ie., money, manpower, networks etc. Finance is the lifeblood of any business and limitation of source of finance can curtail the firm's growth. Hence, it is important to have adequate source of capital, either for one's own or through other sources. Practically, all entrepreneurs are not born rich. Luckily, the present policy initiatives of the Government are pro-entrepreneur that there is no dearth of schemes which offer subsidies, for those who are interested in starting a business. You may remember the mention of such schemes in Chapter 1.

If you are an established business person thinking of expanding your business, manpower will not be a problem because you might possess a customary organizational structure with suitable persons occupying positions. A new entrepreneur may find himself at crossroads in the establishment stage, as to what type of manpower he should look for; whether it should be the top notch professionals with experience of running business (question arising is whether the entrepreneur will be able to afford one) or should he be looking for a fresher who has creative talents for executing a business (question of risking the execution of a new business with an inexperienced person prevails).

Networks are said to be integral part of the business, the talent to create and use networking as a tool for harnessing resources to execute strategies is the success formula for any business. Networking strategies may be looked upon by the entrepreneur to connect to technology based resources, finances, marketing and other functional requirements.

While executing the strategy, one of the aspects that require attention is concurrent monitoring and evaluation. This is relevant in the sense that monitoring and evaluation after execution facilitates only post mortem analysis of why business failed. Rather, a routine inquiry into execution strategies can allow you room for identifying mistaken strategies and even let you make appropriate changes according to changing business environment.

Once you have found answers to questions posed by these steps involved in enterprise building process, one can assume that he possesses the basic skills in starting the business venture. While these questions are only preliminary assessments, a detailed version of each of these have to be significantly addressed while attempting to float an enterprise. A deeper insight and discussion into functional strategies and requirements follow in the ensuing chapters.

Functions of entrepreneurs

Graduating from the broad framework that an entrepreneur needs to have, it is important to have a quick view of the functions of an entrepreneur as a business initiator. A business becomes functional when an entrepreneur envisages an idea, decides to tap its potential in the market, finding ways and means for delineating ideas into practical context by designing and closely pursuing functional strategies intended for the project. Basic functions that an entrepreneur has to pursue are discussed below.

Idea generation: Embarking on an en-cashable business option is the first step in establishing any business. Ideation stage is the most creative and innovative part in the entrepreneurial journey, the stage which involves finding of, analyzing and churning the identified business preposition. While there is no hard and fast rule for identification of what business to be done; environmental scanning, market surveys, brain storming, discussion with established entrepreneurs, trade magazines etc. could help you in arriving at an innovative idea.

Market research: Once you have an idea, the next step is finding out the feasibility of rolling out the idea into a workable business preposition. It can also happen that even great ideas cannot be en-cashed as it fails to fit into the customer's demand frame. Market research will help you in finding out what are the customer expectations and whether your idea has the potential in finding solutions to customer's problems or suiting customer's aspirations. If you have gone through this stage and found that there is a potential market for your proposed product/service, you can proceed along your idea.

Determination of objectives: Once you are convinced about the potential idea, it is time to decide your objectives. An objective says the direction in which you are focused upon and the scope of it. At this point, you have to decide on the nature, type and scope of the business that you are embarking upon.

Deciding on form of enterprise: The forms of business organization include sole proprietorship, partnership, company (one person/public/private), co-operative or Farmer Producer Company. One has to select the convenient form of organization suiting to the requirements of the business option selected.

Raising funds: Finance is the lifeblood of any business; funds have to be tapped from available sources for running your business. This requires meticulous exercise like making a business plan to convince your financiers that you are capable of managing the business and the said business will be viable so that the lender need not be worried about how his money along with his returns would be got back.

Establishment of infrastructure & machinery: Once the adequate funds are available, the project foundation is laid; bring all requisites into a place including fixing of site, buildings, plant and machinery, infrastructure etc suitable to the business you have opted. Plant & machinery may be imported/indigenously manufactured depending on the availability, level of precision and efficiency desired for the machinery.

Recruitment of manpower: After establishing the project layout, attention may be given for recruiting and selecting skilled/semi skilled/unskilled human resources for running the business. The level of expertise of human resources depends on the type of business adopted by an entrepreneur.

Procurement of raw materials: By the time, manpower is set for action on their job after their induction training; the raw materials (if required for manufacturing the product) may be procured from various suppliers so that you are in a position to kick start your business.

Trial run of the project: Once the machine, men and materials are on place, the trial run may be initiated before formally running the project so that early defects and mishaps may be rectified.

Need help for establishing the business?

Is there any place where you can have mentoring and assistance for establishing your enterprise? Of course, the answer is 'yes'. Business incubators are the models wherein you can have your idea turned into full-fledged business. Incubators offer the entrepreneurs with the much required help for refining the business concept, accessing technology, getting through formalities for establishing the business, linking with funding sources, trouble shooting teething problems and other hurdles that may come up in the way of starting the business. The next chapter deals in detail about business incubators and their role in helping potential entrepreneurs in setting up their business.

References

Bhide, A. 1999. The Questions Every Entrepreneur Must Answer, *Harvard Business Review on Entrepreneurship*, Harvard Business School Publishing, Boston, USA, pp: 1-28

Fisher, J. L. & J. V. Koch. 2008. *Born, Not Made: The Entrepreneurial Personality*, Praeger Publishers, 2008.

Kurtako, D. F. & R. M. Hodgetts. 2005. *Entrepreneurship: Theory, Process and Practice*, Thompson Asia Pvt Ltd, Singapore, pp 866.

Mohanthy, S. K. 2007. *Fundamentals of Entrepreneurship*, Prentice Hall of India Private Limited, New Delhi, pp. 8.

www.tata.com

3
Role of ABI for Entrepreneurship Development in Value Addition Sector

Manoj P Samuel, George Ninan and Ravishankar C N

Introduction

Compared to the previous decades, the scope and power of intellectual property rights (IPRs) in agriculture and biotechnology has grown substantially and their international reach has expanded. The IP regime has set up the stage for healthy competition among research centres and industries for developing and seeking novel technologies. However, compared to developed countries, once the technology is created, not much attention is paid in the developing nations on their commercial, policy, environmental, ethical and societal implications. Hence, better techniques are needed for their management, to create policy and educate professionals to commercialize and govern them. Due to the critical role of technology in a competitive environment, strategic technology management is important for farmers and agri-enterprises too. The translation of nascent technologies into new products/ services and their commercialization requires some amount of hand-holding, mentoring and incubation support.

The mission of agri-business incubation is improving the well-being of the poor through the creation of competitive agri-business enterprises by technology development and commercialization. Agri-Business incubation is defined as a process which focuses on nurturing innovative early-stage enterprises that have high growth potential to become competitive agribusinesses by serving, adding value or linking to farm producers.

The major objectives of agri-business incubation initiatives are

- To foster the innovation through creation, development of agri-businesses to benefit the farming community
- To facilitate agro-technology commercialization by promoting and supporting agribusiness ventures.

- To promote successful agribusiness ventures in order to benefit the famers through new markets, products and services

The commercialization including dissemination, transfer and marketing of technology has been evolving as a major pillar that supports the R&D systems. The commercialization process is linked to various activities in the technology management pipeline like protection, valuation, incubation, test marketing, technical and economic feasibility studies, showcasing, licensing and marketing of the technology. Incubation process helps to nascent technology to fully evolve into a business product or service which can compete in real world environment. In a globalized economy, technology licensing and transfer of technology are important factors in strategic alliances and international joint ventures in order to maintain a competitive edge in a market economy.

IPR and its management

Currently, the term 'intellectual property' is reserved for types of property that result from creations of the human mind, the intellect. The most accepted definition is the one advocated by the World Intellectual Property Organization (WIPO). The convention establishing the World Intellectual Property Organization defined the term 'intellectual property' as- "Literary artistic and scientific works; performances of performing artists, phonograms, and broadcasts; inventions in all fields of human endeavour; scientific discoveries; industrial designs; trade-marks, service marks, and commercial names and designations; protection against unfair competition; and all other rights resulting from intellectual activity in the industrial, scientific, literary or artistic fields."

A compilation of the major types of IP assets in agriculture R&D with their qualifying attributes under relevant legislations in India is presented in Table 1. The broad institutional mechanisms, legislative provisions and potential returns to the stakeholders of agri-value chain are also depicted. Considering special nature of use of bio-resources and traditional knowledge (TK) in agriculture, the various provisions and legal mechanisms for protection of these are also enumerated.

Business incubation in Agriculture

Agri-Business Incubators (ABI) open new entry points in the agricultural value chains, which in turn can use to access new markets. They afford leverage through these entry points to accelerate agricultural development and offer the unique potential to develop small and medium-sized enterprises (SME's) which can add value along these chains in ways which other development tools do not offer. There is no single "right way" to perform agribusiness incubation. Rather, the work of agribusiness incubation depends on the state of development of the

Table 1: Broad institutional mechanism (s), legislative provisions and potential returns to stakeholders with respect to IP and related forms of knowledge and resources

S. N.	IP	Legislation	Administration authority	Qualifying attributes	Possible field(s) of application in agricultural sector	Potential stakeholder(s) to benefit
1.	Patent	Patents Act, 1970	Controller General of Patents Designs and Trademarks (CGPDTM)/ Controller of Patents	Novel, non-obvious, capable of industrial application and not fall within the provisions of Section 3 and 4 of the Patents Act, 1970	Agricultural products, processes, value addition	Inventors, traders, economy
2	Design	Design Act, 2000	CGPDTM/Registrar of Designs	New or original; significantly distinguishable from known designs or combination of known designs	Agricultural machinery; post harvest technology products, packaging of processed food/inputs etc.	Industry including SMEs
3	Trademark	Trade Marks Act, 1999	CGPDTM/Registrar of Trademarks	Capable of distinguishing features of goods and services, capable of graphical representation, used or proposed to be used to identify goods/services	Goods and services in agri-business sector	Industry- products and service sector
4	Geographical indication	Geographical Indications (Registration and Protection) Act, 1999	CGPDTM/Registrar of Geographical Indications	Specific geographical origin, possessing qualities, reputation or characteristics that are essentially attributable to that place of origin	Goods, naturally occurring breeds/varieties of commercial value	Communities, traditional practitioners, knowledge holders
5	Copyright	Copyright Act, 1957	Director and Registrar of Copyright	Original expressions of ideas, creations	Software, databases, expert systems, books,	Creators of all works

S. N.	IP	Legislation	Administration authority	Qualifying attributes	Possible field(s) of application in agricultural sector	Potential stakeholder (s) to benefit
6	Integrated circuit design	Semiconductor Integrated Circuits Layout-Design Act, 2000	Registrar, Semiconductor Integrated Circuits Layout-Design Registry	Original; not commercially exploited anywhere in India or in a convention country; inherently distinctive; inherently capable of being distinguishable from any other registered layout-design	Automated machineries/ irrigation systems/ agri-processing	Electronic industry, traders, SMEs
7	Plant varieties	Protection of Plant Varieties and Farmers' Rights Act, 2001	Registrar, Protection of Plant Varieties and Farmers' Rights (PPV&FR) Authority/ Registry	New, distinct, stable and uniform	Seeds/ seedlings/ propagation materials	Plant breeders, farmers, industry
8	Biodiversity	Biological Diversity Act, 2002	Registrar, National Biodiversity Authority (NBA)	Biological resources, herbal remedies, associated traditional knowledge	Access and utilization of biological resources	Knowledge holders, farmers, communities, researchers, etc.
	Traditional knowledge	None	Secretary of the concerned Ministry(ies)	Traditional knowledge/ genetic resources	TK based products and processes	Knowledge holders in communities by sharing of accrued knowledge

agribusiness ecosystem and changes over time as that ecosystem matures and develops. In its earliest phases, incubators demonstrate the viability of new business models and look to create and capture additional value from primary agricultural products. In underdeveloped agricultural economies, incubators help by strengthening and facilitating linkages between enterprises and new commercial opportunities. They open new windows on technologies appropriate to agribusiness enterprises and help agricultural enterprises discover new, potentially more competitive ways of doing business. In subsequent phases of development, incubators operate as network facilitators: they link specialized service providers to agribusinesses and link separate agribusinesses to one another. Finally, in a more advanced state of business development, incubators operate as conduits for the exchange of technology, products, inputs and management methods across national borders.

Agri-business incubation and technology transfer in NARS

The National Agricultural Research System (NARS) in India employs about 4000 researchers in Indian Council of Agricultural Research (ICAR) and almost 15,000 academic faculty members in various State Agricultural Universities (SAUs). In view of changing circumstances and policies, the NARS has initiated steps to strengthen its IP portfolio management and encourage its researchers and academicians to develop and commercialize their innovations for the benefit of farming community. A more pragmatic system for business incubation and promoting start-up companies with respect to agricultural technologies have been evolved in recent times within the National Agricultural Research System (NARS). Generally, agricultural technologies are low-cost technologies and entrepreneurs are not much enthusiastic about it, considering the less purchasing power of the target market.

Since the implementation of the XI Five Year Plan (2007-12) of Government of India, the three-tier IP management mechanism has been established in Indian Council of Agricultural Research (ICAR) towards developing an institutional setup for commercialization of agriculture research products/technologies generated from public research institutions. Accordingly, Institute Technology Management Units (ITMUs) were established in its 95 institutes as a single-window mechanism to showcase the intellectual assets of the institute and pursue matters related to IP management and transfer/commercialization. Five Zonal Technology Management and Business Planning and Development (ZTM&BPD) units were established at the middle-tier, in synergy with the ITMUs, in their respective zones. 12 new BPD units have been initiated in 2013-14 to promote business incubation and technology commercialization. Subsequently, the National Agricultural Innovation Fund (NAIF) has been schematized for the 12th Plan period (2012-17) by the Government of India and establishment of Agri-Business Incubation (ABI) Units in 27 Agricultural research institutes and promotion of Grass-root Innovations are the highlights of the scheme. Under the new initiative, sector wise Zonal Technology Management Centres (ZTMC) coordinate the technology incubation, protection and commercialization activities. Apart from these, Technology Business Incubators (TBI) supported by Department of Science & Technology (DST) has been set up in three NARS institutions and incubation and innovation centres are established at different State Agricultural Universities. Support and services needed by bigger firms and investors for technology transfer as well as for incubation and funding can be addressed through the new flexible business innovation-incubation centres like AgrInnovate India and Technology Business Incubator under the NARS itself (Fig. 1).

The requirement of incubation support by the bigger firms may also be met by these institutional innovations. Provisions were also made to protect the interest of farming community to a larger extent. The established mechanism helps to answer the questions, which may arise from the society on the righteousness and ethical issues of commercializing the public funded research outputs.

The Agribusiness Incubator Program under NARS seeks to provide business consulting services to agriculture-related businesses and helps to develop a strategic business plan. The new initiatives by the Govt. of India as well as ICAR have encouraged start-up companies in agriculture, especially by attracting

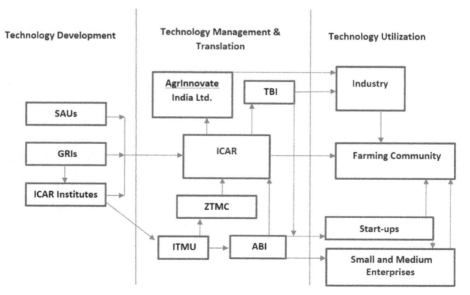

Fig. 1: Institutional framework for Tech transfer and commercialization

rural youth to agri-entrepreneurship. Apart from guidance and consultancy services, the new initiatives also assist in making venture capital funds available to the start-ups. The local communities can also be involved in developing business ideas and entities with respect to agriculture.

Potential institutional innovations for strengthening Agri-Incubation and commercialization capacity of NARS in India

A novel approach should be designed to encourage start-up companies in agriculture, especially by attracting rural youth to agri-entrepreneurship. Apart from guidance and consultancy services, the new initiative should also assist in making venture capital funds available to the start-ups. The local communities can also be involved in developing business ideas and entities with respect to agriculture.

The development processes in the suggestive framework (Fig.2) for the Agriculture Business Incubation (ABI) involve scouting of the technology, assessment and valuation. The technology management services focus on the protection of the developed technologies having a commercial value. The technology generation cycle is the phase where product prototype developed out of the technology innovation undergoes continuous transformation leading into the final product development. The process such as innovation process; technology generation process; and agriculture business incubation are individual entities but complete a cycle of a business. Combining all these processes in a framework, a holistic approach for fostering innovation and incubation eco-system has been envisaged. Through this framework, the role of the individuals or public and private players at various levels and at various places are defined in the process of innovation of various technologies and products.

The nodal centre, which can act as a networking platform of technology managers in SAUs and ICAR institutes in line of a registered society will be helpful in networking relations and exchange of ideas and information related to IP management in agriculture. Further, it can be extended by incorporating other areas of scientific organizations, institute of technologies, engineering colleges, law and business schools and traditional universities. Such a platform can be linked to similar organizations in other countries like Association of University Technology Managers (AUTM) in USA in order to explore the possibility of global technology transfer and commercialization. This initiative will also aid in updating with recent trends in IP regime, new changes in IP laws in a national and international perspective. The platform can also be extended to private companies to foster public-private partnerships.

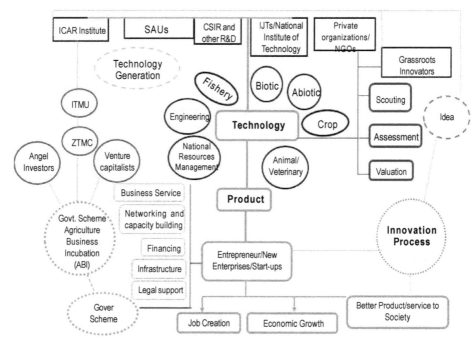

Fig. 2: Conceptualized framework for agri-innovation-incubation process

The nodal centre can bridge the gap between research institutes, industry, society, and the Government. It can play a proactive role in framing technology transfer and commercialization policy in coordination with Central and State Governments, related agencies, business houses and other players in the industry. Nodal centre can be mooted in all research councils/organizations like CSIR, ICAR, ICMR etc. and all can be pooled together to form a National level umbrella consortium under Government of India. The Consortia is envisaged to facilitate the convergence and effective deliverance of all schemes with respect to innovation, incubation and commercialization and provide necessary information to the Government and policy makers and also can play an advisory role.

Agricultural technology management and IPR- emerging opportunities and challenges

Intellectual property rights could play a significant role in encouraging innovation, product development and technical change. Developing countries like India tends to have IPR systems that favour information diffusion through low-cost imitation of foreign products and technologies. This policy stance suggests that prospects for domestic invention and innovation are insufficiently developed to warrant protection. However, an inadequate IPR system could stifle technical change even at low levels of economic development. This is because much invention and product innovation are aimed at local markets and could benefit from domestic protection of patents, utility models, and trade secrets. Moreover, IPR systems could help reward creativity and risk-taking among new enterprises and entrepreneurs. Countries that retain weak standards could remain dependent on technologically inefficient firms that rely on counterfeiting and imitation. It is therefore necessary that a development oriented country like India must have strong IPR legislation and policies and strive to create awareness among industry, academia, students, farmers and public on the IP regime and precautions to be taken to protect their intellectual assets. India is rich in its indigenous technical knowledge and therefore, the avenues to cash upon the traditional wisdom should be opened to the grass root innovators and farmers. A general awareness of the global scenario and clear understanding of the modalities under the IPR is inevitable to the above mentioned stakeholders and it can be cultivated only through educational programmes at the degree level and above especially to the students of agriculture, law, engineering and management.

Inventive firms in developed economies tend to orient their research programmes toward products and technologies for which they expect a large global demand and that may be protected by IPR. This means that a disproportionately small amount of global R&D is focused on the needs of developing economies with low income and weak IPR protection. The efforts to strengthen IP protection in developing countries like India could induce greater R&D aimed at meeting the particular needs of the country. The evidences suggest that IPR protection could generate more international economic activity and greater indigenous innovation, but such effects would be conditional on circumstances. These circumstances vary widely across countries and the positive impacts of IPR should be stronger in countries with appropriate complementary endowments and policies. Countries face the challenge of ensuring that their new policy regimes become pro-active mechanisms for promoting beneficial technical change, innovation, and consumer gains. Educating all stake holders along with policy makers on the dynamic environment of the IPR regime is a vital pre-requisite for conceiving and enforcing strong IP legislations. Apart from encouraging their innovativeness and accelerating returns for re-investments, the stakeholders should also be taught to extract profit from their innovations on traditional or modern technologies using the means provided under IPR by commercializing them through licensing or similar agreements on an international arena.

The lack of formal IP education makes future agriculture professionals incompetent in the face of global business and technological challenges and therefore, a well-structured and comprehensive academic programme in IP and technology management should be included as part of curriculum at the university level. The future global economies will largely be governed by climate change/GHG emission approach, carbon trading, environmental issues and sustainable livelihood based food and water policies. Hence, IPR and technology management educational programmes should also be directed towards these issues and related socio-economic factors. The factors such as changes in global and local businesses, dynamics of supply and value chain systems, advances in technology management protocols, change in preference of consumers and industry should be considered while formulating education policies with respect to IPR in agriculture.

Business incubation initiatives in fish processing sector

Fisheries sector with its important role played in the socio-economic development of the country has become a powerful income and employment generator, and stimulates the growth of a number of subsidiary small, medium and large scale industries. In order to translate the research results arising from the field of fisheries and other agricultural sectors, ICAR have set up an innovation based Business Incubation Centre (BIC) at the ICAR-Central Institute of Fisheries Technology (CIFT), Cochin (Fig.3). BIC is managed by Zonal Technology Management – Business Planning and Development (ZTM-BPD) Unit and aims at establishment of food business enterprises through IPR enabled ICAR technologies.

BIC supports operations on business projects as a measure of enhancing the foundation for new technology based industries and establishing a knowledge-based economy. It focuses on finding new ways of doing business in fisheries and allied agricultural fields by finding doors to unexplored markets. The Centre helps prospective entrepreneurs, by providing pro-active and value-added business support in terms of technical consultancy, infrastructure facility, experts' guidance and training to develop technology based business ideas and establish sustainable enterprises. It acts as a platform for the speedy commercialization of the ICAR technologies, through an interfacing and networking mechanism between research institutions, industries and financial institutions. The Incubator at ICAR-CIFT differs from traditional Business Incubators as it is tailored specifically for technology based industries and is operational at an area with a high concentration of fish production. This industry-specific incubator also allows new firms to tap into local knowledge and business networks that are already in place. BIC offers their services to industries not only in Cochin, but also all over India through virtual incubation. Beyond promoting business growth, the Centre is also trying to bring its benefits to all the fisheries communities in India.

This unique Business Incubator is now known as a "One Stop Shop", where entrepreneurs can receive pro-active, value-added support in terms of technical consultancy, and access to critical tools such as entrepreneur ready technologies, vast infrastructure and other resources that may otherwise be unaffordable, inaccessible or unknown. With the aim of transforming the incubator into a symbol of entrepreneurship and innovation, the ZTM-BPD Unit has created an environment for accessing timely scientific and technical assistance and support required for establishment of technology based business ventures. The activities of the ZTM-BPD Unit focuses on finding creative and innovative ways for linking public sector resources and private sector initiatives within and across regional and national boundaries for promoting economic growth. The Centre uses the right expertise in relevant fields to identify and analyse the constraints and barriers hindering the growth of a business, and devise appropriate strategies. It explores the various structures and strategies to help small enterprises to grow and ensure a promising future in the global market. It fosters corporate and community collaborative efforts, while nurturing positive government-research-business relationships.

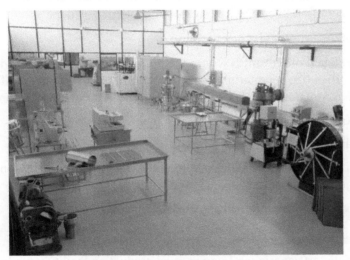

Fig. 3: Pilot Plant at CIFT for Business Incubation

Process of incubation

The Centre regularly conducts industry interface and technology promotional programmes for sensitization of entrepreneurs and to identify interested potential candidates for physical and virtual incubation. The clients at BIC get the privilege of meeting scientists, business manager and business associates directly, to discuss and finalise the strategies to be adopted to take the business forward. It is also the peer-to-peer relationships that develop within the incubator, that ensures the delivery of basic services such as how to actually incorporate a business; what are the legal issues; how to take intellectual property protection; how to do basic accounting and cash flow; how to do business presentations etc.

The residency period for direct incubatees is normally for two years, extendable by another year in special cases, depending on the progress of incubation. As the business venture becomes mature enough, the concessions and the facilities provided to the incubatee companies will be gradually withdrawn. Each incubatee of the unit will have to pay to the Institute a charge for utilization of space, at a rate concessional to the benchmark rate which is the prevailing market rent realizable. Incubatee mentoring will continue in virtual mode after graduation, on need basis.

Services and facilities offered by ICAR-CIFT business incubator

The Centre through its business support services provides links to supporting industries; upgrade technical/managerial skills; provide scientific/technical know-how; assist in market analysis, brand creation and initial test marketing; protect IP assets; and find potential investors and strategic partners.

Incubation facilities under one roof are:

- Furnished office suites within the premises of ICAR-CIFT, with shared facilities like secretarial assistance, computing, copying, conferencing, video conferencing, and broad band internet and communication services.
- Pilot level production lines
- Culinary facility
- Access to modern laboratory facilities for product testing and quality control
- Access to well-equipped physical and digital libraries

Pilot level production lines

A state-of-the-art generic semi-commercial production facility is made available to incubating entrepreneurs for developing value added products from fish. BIC provides access to these facilities along with support of manpower, and assists the entrepreneurs in production and testing of new product formulations. For the tenants, the pilot plant is an ideal testing arena to determine the commercial viability of new products. The plant also serves as a process lab, a place to see how processing equipment impacts food products under varying conditions. There are production lines for pre-processing, cooking, retort pouch processing, canning, sausage production, extruded products, chitin & chitosan production, smoking, curing & drying, breading & battering and product packaging. By providing access to these resources, the Centre greatly reduces one of the major barriers to the commercialization of institute technologies by smaller firms - the high capital cost of intermediate or large scale process equipment.

Business services

The business oriented services offered by BIC include assistance in complying with business regulations and licensing procedures, financing, information services, marketing, and tailor-made services designed for the various tenant enterprises. Incubator clients can also gain special advantage in terms of tax savings through special regulations for Business Incubators. BIC also offers a wide variety of services, with the help of strong associations throughout the Business Incubation Network.

Success stories of business incubation programme at ICAR-CIFT

The Central Institute of Fisheries Technology established a successful pathway in the areas of business incubation and igniting start-ups in food processing, value-addition and packaging sector. The ITMU, ZTMC and ABI units operating in the institute cater the needs of young entrepreneurs especially in micro-small and medium sectors. It also operates in tandem with other ABIs and ITMUs in the fisheries domain and offer mentoring and consultancy services. Some of the success stories of CIFT Business Incubation centre, especially in value addition sector are depicted below:

Chitin and chitosan

Chitosan is a natural product derived from the polysaccharide chitin. It consists of units of amino sugar D-glucosamine. Chitosan is not digestible and has the unique ability to attach itself to lipids or fats. There are no calories in chitosan and it traps the fat and prevents its absorption in the digestive tract. Matsyafed, (Kerala State Cooperative Federation for Fisheries Development Ltd.) in technical collaboration with Central Institute of Fisheries Technology, Cochin has developed CHITONE capsules, a natural chitosan product that can be used to counter obesity/overweight and high blood cholesterol level. The over-the-counter (OTC) chitone capsules hit the Indian markets during December 2009. A chitosan plant has been established by Matsyafed at Neendakara in Kollam District, Kerala for the commercial production of chitone. The plant has the capacity to produce 1.5 lakh capsules a day. The chitosan used for commercial production of chitone capsules is extracted from exoskeleton of fresh marine prawns, crabs and lobsters.

The technology was again commercialized to Uniloids Biosciences Pvt. Ltd. Hyderabad, who is specialized in the domain of bio fertilizers and respective chemicals. The company was given the technology know-how and training to convert the seafood process waste to chitin and chitosan using the scientific methods developed at CIFT. Uniloids Biosciences is a registered incubatee under the ZTM-BPD Unit, an Agribusiness Incubation Centre established at CIFT. The company is provided business support services through the ZTM-BPD Unit and technical support through the Fish Processing and Quality Assurance & Management Divisions at CIFT. Uniloids is successfully manufacturing, supplying and exporting chitin and chitosan to major market players in this field.

India's first integrated- zero waste agri-business venture, pioneered by CIFT incubatee

The Central Institute of Fisheries Technology (CIFT), Cochin under Indian Council of Agricultural Research (ICAR) has set a model for the public private partnership through the establishment of India's first inland fish processing facility in the village of Bhutana, District of Karnal, Haryana. Mr. Sultan Singh is the man behind the establishment of the "Sultan Singh's Fish Seed Farm", "Sultan Singh's Food Court" and the processing unit for the production of value added products from fish. Mr. Sultan Singh is a registered incubatee under ZTM-BPD Unit, South Zone. The processing unit at Karnal was set up in technical collaboration with the Fish Processing and Quality Assurance & Management division at CIFT, Cochin. He is the first incubatee from CIFT, to establish a successful business venture in the field of inland fisheries in India. Scientists from CIFT have provided technical guidance in setting up the zero waste fish processing unit and have imparted training in the production of fish based value added products. The plant is expected to improve the economic status of hundreds of families engaged in fish farming in the village ponds, and other entrepreneurs. The products like fish nuggets, burger, fingers, balls etc are being prepared and marketed under the brand name "Fish Bite". The Unit is designed in such a way that even the waste from fish processing would be converted into fish feed, thereby setting a fine example of zero waste agriculture.

Ready-to-cook fish product chain 'Meenootty'

Baigai Marine Foods, a Cochin based company, incubated at Business Incubation Centre (ZTM-BPD Unit), Central Institute of Fisheries Technology, Cochin, has hit Kerala market with an innovative concept of establishing a retail marketing network of chilled and packed fish products. The whole idea is to bring fish to the customer's doorsteps in a ready-to-cook form. The product line "Meenootty" attains great importance in today's daily life as the number of seafood consumers in India is showing an increasing trend. This is mainly due to the recognition of the nutritional value of fish. But, as one kind of perishable and short shelf life goods, fishes are easy to deteriorate and the process is accelerated with increasing temperature owing to a number of factors such as microbial metabolism, oxidative reaction and enzymatic activity. Consequently, economic value and use value of fish is seriously affected. Baigai Marine Foods launched "Meenootty" by giving importance to the scientific interventions in quality assessment and packaging, and organised business model designed by ZTM-BPD Unit, CIFT. The product is processed and packed in par with the natural conditions preserving their nutritional qualities and freshness. The whole concept is to make available clean and fresh fish to every house, like milk products in the market.

Conclusion

In recent years, many institutional innovations and policy changes within the National Agricultural Research System have catalyzed agricultural technology commercialization and business incubation processes. To further increase Indian industrial competitiveness, new institutional innovations have to be encouraged for agri-business incubation and commercialization. Translating research into technologies and then to product and services requires a coordinated and concerted effort by all stakeholders including the Government, academia, research organization, industry, business houses and the public.

Though, there are many agencies, schemes and government departments in the country to act as support mechanisms for IP protection and subsequent commercialization, the benefits are not really get extended to the needy entrepreneurs, especially in case of small and medium scale agri-businesses. Hence, an effective umbrella structure should be conceived and established which ensures the deliverance of governmental schemes and financial grants to the appropriate agri-enterprises and start-ups.

The National level umbrella consortium can be mooted by the Govt. of India for coordinating and converging the individual initiatives to an integrated and focused effort. A technology transfer protocol for forward integration with the Government machinery, policy makers and other clients and the backward integration with the framers, research institutes, NGOs and other organizations such as IIMs, IITs and business houses have to be designed with clearly defined channels of communication and data flow.

Partnerships should be developed among the research producers, users, and funders both at nodal centre and consortia levels. The scope of public-private partnerships in agriculture and biotechnology in the areas of technology development, protection, transfer and commercialization has to be explored.

Apart from this, a concerted effort of all public institutions under various platforms in India such as Department of Science & Technology (DST), Council for Scientific and Industrial Research (CSIR), Department of Bio-technology (DBT), ICAR, Ministry of Micro, Small and Medium Enterprises etc. should be ensured for making sure of effective flow of information, timely consultancy services and speedy delivery mechanisms to the grass-root level agripreneurs. Effective communication, coordination and cooperation among the various nodal centres, umbrella consortium and the industry are inevitable for the successful implementation of the schemes.

References

Castle, D. *et al.* 2010. Knowledge Management and the Contextualisation of Intellectual Property Rights in Innovation Systems, *SCRIPTed*, 7(1). 32.

Dahlman, C. 2005. *India and the Knowledge Economy: Leveraging Strengths and Opportunities* (The World Bank, New York). Available at: https://openknowledge.worldbank.org/handle/10986/8565.

Darrell, M. W. 2012. Improving University Technology Transfer and Commercialization. *Issues in Technology Innovation, 20.* Center for Technology Innovations, Brookings. Available at: https://www.brookings.edu/wp-content/uploads/2016/06/DarrellUniversity-Tech-Transfer.pdf.

Fikkert, K. A. 2005. Netherlands- Judgment on Essentially Derived Varieties (EDVs) in the First Instance, *Plant Variety Protection*, 99. 11-12.

Geographical indications: Its evolving contours, *WTC Study Report*, (MVIRDC World Trade Centre, Mumbai and NMIMS University), 2009, http://www.iips.ac.in/main_book.pdf (12 September 2013).

Kochhar, S. 2008. Institution and Capacity Building for the Evolution of IPR Regime in India: Protection of Plant Varieties and Farmers Rights, *Journal of Intellectual Property Rights*, 13 (1). 51-56.

Making the Indian Higher Education System Future Ready: *Report of FICCI Higher Education Summit* (Federation of Indian Chambers of Commerce and Industry, New Delhi), 2009.

Manoj, P. S., R. Kalpana Sastry & R. Venkattakumar. 2014. Status and Prospects of IP Regime in India: Implications for Agricultural Education. *Journal of Intellectual Property Rights*, 19. pp.189-201.

Sastry, K. & A. Srivastava. 2013. *Indigenous Traditional Knowledge for Promotion of Sustainable Agriculture*, edited by V. S. Babu., K. Suman Chandra & S. M. Ilyas (NIRD, Hyderabad, India), pp.145-160.

United States Patent and Trademark Office 2010–2015 Strategic Plan. Available at: http://www.uspto.gov/about/stratplan/USPTO_2010-2015_Strategic_Plan.pdf

WIPO 2012. IP *Facts and Figures: WIPO Economics and Statistics Series* (World Intellectual Property Organization, Geneva), p. 7.

4

Trends in Food Processing – An Opportunity to Entrepreneur

Sudheer K P, Saranya S, Ranasalva N & Seema B R

Introduction

The food processing industry is a sunrise sector and contributes around 14% of Gross Domestic Product, 13% of India's exports and 6% of total industrial investment. The Indian food retail market is expected to reach Rs 61 lakh crore by 2020 and the food processing industry accounts for 32% of the country's total food market. This is considered as one of the largest industries in India and is ranked fifth in terms of production, consumption, export and expected growth.

The drift towards urbanisation and busier way of life, pave path to tremendous bounce in the food market. Nowadays consumer's approach to being fit is not about going to gym, but also consulting dietitians and adopting an adjusted eating routine. Consumer's behaviour concerning food consumption has undergone significant changes as a result of the increased awareness about different value added food products, brands, and cuisines, and also their greater willingness to experiment and ability to pay. It leads to the concept of food security in to nutritional security. Companies are offering a wider range of products to the consumers as a result of their investments into product innovation, research and development. At present, various value added products with versatile colour, shape, ingredients etc are available in worldwide market. Among this, some food items like processed fruits and vegetables and dairy products have witnessed a momentous growth in export also. Recent surveys indicated that for companies manufacturing processed foods and beverages, rise in the obesity rate is a great challenge. Hence, efforts were made to develop commercially viable, new process protocols to design healthy products from underutilised food commodities.

In India, a lion share of available food is consumed in fresh form and only a small quantity is processed for value addition. Huge quantities of underutilised food commodities with enormous potential to fetch additional income to the farming community are wasted every year. Promotion of value addition technologies and agripreneurship reduce wastage of foods and offer safe, nutritious and healthy foods with better income to the farming community and thus ensure food as well as nutritional security.

Increased literacy and rising per capita income have induced the Indian customers to increase their expenditure on value added foods, which has higher shelf life, greater nutritive value and easy to cook. This has led to change consumer preferences towards processed foods. Adoption of innovative technologies in the field of food processing for the utilisation of available raw materials to produce safe and healthy products with higher market value is the need of the hour. Government agencies like agribusiness incubators had already developed so many green technologies and value added products in the field of food processing which can fetch higher market to several underutilised food crops. These incubators support economic development strategies through innovation and application of technology, and provide a mechanism for technology transfer. Some of the innovative technologies for the production of safe and healthy value added products developed under agri business incubator of Kerala Agricultural University are briefed below.

Technology for the production of tasty Ready-To-Eat snacks

A "Ready-To-Eat" food product is the one which needs minimum processing procedures on the part of consumer before it is good enough for consumption and is ready-to-eat as soon as the pack is opened. Nowadays, many consumers do not have time to prepare traditional meal and even lack knowledge of how to cook. They also want to relax in the comfort of their own home rather than to spend time at a full service restaurant. Presently, snack foods play a very important role in the diet of modern consumers and the production and consumption of expanded RTE (Ready-To-Eat) products prepared through extrusion cooking has notably increasing worldwide. Extrusion technology plays a pivotal role in the snack and ready to eat breakfast food industry.

Several RTE products are available in the Indian market. Kurkure type extruded products are becoming popular day by day due to change in the food habits and convenience to use. Such products are more palatable and acceptable to the modern day consumers. Extrusion technology is very useful to retain the nutrients compared to other thermal processing methods (Moscicki *et al.*, 2003). It offers several advantages, like faster processing time and reduction in energy consumption over other cooking processes and thus lowering the cost of production. Considering the nutritional and health benefits of ragi, corn, rice and yam, nutritious, convenient and safe to consume RTE snack food was developed at the agribusiness incubator attached to Kerala Agricultural University. The detailed process flow chart for the production of RTE snack is given in Fig.1.

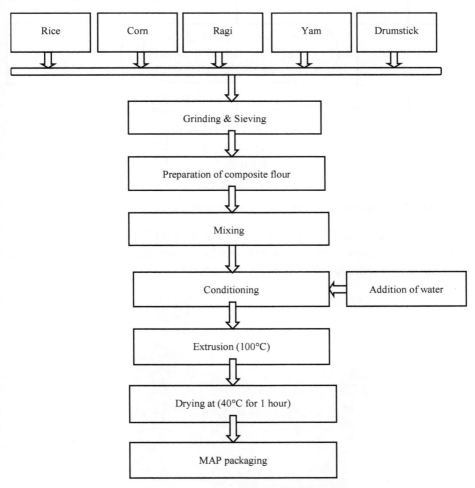

Fig. 1: Process flow chart of RTE product

Standardisation of the extruded snack food was conducted with different combinations of raw materials and at varying temperatures. The product was evaluated in agri incubation laboratory and the best combination was found to be the one prepared with 60% corn, 15% elephant yam, 20% purple yam, and 5% drumstick prepared at 100°C. The proximate composition and the physical and textural properties of this combination are presented in Fig.2.

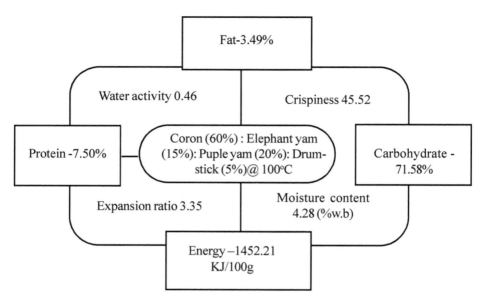

Fig. 2: Physical-textural properties and proximate components of RTE food

Machineries required for large scale production of RTE snacks at the industrial level with their approximate cost and popular manufactures are listed in Table. 1. Twin screw extruder machine is represented in Fig.3.

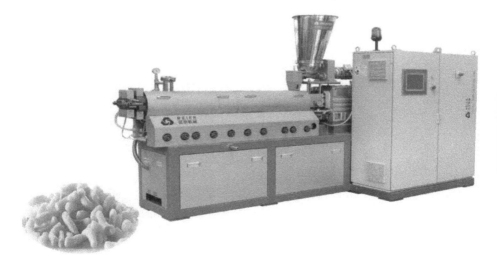

Fig. 3: Twin screw extruder machine

Table 1: Machineries required for the industrial production of RTE products

Machineries	Expected cost (Rs.)	Manufacturing companies
Cleaning unit	30,000	*Process engineers and associates, Koyka electronics etc*
Slicer	30,000	*JAS enterprises, Thomas enterprises, Ponmani enterprises etc*
Cabinet drier (SS)	2,00,000	*Sarah's techno consultancy, Marel engineering & food processing machinery, Shree narayana machines, Viswakarma traders etc*
Pulveriser	1,00,000	*Pilot smith, Sarah's techno consultancy, Modern industry, Pharma tech international, Sri Krishna etc*
Mixer	50,000	*Sarah's techno consultancy, Pilot smith, perfect engineering works, Star tech engineering & construction etc*
Twin screw extruder	18,00,000	*Basic Technologies, Gungunwala food equipment pvt.ltd, Vikas machinery, Micro powder tech etc*
Coating machine (100 kg/batch)	1,00,000	*Prism pharma machinery, Desain engineering, Hi-Tech machineries etc*
Vacuum packaging machine	2,00,000	*Unitek packaging pvt ltd, Royal pack industries, Sevana electrical appliances etc*
Building cost	2000 sqft @ Rs. 1500/-	*NB: Floor area will vary according to the requirement*

Technology for the production of healthy vacuum fried chips

Today's consumers are more concerned on healthy snacks with low fat content. Though, deep fried products are preferred by all age groups, excess consumption may lead to several health problems like obesity, cancer, heart diseases etc. Frying under atmospheric conditions absorb more oil and degrade the quality of oil to a considerable level. Hence, the alternate innovative technology to conserve the quality of oil and to reduce oil absorption and retain the nutritional quality of the fried product is through vacuum frying.

During vacuum frying, the sample is heated under a negative pressure that lowers the boiling point of oil and water in the sample (Troncoso *et al.*, 2009). Therefore, the unbound water in the fried food is rapidly removed when the oil temperature reaches the boiling point of water. Moreover, the absence of air during frying inhibits lipid oxidation and enzymatic browning and thus preserve the colour and nutrients in the sample. The formation of acrylamide, a carcinogenic agent is also reduced to a negligible amount. Moreover, the oil used for vacuum frying can be reused several times without change in the quality when compared to atmospheric frying.

Fried banana snack is a flagship product of Nendran banana which contributes a major share in the snack market. Both raw and ripe banana chips were produced in the agri incubation laboratory using vacuum frying system. The raw banana chips were produced without removing the peel, so as to enrich the fibre content of banana chips. The flow chart for vacuum frying of banana chips is presented in Fig.4. Three D view of vacuum frying system is depicted in Fig. 5.

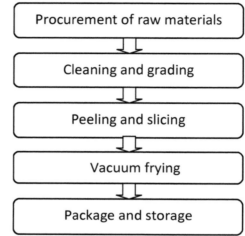

Fig. 4: Process flow chart of vacuum frying of banana chips

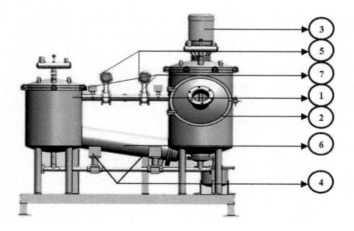

1. Oil storage chamber   2. Frying chamber   3. De-oiling motor
4. Oil flow control   5. Vacuum valve   6. Condenser
7. Nitrogen flow valve

Fig. 5: Three D view of vacuum frying system

The main advantage of vacuum fried banana chips is the retention of colour (Fig.6), since frying was done under reduced temperature and pressure. The vacuum fried product absorbed very less oil compared to atmospheric fried banana chips and maintained the quality after repeated frying.

Fig. 6: Atmospheric fried banana chips & vacuum fried banana chips

The quality of edible oil can be determined by the percentage of total polar molecules (TPM) using TESTO 270 device and maximum allowable limit for the total polar molecule is 25%. The rice bran oil which was used in the laboratory for vacuum and atmospheric frying was tested for the TPM after each batch of frying. Vacuum fried oil showed very gradual rise in TPM and was to the safe level after 50 batches of frying while the atmospheric fried oil showed a rapid rise in TPM percentage and it reached 25% after 8 batches of frying (Fig.7). The low value of TPM in vacuum frying is due to low temperature and low pressure conditions applied for frying when compared to atmospheric frying (Ranasalva and Sudheer, 2016). Similarly, the TOTOX (total oxidative) value for vacuum frying is less compared to atmospheric frying (Fig.8). Hence, the oil used for vacuum frying can be reused more than 50 batches without any degradation in oil quality. Correspondingly, the quality of chips fried in vacuum frying was also found to be superior when compared to atmospheric frying (Table. 2). Hence, it can be concluded that vacuum frying is a promising technology to maintain the quality of fried products and frying oil. Banana chips produced using vacuum frying technology provides a bench mark product with good market value.

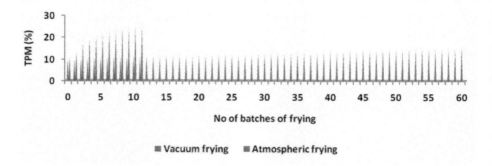

Fig. 7: Changes in TPM value in frying oil

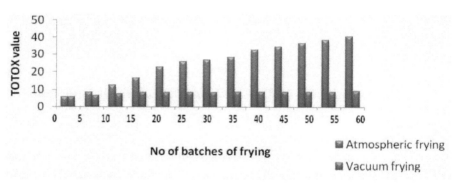

Fig. 8: Total oxidation value of rice bran oil used for vacuum and atmospheric frying

Table 2: Quality attributes of atmospheric and vacuum fried chips

Quality parameters	Atmospheric frying	Vacuum frying
Crispness (N)	3.98	3.54
Colour values L*(ranges from 0-black to 100-white)	43.32	71.57
a*- ranges from -60 (Greenness) to +60 (Redness)	10.54	9.42
b*- ranges from -60 (blueness) to +60 (yellowness)	35.54	48.35
Oil content (%)	29.3	13.2

The shelf life of vacuum fried banana chips was found to be more than 6 months when packed under active packaging (Nitrogen fill package MAP). This technology can also be used for the production of ripened jackfruit, bitter gourd, carrot, beet root, ladies finger etc (Ranasalva and Sudheer, 2017). Various machineries for the production of vacuum fried banana chips are listed in Table 3.

Table 3: Machineries for industrial production of vacuum fried chips

Machineries	Expected cost (Rs.)	Manufacturing companies
Cleaning unit	30,000	*Process engineers and associates, Koyka electronics etc*
Slicer	30,000	*JAS enterprises, Thomas enterprises, Ponmani enterprises etc*
Vacuum frying machine (Capacity- 25 kg/hr)	18,00,000	*Locus exim, Future Tech Foods India ltd.*
Cooling tray	10,000	*Sarah's Techno*
Vacuum packaging machine	2,00,000	*Unitek packaging pvt ltd, Royal pack industries, Sevana electrical appliances etc*
Building cost	2000 sqft @ Rs. 1500/-	*NB: Floor area will vary according to the requirement*

Technology for the production of Intermediate moisture foods (IMF)

Osmotic dehydration is one of the food preservation techniques used for partial removal of water from fruits and vegetables by immersing in aqueous solutions of high osmotic pressure *viz.,* sugar and salts (Pandharipande *et al.,* 2012). As a cost saving drying technology, osmotic dehydration is receiving much attention in the food industry. Preparation of osmotic dehydrated fruits and vegetables involves various unit operations *viz.,* cleaning, grading, peeling, slicing, blanching, and soaking of sliced fruits/vegetables in sugar syrup, drying and packaging. Various perishable fruits such as banana, seasonal fruits like jackfruit and vitamin rich gooseberries etc can be preserved by this method and add value to the locally available fruits and vegetables. Traditional practice of drying fruits and vegetables increase the chances of shrinkage and deteriorate colour. Apart from these drawbacks, spoilage due to insect attack and non uniform drying are other drawbacks of traditional drying. Hence, a process protocol has been developed and standardised under agribuisness incubator at Tavanur to solve these hurdles and to ensure prolonged shelf life and quality retention of final product. The various steps involved in osmo-vac dried nendran banana are charted in Fig.9 and the product is depicted in Fig.10.

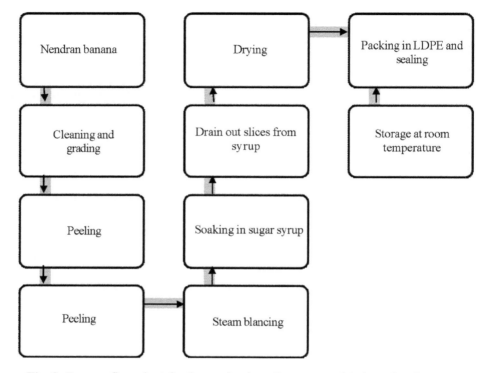

Fig. 9: Process flow chart for the production of osmo–vac dried nendran banana

To produce Intermediate Moisture Foods (IMF), selected fruits/vegetables are cleaned to remove undesirable foreign materials. Then, they are graded based on their maturity and ripening. Cleaned fruits/vegetables are sliced in circular or longitudinal shapes using a slicer so as to get uniform thickness. Sliced fruits/vegetables are then blanched before soaking in sugar syrup. Blanching is carried out mainly to retain the colour of fruits/vegetables after drying and to prevent enzymatic activity. According to the commodity, blanching process and blanching time will vary. For ripe banana, steam blanching is done for 2 minutes, and subsequently the blanched slices are soaked in sugar syrup.

A blancher cum drier or a vacuum dryer is utilized for drying operation. Since, vacuum drying is a low temperature process; the quality of the product is maintained even after drying. From an industrial standpoint, blancher cum drier is more economical and less time consuming

Fig. 10: Osmo–vac dried intermediate vacuum dried nendran banana

since both blanching and drying are done within the same machine. Dried slices are cooled to room temperature and packed in attractive LDPE film. This can be stored up to 6 months under room temperature. Presently, so many SHGs and women entrepreneurs are successfully running their dry fruit industry in a profitable manner. Some of the famous dry fruit industries like shipra foods, suma foods etc are the growing industries in Kerala. Machineries required for the industrial production of IMF are represented in Table. 4

Table 4: Machineries for industrial production of intermediate moisture fruits/ vegetables

Machineries	Expected cost (Rs.)	Manufacturing companies
Cleaning unit	30,000	*Process masters, Revlon industries*
Slicer	30,000	*JAS enterprises, Thomas enterprises Ponmani enterprises, Balakrishna Engineering etc*
Blancher cum drier (Vacuum dryer –optional)	2,00,000	*Sarah's Techno*
Sealing and packaging machine	50,000	*Unitek packaging pvt ltd, Royal pack industries, Sevana electrical appliances etc*
Building cost	2000 sqft @ Rs. 1500/-	*NB: Floor area will vary according to the requirement*

Microencapsulation technology for the production of healthy ready to drink mix formulations

Our custom of traditional familial foods is gradually shifting to industrial foods and at present pre-cooked/fast foods represents a large part of our consumption. Adoption of powdered mixes and recomposed powders has fundamentally modified the storage period and handling of the foodstuffs. From a commercial perspective, converting food products into powdered form will make it much simpler and reduce storage volume and transportation cost. Consequently, it is necessary to supply food powders by retaining the inherent aroma, nutrients and other properties. Microencapsulation by spray drying can be used to develop nutraceutical products from commodities with high medicinal and nutritive value. Microencapsulation can improve the retention of nutrients in food materials. It also offers protection of sensitive food components against nutritional loss and also preserves flavour by coating tiny droplets with a suitable wall material. Spray drying will assist in converting the liquid slurry to powdered form which reduce the storage space, transportation cost and prolong shelf life.

Microencapsulated powders were developed at agri incubator from banana pseudo stem, horse gram, milk, whey and kokum. These are healthy ready to drink formulations that can compete with the fruit powders available in the market. To enhance the organoleptic properties, natural flavourings like ginger juice, cardamom powder, mint extract etc were incorporated in the powder. The detailed process flow chart for the production of microencapsulated powder is depicted in Fig. 11. A schematic diagram of spray drying process is given in Fig.12.

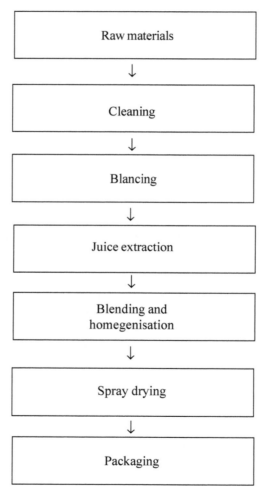

Fig. 11: Process flow chart for the production of microencapsulated healthy ready to drink formulations

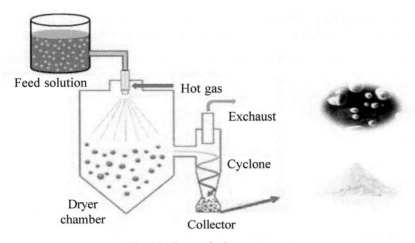

Fig. 12: Spray drying process

Three powder based products were developed from banana pseudostem juice by spray drying technology (Product i, ii, and iii in Table 5). The combinations comprises of (i); pseudostem juice-sugar with ginger extract (Saranya *et al.*, 2016), (ii); blend of banana pseudostem and horse gram with ginger extract (Saranya and Sudheer, 2017$_a$), and (iii) banana pseudostem juice fortified with milk, horse gram extract, and cardamom flavour (Saranya and Sudheer, 2017$_b$). The quality parameters of these powders were determined by standard procedures and are shown in Table 5. The operating/process parameters were optimised for each product. In general, the standardized operating conditions for these spray dried products were spray drying temperature of 180- 185°C, feed pump speed of 15 rpm, blower speed of 1800 rpm, and an atomising pressure of 2 bar. Spray dried ready to drink powders were also developed from kokum juice and a combination of whey and melon juice (product iv and v), with maltodextrin as carrier agent. The important quality parameters of the spray dried product from these studies are given in Table 5.

Table 5: Quality parameters of standardised spray dried fruit juice powders

Powder	Moisture content (% w.b)	Water activity	Total soluble solids (°B)	Wettability (s)
i) 15% S + 25% MD + 56% PS	3.240	0.313	16	80
ii) 25% MD + 30% HSE + 43% PS	4.210	0.352	23	563
iii) 50% milk + 30% HSE+ 20% PS	3.960	0.263	19	603
iv) 45% W+15% WM+25% MD+ 15% S	4.000	0.160	22	37
v) 75% KKM+25% MD	5.000	0.400	19	26

S-Sugar, MD-Maltodextrin, PS- banana Pseudostem juice, HSE- Horse gram extract, W- Whey, WM-Water melon juice, KKM-Kokum juice

Microencapsulated powders can be formulated with the help of a series of machineries. The details of machineries are given in Table 6.

Table 6: Machineries for industrial production of healthy ready to drink mix formulations

Machineries	Expected cost (Rs.)	Manufacturing companies
Cleaning unit	30,000	*Process engineers and associates, Koyka electronics etc*
Blancher	70,000	*Master proseco, Roytech Engineers, Amit Finisher's, Sarah's Techno*
Juicer cum filter	15,000	
Homogeniser	2,00,000	*Diamond Engineering works, Create industries, Select best solutions*
Spray dryer	18,00,000	*SMST, Excel plants & equipment pvt ltd., Spray Tech systems*
Packaging unit	50,000	*Unitek packaging pvt ltd, Royal pack industries, Sevana electrical appliances etc*
Building cost	2000 sqft @ Rs. 1500/-	*NB: Floor area will vary according to the requirement*

Technology for the production of nutraceutical pasta

Ready to cook food items are popular in developed countries because of its versatility in taste, convenience, ease of preparation and appetizing nature. Subsequently, these ready to cook products have flooded the market and emerged as the most popular product as it is cheaper, and fitting to the current life style. Major component associated with pasta production at present is maida, a rich source of gluten. Since, gluten is allergic to some people, gluten free pastas were developed using ragi, corn, purple yam, elephant foot yam and drumstick as pasta flour to enhance the medicinal as well as nutritional value of final product.

An industrial model Pasta Machine is used for preparing cold extruded products (Fig.13). Different types of "die" could be attached to produce pasta of various shapes as per the requirement. The dough to be extruded is prepared just prior to extrusion with the aid of an industrial mixer. Once the dough of required consistency is ready, extrusion is carried out with greater ease and desired shape. A pasta cutter

Fig. 13: Industrial pasta maker

blade, optionally attached at the outlet of the "die", cut the extruded pasta to desired size. The pasta products were manufactured by following the procedure advocated by pasta machine manufacturer.

The process flow chart for production of nutraceutical pasta is detailed in Fig.14. A formulation with minimum cooking time, minimum moisture content, maximum expansion, and maximum swelling power was standardised at agri incubator for commercial production. The ingredients in the standardised pasta included 25% ragi, 20% corn, 25% atta, 10% elephant yam, 15% purple yam, 3% drumstick and 2% gaur gum. These pasta products can be stored up to six months without any spoilage (Seema *et al.*, 2016). The physical properties of developed pasta are represented in Fig. 15.

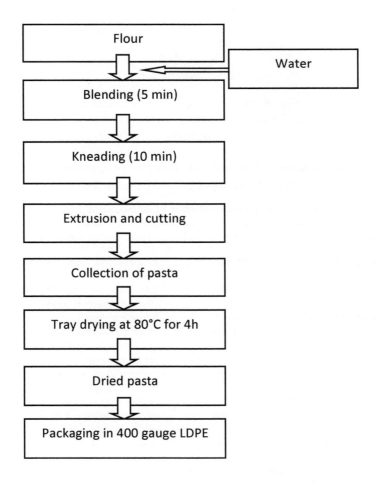

Fig. 14: Process flow chart for production of nutraceutical pasta

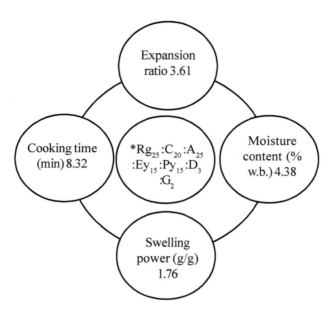

Fig. 15: Physical properties of nutraceutical pasta
(*Rg-Ragi flour :C-Corn flour :A-Atta:Ey-Elephant yam flour:Py-Purple yam flour:D-Drumstick pulp:G-Gaur gum)

Different machineries associated with industrial production of nutraceutical pasta with their expected cost and suppliers are listed in Table 7.

Table.7: Machineries for industrial production of nutraceutical pasta

Machineries	Expected cost (Rs.)	Manufacturing companies
Cleaning unit	30,000	*Process engineers and associates, Koyka electronics etc*
Slicer	30,000	*JAS enterprises, Thomas enterprises, Ponmani enterprises etc*
Blancher cum drier	2,00,000	*Sarah's techno*
Pulvariser	1,00,000	*Pilot smith, Sarah's Techno consultancy, Modern industry, Pharma tech international, Sri Krishna etc*
Blender	50,000	*Sarah's Techno consultancy, Pilot smith, Perfect engineering works, Star tech engineering & construction etc*
Pasta maker	9,00,000	*Rising industries, S.K industries, Grace food processing & packaging machinery,*
packaging machine	10,000	*Unitek packaging pvt ltd, Royal pack industries, Sevana electrical appliances etc*
Building cost	2000 sqft @ Rs. 1500/-	*NB: Floor area will vary according to the requirement*

Retort processing technology for shelf - stable convenience food

Life style and food habits have undergone drastic changes in India during the last decades. Busy life, dual income, innovative technologies and nuclear families etc., increased the demand of ready to cook and ready to eat products. Modern life style often demand the adoption of fast food culture either due to the time constraints or due to the dislike towards time consuming cooking practices. Hence, it is imperative to design a safe alternative to fast food which could be both nutritious and shelf stable. The retorting or sterilization process ensures the stability of the Ready-to-Eat foods in retort pouches on the shelf and at room temperature. Retort thermal processing is the procedure used to retortable containers in a chamber with steam valves that allow for precise temperature control in order to destroy bacteria and spores. The injection of steam under pressure allows the temperature to exceed the boiling point of water inside each pouch within the chamber (Praveena *et al.*, 2016). Retort processing will assist in the destruction of all viable micro organisms especially the heat resistant pathogenic bacteria *Clostridium botulinum* to produce commercially sterile product. Flexible laminated pouch also known as retort pouches are used for retort processing, which can withstand thermal processing temperatures and combines the advantages of metal cans and plastic packages. The multi-layer (4 layered) structure of retort pouch act as a good barrier for gas, moisture etc and thus prolong storage life.

Retort processing include filling of specific quantity food product in retort pouches, which can be achieved by using a filling machine. The pouches can be heat sealed using seal jaws and it should arrange in stackable pallets in retort chamber before retorting. After reaching the pre-set time, temperature and pressure, sterilisation of product can be done by injecting steam or water into the retort. The detailed process flow chart for retort processing of Ramasseri idli is given in Fig.16.

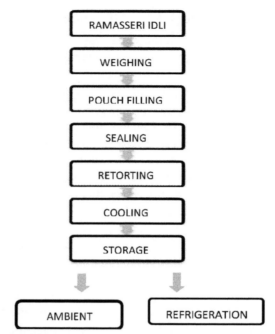

Fig. 16: Process flow chart of retort processed Ramasseri idli

The processing parameters for thermal processing can be determined by conducting heat penetration tests which is used to determine the heating rates of a specific food product in a given container under particular conditions. Thermal heating characteristics (F value and temperature profile during retorting) of Ramasseri idli processed at 100°C are given in Fig.17.

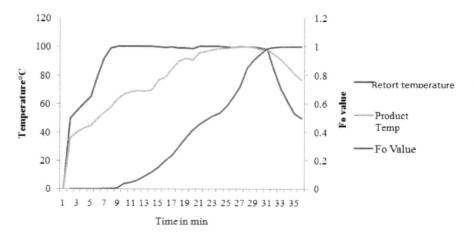

Fig. 17: Heat penetration characteristics for sterilisation (100°C)

Wastage of seasonally available crops like jackfruit, traditional foods (Ramasseri idli), can be preserved and can be made available at off season by adopting retort pouch processing (Fig.18 &19). Since, the cost of retorting is lower than canning; this technology can be adopted for preserving and marketing high value products with economic feasibility. Mr.Subhash Koroth is one of the success entrepreneurs in the field of retort processing. His company named Artocarpus foods private limited situated at KINFRA, Kannur is one of the leading companies in export of retorted tender jackfruit, jackfruit pulp etc. It is India's first full-fledged industry exclusively for jackfruit processing with an investment around 1.3 crore. Products from the industry are sold under the brand name of "HEBON" with a punch line of "Jack of All Tastes". The company has 8,000 square feet of built-up area and employs 30 people. Presently, company process about 800 tons of jackfruit per annum and it is expected to touch 10,000 tons by 2018. Estimated cost for various machineries in retort food processed line is given in Table. 8.

Fig. 18: Retorted jack pulp   Fig.19 Retort machine

Table 8: Machineries for industrial production of retort processed foods

Machineries	Expected cost (Rs.)	Manufacturing companies
Cleaning unit	30,000	*Process engineers and associates, Koyka electronics etc*
Slicer	30,000	*JAS enterprises, Thomas enterprises, Ponmani enterprises etc*
Blancher	70,000	*Master proseco, Roytech Engineers, Amit Finisher's, Sarah's Techno*
Filling and sealing machine	1,00,000	Standard pack packaging machines, Packmach solutions, For-bro engineers
Retort unit	25,00,000	
Building cost	2000 sqft @ Rs. 1500/-	*NB: Floor area will vary according to the requirement*

Conclusion

Entrepreneurship has a prominent role in the economic development of the country. Developing nations like India need a full hand support in the field of entrepreneurship to foster the per capita income of the farming community. Entrepreneurship can boost the supply of manufactured goods, promotion of capital formation, development of local enterprises and skills and creation of employment opportunities. Selection of appropriate entrepreneurial sector plays an immense role in creating success rate and better utilisation of available resources. Enterprises/entrepreneurs should follow the fluctuating consumer styles and provide high quality products with desired nutritional and sensorial qualities at reasonable cost. Compared to other sectors, food processing sector is a blooming sector having ample scope for entrepreneurship opportunities. Since, the demand of processed food is rising day by day, the production and supply of more products is also essential. Currently, our food market is ruled by multinational food companies and our diets are filled with unhealthy junk foods. This increases the chances of various diseases among different age groups especially youngsters who are addicted to fast food culture. Hence, the production of healthy products which are convenient and nutritional should be promoted to solve these hurdles. To fetch the market value and consumer acceptance, the products should be superior in quality as well as versatile in form, colour, attractiveness etc. Technological as well as mechanical support is mandatory for the production of better and safe foods. In this concept, Government has established agri-preneurship supporting centres viz., agri incubators all over the country. Systematic research, monitoring and support of young entrepreneurs, technology development and transfer of developed technology are major activities of these incubators. Various technologies developed under the agri incubator at KCAET, Tavanur, under Kerala Agricultural University, suitable for marketing are discussed in this chapter. We can hope that these technologies and agribuisness incubators will fill more colours to the food processing sectors, and farming community to boost the economy of our country.

References

Moscicki, L., M. Mitrus., A. Wojtowicz & T. Oniszczuk. 2003. Application of Extrusion Cooking for Processing of Thermoplastic Starch. *Food Res. Int.* 47: 291-299.

Pandharipande, S. L., P. Saurav & S. S. Ankit. 2012. Modeling of Osmotic Dehydration Kinetics of Banana Slices using Artificial Neural Network, *Int.J. Computer Appl.* 48 (3): 26- 31.

Praveena, N., K. P. Sudheer & N. Ranasalva. 2016. Development and Quality Evaluation of Retort Pouch Packaged Thermally Processed Tender Jackfruit, In: *Proceedings of 28th Kerala Science Congress*, held at Calicut during 28th to 31st January, 2016.

Ranasalva, N. & K. P. Sudheer. 2016. Production of Traditional Low Fat Snack using Novel Green Technology – Vacuum Frying System. *International Conference on Science and Technology for National Development at KUFOS*, Kochi on 25th and 26th October, 2016.

Ranasalva, N. & K. P. Sudheer. 2017. Comparison of Atmospheric and Vacuum Deep Fat Frying on Product Quality for Selected Vegetables. *National Seminar on Biodiversity Conservation and Farming Systems for Wetland Ecology* held at Regional Agricultural Research Station, Kumarakom on 22-23rd February 2017.

Saranya, S. & K. P. Sudheer. 2017[a]. Process Protocol Standardisation and Quality Assessment of Nutraceutical Powder from Banana Pseudostem Juice. In: *Abstract, 29th Kerala Science Congress*, held at Marthoma College, on 28th to 30th Jan. 2017.

Saranya, S. & K. P. Sudheer. 2017[b]. Development and Quality Assessment of Milk Fortified Banana Pseudostem Juice. In: *Abstract, National Seminar on Biodiversity Conservation and Farming System for Wetland Ecology*, held at Regional Agricultural Research Station, Kumarakom, on 22nd to 23rd Feb 2017.

Saranya, S., K. P. Sudheer., N. Ranasalva & C. Nithya. 2016. Effect of Process Parameters on Physical Properties of Spray Dried Banana Pseudostem Juice Powder. *Advances in Life Sciences 5(17): 6768-6773.*

Seema, B. R., K. P. Sudheer., N. Ranasalva., T. Vimitha & K.B. Sankalpa. 2016. Effect of Storage on Cooking Qualities of Millet Fortified Pasta Products. *Advances in Life Sciences* 5(17): *6658-6662.*

Troncoso, E., F. Pedreschi & R. N. Zuniga. 2009. Comparative Study of Physical and Sensory Properties of Pre-treated Potato Slices during Vacuum and Atmospheric Frying. *Food Sci. & Technol.* 42:187-195.

5

Rice Milling Sector– A Prospective Avenue for Entrepreneurs

Sudheer K P & Ravindra Naik

Introduction

Rice (*Oryza sativa* L.) is a staple food of over half the world's people and is grown on approximately 146 million hectares, more than 10 per cent of total available land. Ninety seven per cent of the world's rice is grown by less developed countries, mostly in Asia. Post harvest losses are considered to be contributed by storage alone, and not much attention was paid to losses occurring during harvesting, transport, threshing, drying, milling and pre-milling treatments. Quantitative loss of food grains was the main concern of the producers and the consumers. Qualitative losses at various stages were not fully recognized, but systematic studies by different organizations revealed that these losses are enormous and should be prevented. These losses cumulatively could be as high as 40 per cent and are normally accompanied by loss of quality also. Quality loss of products can take place at all stages but more particularly during drying or curing of harvested crops, storage, and milling and pre-milling treatments. Post-harvest attention is essential if the grains of higher production are to be fully exploited. Also the quality loss occurring in food grains can be extremely serious if the farming community is not aware of them.

Grain structure, composition and consumers' criteria for quality

The rice grain (rough rice or paddy) consists of an outer protective covering, the hull, and the rice caryopsis or fruit (brown, cargo, dehulled or dehusked rice). Brown rice consists of the outer layers of pericarp, seed-coat and nucellus; the germ or embryo; and the endosperm. The endosperm consists of the aleurone layer and the endosperm proper, consisting of the sub aleurone layer and the starchy or inner endosperm. The aleurone layer encloses the embryo. Pigment is confined to the pericarp.

The grain contains a large centrally located starchy endosperm, which is also rich in protein. The seeds are covered with protective layers of hull and bran. The hull is coloured and may contain tannins and is largely indigestible. The grains also contain certain germ which is high in oil and is enzymatically active and under certain conditions may produce rancidity in the grain. Therefore, hulls, bran and germ are removed in milling operation. One hundred kg of clean bold paddy would yield about 20 – 22 kg husk, 4 kg bran, 2 kg germ and about 72-74 kg of white rice, if nothing were lost in the process of conversion of paddy into white rice.

Harvesting stage for quality and quantity

Farmers in general are not well aware of the advantages of early harvesting of field crops. Conventionally, most farmers in the mono-culture system of paddy cultivation harvest the paddy crop at about 16 per cent moisture content, and suffer a loss of 10 to 15 per cent of the expected field yield. A definite correlation between grain moisture and days after flowering has been established through field studies. Normally, about a week would pass during which the grains are at moisture content between 20 and 24 per cent. If the crop is harvested during that period, the losses are considerably reduced and the field yields are comparatively high. Maximum yields were actually recorded from the harvest carried out on the thirty-first and thirty-second day after flowering, all field yields being corrected to a fixed moisture content of 14 per cent. An appropriate period for harvesting paddy crops may therefore lie between the twenty-eighth and thirty-sixth day after flowering. This provides about nine days to complete the operation without sacrifice of either the yield or quality of the grain. If the crop is over-dried, sun-cracks develop on the kernel and cause breakage during threshing and milling.

Cleaning and Grading

The crop after harvesting and threshing contain organic and inorganic impurities like straw, chaff, weed seed, stones, mud etc. Grains with impurities fetch lower price. Presence of foreign materials increases the bulk; hence increase the cost of handling and transportation. It may cause wear and tear to the processing machineries. Cleaning and grading are the first and the most important operation undertaken to remove the foreign and undesirable materials and separate the grains into various fractions.

Cleaning generally refers to the removal of foreign and undesirable materials from desired grains. Grading refers to the classification of cleaned product into various quality fractions depending upon various commercial values and other usage.

Fig. 1: Air screen cleaner

The grade factors further depend upon.

- Physical characteristics like size, shape, moisture, content, colour etc.
- Chemical characteristics like odour, free fatty acid etc.
- Biological factors like germination, vigor index, insect damage etc.

Equipment used for cleaning (Fig. 1) and grading are classified based on the following properties of the materials (Kachru *et al.*, 1986).

S.No.	Properties	Equipment used
1.	Size	Screen cleaner, Air Screen cleaner
2.	Shape	Disk separator
3.	Specific gravity	Specific gravity separator/destoner
4.	Surface roughness	Inclined draper
5.	Aerodynamic property	Cyclone separator
6.	Magnetic property	Magnetic separator
7.	Colour	Colour separator

Importance of grain grading

Grain grading is a set of standard procedures and methods in quality determination which is essential in marketing, quality assurance operations and in the varietal improvement programme of the country and other research projects involving paddy and milled rice.

Grades and marketing

Grading is necessary in the development of quality standards that define the relationship between grades and prices in the assessment of the value of grains. Official standards are important in the marketing process, because they furnish the means of describing variations in quality and condition. They also provide a basis for merchandising contracts, for quoting prices, for loans on product in storage and for sorting and blending by producers to meet market requirements. Grading then provides for an orderly marketing and trading system. When grades and prices are defined, the farmers become virtually interested in producing better crops because with grading they are assured that their return are based on the quality of their produce. This is supportive of the quality assurance programme of the agency.

Grades and quality assurance

With better crops procured, quality assurance in the other post harvest operations becomes more manageable. With grades and quality standards, quality evaluation or quality assessment operations aimed at preventing quality deterioration and reducing post harvest losses becomes more uniform. Grading is conducted at regular intervals in the various stages of post harvest operations as a means of quality monitoring. It becomes a basis for comparison between grain quality, particularly of stocks before and after long storage, thus, as basis for remedial measures to be undertaken.

Quality has become one of the dominant factors for consideration in rice industry and the first step towards achievement of quality rice is grading. Setting-up modern post-harvest facilities alone cannot solve the quality problem completely. Grading is particularly desirable prior to milling as it offers the following advantages: i) immature grains are separated ii) more precise adjustment of the huller is possible, which minimizes breakages, and iii) independent milling of graded lots is possible.

Paddy grading

The present grading system is of two types: Field grading during procurement and laboratory grading after procurement.

Field grading during procurement is a very important phase of the agency's post-production operations. At this stage, all grains being procured from individual farmers, corporations or cooperatives are inspected and analyzed to establish the quality of the commodity. By doing so, the commercial value of the commodity is assessed as well as its fitness for processing, storage or distribution.

During procurement of paddy, moisture content and purity are quality factors which form part of the basis of payment and are therefore properly determined before the grain is procured. At the buying stations, farmers' produce are sampled randomly by means of a grain probe and the gathered sample is then inspected. Purity, damaged and discoloured kernels are determined by ocular inspection by quality assurance officers, while moisture content is determined by means of a calibrated moisture meter.

Paddy drying

Paddy drying is an essential part of post harvest handling. If paddy is too wet then storage period will be very short due to pest and microbial activity, resulting in heat accumulation and it will made rice turn yellow. In worse case mould will develop. Paddy drying determines storage time, milling outturn and appearance of rice. Different millers have different techniques which determine the competitiveness of each mill. Generally, paddy and rice will store longer at lower temperature. The less temperature change during drying and slower drying will result in better milling outturn and rice appearance.

Usually 1-2% moisture is removed in a single pass during paddy drying. However, care should be taken while drying between 16% and 14% moisture content. In this moisture range, not more than 1% moisture should be removed per hour. This is done in two steps.

- One per cent of moisture is removed to bring it down to 15%. After removal of this one per cent, allow grains to cool/temper for at least 8 hours.

- Another one per cent of moisture is removed to bring the moisture content down to 14%. The grain is allowed to cool/temper for at least 8 hours before milling.

In ancient time, farmers always leave mature paddy on plant in field for few days to dry but this will lower milling yield and harvesting yield. Also leaving rice in field may not be practical for some area which has rainy weather or rice is planted off season, so dryer is needed. Though, many dryers are available, the most common one is LSU drier.

LSU (Louisiana State University) dryer

It was introduced by Louisiana State University. It has a form of large rectangular bin with many inverted V shape fins connecting to opposite bin walls. Each inverted V fins has triangle opening on one wall and closed on the other. This opening is alternated from one fin to the adjacent fin. The working principle is to blow air into one wall side. Air will flow into fin opening, flow through paddy along the bottom of fin and to the adjacent fin and out through the opening of the other wall. Distance between adjacent fins is between few inches to one foot. Paddy is fed to dryer by elevator and out by gravity down flow. Normally, hot dry air is used for blowing. Paddy in LSU dryer will mix well while flow down the bin, so paddy tends to dry evenly. One disadvantage is the complicated bin design and construction.

Compared with batch-in-bin dryers and re-circulating batch dryers, continuous-flow dryers offer the largest drying capacity. When large volume of wet grain is to be dried in a single site these are the types to be considered first. They are most commonly used in a multi-pass drying operation, but because of the large throughputs, operating costs per ton are lower than the larger batch-in-bin dryers and re-circulating dryers.

In a multi-pass drying system, continuous-flow dryers are used in association with tempering bins. During each pass through the dryer, the grain is dried for 15-30 minutes with a reduction in moisture content of 1-3%. Drying at this rate sets up moisture gradients within the individual grains. After each pass, the grain is held in a tempering bin where the moisture within the kernel equalises as moisture diffuses from the interior of each kernel to the surface. The combination of rapid drying and tempering is repeated until the desired moisture content is attained. Using this procedure the actual residence time of the grain within the continuous-flow dryer is of the order of 2-3 hours to effect a 10% reduction in moisture. Selection of the number of passes is a compromise between the dryer efficiency (ie fewer passes), and grain quality (ie longer drying time). Tempering periods are usually 24 hours in duration. The tempering bins may be aerated with ambient air to cool the grain with some slight moisture removal.

It is vital that the operation of drying with tempering is carefully planned and managed to ensure maximum throughput and efficiency.

Parboiling

In this process, paddy is soaked and the wet paddy is heated and then dried. The structure of paddy grain show that the endosperm which covers the major volume of rice grain, is mainly composed of polygonal spaces filled with air and moisture. The presence of voids and the fissures and/or cracks, developed during maturity, causes the breakage of rice during milling. There are three main steps in parboiling viz., soaking, cooking and drying. During soaking of paddy, water penetrates into starch granules and results in swelling of grains. Theoretically soaking of paddy can be done at or below its gelatinization temperature. The lower the temperature used, slow is the process of soaking and vice – versa. Soaking period can be reduced by subjecting the paddy to vacuum for a few minutes before soaking and /or soaking under pressure in hot water. In heating, the energy weakens the granule structure and more surfaces becomes available for water absorption and results in irreversible granule swelling. This phenomenon is called gelatinization of starch. During gelatanization of starch, this will fill the voids and cement the fissures and the cracks. Thus, during the parboiling process, crystalline form of starch is changed into amorphous one, due to the irreversible swelling and fusion of starch. Heat of gelatinization of starch is supplied by saturated stem. Parboiled paddy may be dried in the shade or in the sun or with hot air. Shade drying takes longer time

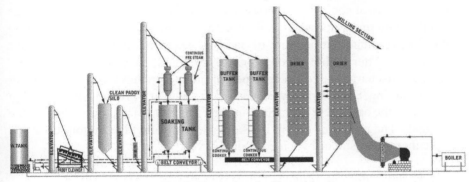

Fig. 2: Typical parboiling unit

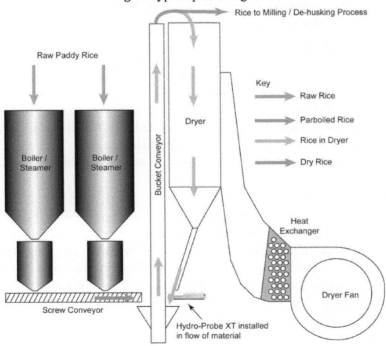

Fig. 3: Shematic diagram of tempering process

but gives excellent milling qualities. Rapid drying in sun or with hot air causes higher breakage during milling. The most convenient practice is to dry in two passes with a tempering period in the moisture range of 15-19% (wet basis) (Fig 2 &3)

CFTRI method

This process was developed at Central Food Technological Research Institute, Mysore. In this process, parboiling tanks are filled with clean water and heated to a temperature of about 85°C by passing steam through the coils placed inside the tank. Sometimes hot water is pumped from other sources into parboiling tanks. The resultant temperature of paddy – water mixture in the tank stays around 70°C. After soaking paddy for 3 to 3.5 hours, the water is drained out. The water discharge valve is kept open in order to remove condensed water during steaming. Soaked paddy is exposed to steam at a pressure of about 4kg/cm² through the open stem coils. Soaking and steaming of paddy are done in the same tank. The parboiled paddy is taken out by opening the bottom door and dried either under sun or by a mechanical dryer.

Pressure parboiling method

This method of parboiling was developed at Tiruvarur in Tamil Nadu. The parboiling is achieved by penetration of moisture into the paddy in the form of water vapour under pressure. This results in gelatinization of starch of the kernel. The paddy is soaked for 40 minutes at 85-90°C. Thereafter, it is steamed under pressure for 18 minutes. The water vapour which penetrates the kernel drives out entrapped air. It is reported that the whole process is completed in 1 to 1.5 hours. The rice obtained by this method has a pleasing and slightly yellowish uniform colour. Reduced soaking period of paddy is the main advantage of this method. It was also observed that such parboiled paddy has better shelling efficiency, has more fat in bran and increased storage life of rice grain.

Drying of parboiled paddy

After parboiling, paddy contains about 35-45% moisture. During the parboiling process the starch is gelatinized which confers quite different drying properties to that of field paddy. It has been shown that in drying parboiled paddy, significant damage (ie kernel cracking) does not occur until moisture content falls to 16%, regardless of drying method or rate of drying. Cracking then occurs some time after grain has cooled. The recommended drying procedure is to dry parboiled paddy to 16-18% moisture as fast as facilities permit, temper it for four hours if warm or eight hours if cooled, and then dry in a second operation to 14% moisture.

Soft Drying System (SDS)

The Soft Drying System (SDS) is a novel grain-treatment/process done to retain the natural rice flavour. Many country Elevator (Large size of paddy receiving, drying, processing and complex) in SDS are already in service in Japan as recognized and established technology (Fig.4).

Advantages and features

- Low running cost
- SDS with husk furnace type does not need fossil fuel for producing heated air unlike conventional paddy drying
- Lower electrical consumption as blowers/mechanical dryer are not required
- Storage drying in mixing tank does not need skilled labour
- Paddy dried in SDS is free from cracking which is caused by excessive drying at sun drying or improper handing at mechanical dryers because moisture contained in raw paddy is gradually transferred to husk without heat stress
- Paddy is not moved while drying. Husk dryer with husk furnace for drying moistened husk does not need any fossil fuel and thus saves energy
- In SDS, heated air is used only for drying husks unlike a conventional dryer which requires a special attention for drying time and temperature to avoid cracking. Thus SDS is easy to operate

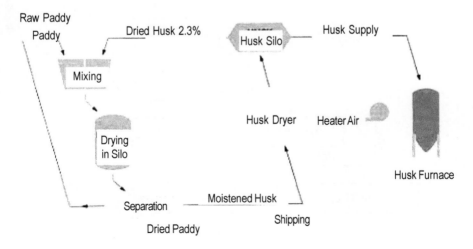

Fig. 4: Schematic of soft drying of paddy

- Rice dried by SDS, is naturally dried or dried under normal temperature and hence retains all the original taste of rice, which could have been lost when it comes in contact with the heated air.

Rice milling

Rice milling is the removal or separation of husk and bran to produce edible portion (endosperm) for consumption by application of external force. Recovering the maximum amount of endosperm or edible portion of paddy grain with none or minimum brokens is the main objective of rice milling. Two basic operations involved are removing the outer covering called husk, or hull and removing the seed coat called bran. The former is called dehusking or dehulling while the latter, is polishing or whitening process. This process has to be accomplished with care to prevent excessive breakage of kernel and improve recovery of paddy. Actual milling process, however, removes also the germ and a portion of endosperm as broken or powdery materials reducing the quantity of grains recovered in the process. The extent of losses on edible portion of grain during milling depends on so many factors as variety of paddy, condition of paddy during milling, degree of milling required, type of rice mill used, the operators, insect infestation and others (Afzalinia *et al.,* 2004). The end product of milling operation is the husk or hull, milled rice or the edible portion, germ, bran and the brokens. Different methods of accomplishing these two operations range from traditional hand pounding using pestle and mortar to high capacity sophisticated milling systems (Fig.5)

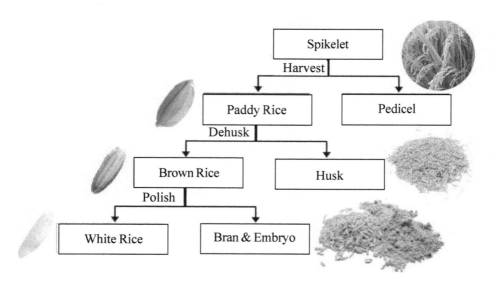

Fig. 5: Rice Milling Flow Chart

Depending on the type of rice mill used, the by-products coming out from the mill will be mixed or separated. Milling is usually done when paddy is dry (about 14% (db) moisture content). Wet soft grain will be powdered. Very dry brittle grain will break and produce broken and powdery materials during milling operations. Some of the rice milling equipment is described below

Single huller

It is also called Engleberg Huller. Most of the other operations, except milling proper, are done invariably. The huller does the job of husk removal and bran removal simultaneously. Though the output/efficiency is less for this huller, most of the traditional rice milling units makes use of this machine. The working element of this machine is made of cast iron, which bears very heavily on rice grains thereby breaking them to pieces which are liable to be lost along with husk. The hullers even though cheaper in cost, wastage (broken) are more and have high labour requirement. It has hollow cylinder containing a rapidly rotating fluted roller. The material is fed axially by a screw conveyor located at the point where paddy enters the hollow cylinder. Husking takes places as a result of shearing action produced on the grain by movement of roller flutes against a stationary blade. A rotating steel roller inside a screen cylinder provides pressure and friction among the grains and effect simultaneous dehulling and whitening or polishing of kernels. The impact force in the huller is absent. Husk, bran, and some broken rice pass through slotted screen in the lower half of the cylinder while the milled rice is discharged at the terminal end. In the first pass, 80-90% paddy is shelled and some polishing also occurs. The average milled rice recovery is 65% and the breakage is high. Besides, it has the largest power equipment per ton of paddy milled than other type of rice mills. Its capacity ranges from 200 to 250 kg/hr. Its average H.P. requirement is 15.

Battery of hullers

These mills consist of a battery of hullers. Additional facilities like, sieves for cleaning paddy, reciprocating sieves for removing broken, aspirators to remove husk and bran etc are also provided. Sometimes, two hullers may be operated in sequence to perform milling, so that better yield and separation of by-products are achieved. This is more suitable for parboiled paddy as extend of broken if used in raw rice is more.

Under runner disc sheller

The under runner disk sheller, often referred to as a disc sheller, consists of two horizontal iron discs partly coated with an abrasive layer. It comprises of two cast iron discs coated with an abrasive composition materials, usually with emery grits. The upper disc is stationary while the lower one rotates. Paddy is fed centrally and the grain contact with the two discs cause husking. Under centrifugal pressure and friction of disc, most of the grains are dehusked. Therefore, the breaking is less and white rice yield is also more of the order of 66-68 per cent with 10 to 15 per cent brokens. The main advantages of the disc sheller are, it is very simple to operate and has low operating cost. The abrasive coating is made at site with inexpensive materials. The clearance between discs could be adjusted for most of the variety of paddy. There is an increase of 2 to 4 per cent in rice output over conventional huller. The main disadvantages are grain breakage and abrasions caused to outer bran layers. The out turn of rice from this type of mill is higher by 1-2% over huller mill.

Sheller-cum-huller mill

This operates with the combination of emery sheller or dehusker and is used for dehusking to obtain brown rice. A huller is used for polishing the dehusked (brown) rice. After cleaning paddy in a sifter, the husk is aspirated. The mixture is then cleaned and aspirated to remove grain and small brokens from head rice. The outturn of rice from this mill is higher by 1-2 per cent over the huller mill for raw paddy.

Sheller-cum-cone polisher mill

This consists of cleaner, disc sheller, husk aspirator, paddy separator, cone polisher and broken rice separator. This mill gives more outturn of rice than hullers by at least 3 per cent (for raw rice). This gives a higher head rice recovery. There is a control over the degree of polishing by proper adjustment. Here, the bran and husk are separated and rice obtained is free from admixtures.

Centrifugal shellers

Here, paddy is under centrifugal force by means of a rotating impeller. The equipment has high capacity and simple constitutional features as there is only one moving part i.e. an impeller. The paddy is fed to the center of the rotor, the centrifugal force acting on the fed paddy throws it towards the casing lined with rubber with great force for the shelling process to occur. Here, the heating of grains does not take place as there is no shearing action involved. Polishing of brown rice obtained is done by a separate polisher. The centrifugal shellers are available which have the provision of winnower to remove the husk.

Modern rubber-roller rice mill

It is basically a sheller mill, with rubber rolls (Fig.6). In addition, it has all other secondary systems needed for good rice milling like pre-cleaners, paddy separators and polishers. The greatest advantage of rubber roller over cast iron cylinder or the emery roller is that due to their compressible nature, the grain is handled gently during the process of dehulling. The functional coefficient between paddy grain surface and rubber is lower than that of paddy and steel. This facilitates easy dehulling. Two rubber rollers rotate in opposite direction at differential speed to take off the aspiration. The shelling components comprise two closely spaced rubber rollers, rotating in opposite directions and at different speeds. One roll moves about 25 per cent faster

Fig. 6: Rubber roll sheller

than the other. The difference in peripheral speeds subjects paddy grains falling between the rolls to a shearing action that strips off the husk. Its shelling efficiency is about 85% and can be improved up to 95% by feeding graded paddy.

After blowing off husk, unshelled paddy is separated in paddy separators and recycled through sheller. Shelled paddy is then polished in a polisher. In this, neither bran layers of the brown rice get damaged nor do sound kernels break during dehusking operation. It is considered as the most technologically advanced method of rice milling. It gives at least 5% more recovery of rice than in a huller mill. Since, entire paddy is not dehusked, paddy must be separated. This is accomplished by a paddy separator. Brown rice (rice with bran coating on) is then fed to rice whitener where bran is removed. Rubber roller mill gives better quality rice and bran which fetch more price in the market

Compared with disc sheller, rubber roll husker has the advantage of reducing grain breakage, loss of small brokens, and risk of damage to grain and operation of machine by unskilled operators. It does not remove germ and therefore primary sieving operation of brown rice can be avoided. Its hulling efficiency is high. Rubber rolls will wear fast due to abrasive nature of husk and constant friction. Hence, this component needs regular replacement, which incurs additional expense. This adds to additional financial burden to the miller and adds to cost of milling. Main reason for the short life of Indian rubber is due to the lack of proper compounding ingredients in the rubber. A new combination of thermoplastic polyurethane rollers can be used as a replacement to rubber rollers.

These rollers have increased life and have reported to give increased head rice recovery with reduced percentage of broken. These materials are accepted by the United States Department of Agriculture (USDA) for use in processing areas for contact with food products.

Modern rice milling process

Operations and equipment

The purpose of rice milling is essentially to remove husk and bran from dried paddy to produce polished rice. Overall milling operation in a modern rice mill involves series of unit operations from handling of paddy coming into milling system to turning it into polished rice (Dhankhar, 2014). The unit operations are cleaning, dehusking, husk separation, paddy-brown rice separation, polishing, grading, weighing and bagging. Material transfer is carried out by means of material handling equipment such as elevators. They do transfer these products from one machine to another for each successive operation (Fig.7).

Various unit operations are given below

Cleaning	This is a unit operation in which foreign materials such as sand, stones, straw, weed seeds and pieces of iron are removed from paddy.
Dehusking	It is a unit operation in which husk is removed from paddy with a minimum stress to the grain.
Husk separation	It is a unit operation where in husk is removed from the mixture obtained after dehusking. This is usually carried out by winnowing process
Paddy separation	It is a unit operation where in brown rice is separated from remaining unhusked paddy. Unhusked paddy is returned for dehusking.
Polishing	It is a unit operation where in part of or all the bran layers are removed from brown rice to produce polished rice.
Bran aspirator	It is a unit operation which removes bran adhering to rice kernel.
Grading	It is a unit operation where in broken rice from head rice are separated into different sizes based on size.
Handling equipment	It is a unit operation where in paddy and rice is conveyed into various processing units.

Removal of bran

This process is called as polishing or whitening. During this process the silver skin and bran layers of brown rice is removed. Some amount of polishing is essential for easy cooking and storage; although excessive polishing reduces nutritive value of rice. This is obtained by abrasion and friction. In the polishing process, friction takes places between brown rice and an abrasive surface, mostly an abrasive stone. In the friction process, friction is caused between grains to peel away bran from outer surface.

The available rice polishing/whitening machinery available in the processing industry can be classified into following groups

- Vertical abrasive whitening cone polisher
- Horizontal abrasive whitening machine
- Horizontal jet pearler

Fig. 7: A typical modern rice mill

Vertical abrasive whitening cone

This machine basically consists of a vertical truncated cone shaped cast iron cylinder with an abrasive coating. Cone is fixed on a vertical shaft that rotates either clockwise or counter clockwise inside a wire screen having a mesh size, as per the paddy variety to be polished. Brown rice is fed into the polisher through a hopper. Feeding of brown rice is adjusted so that brown rice is uniformly distributed on cone surface. Centrifugal force generated by rotation of cone assist in the movement of brown rice between cone and wire mesh. Rubber brakes are provided between wire screen and abrasive cone at regular interval which restricts movement of rice; thereby required pressure is applied. This friction removes bran layer. Partly or fully whitened rice is collected at the discharge outlet. For achieving best performance of machine, it should run half empty. It is recommended that bran layer be removed in three or even more passes. Multi passes produce lower broken and thus leads to higher head rice recovery. Provision of air aspiration through whitener reduces breakage, as heat and dust is removed out of the polisher. In some cases, cone is made of steel and is covered with wood on which leather strips are nailed to reduce broken.

Fig. 8: Horizontal Polisher

Horizontal abrasive whitener

The machine consists of an abrasive (carborundum) roll/disc operating inside a cylindrical perforated metal screen which is horizontally mounted to a steel shaft (Fig.8). The emery is covered by screen cylinder. A uniform gap is provided between the roll/disc and screen where in brown rice is fed through a small screw conveyor. The emery roll rotates at an optimum speed whereas screen cylinder remains stationary. Polishing is obtained by friction between roll, screen and rice grain. At the delivery end of polishing chamber, adjustable weight is placed which helps in regulating the retaining time of brown rice in the polisher, thereby regulating the degree of polish or the whitening index. Brakes, which can be adjusted, are fitted to screen cylinder, which also helps in regulating the degree of polish. Bran removed from rice comes out of the perforated screen and is aspirated out by a blower. Here also, as like in vertical cone polisher, to reduce the brokens, combination of wooden shaft with leather strips is used on moving shaft.

Jet pearler

The jet pearler is used to remove final part of bran layer of brown rice. It also simultaneously cools grain to ambient temperature. It consists mainly of a horizontal partly hollow perforated shaft on which a cast steel cylinder with friction ridges is clamped. The cylinder has a long opening which allows passage for air. This cylinder runs inside a chamber which has screens with slotted perforations. The screen is placed in two halves so that gap between cylinder and screen can be adjusted as per the requirement. Brown rice is fed into machine, in between shaft and screen. A screw controls the clearance between two halves of the screen. Rice produced by this machine is almost free from adhering bran, and cool in nature. Such kind of polishers is best suited for short grain varieties than for medium and long grain varieties.

Recent trends in polishing

Humidified polishing techniques give higher head rice recovery with lower brokens and better appearance to milled rice (Sudheer, 2008$_a$). During this process, brown rice is conditioned with humidified air using air coolers and humidifiers. In 'water milling', required quantity of atmoised water is passed through hollow shaft of the final friction polishers. This process softens kernel surface which helps in easy removal of bran without application of excess force. Air conditioning systems with instrumentation for control of temperature and humidity are used for higher performances. Automation of rice mills for precision control and maximum output is possible through computer or microprocessor based control system.

Pneumatic bran separator

Bran obtained from the polishers is sieved in series of vibratory sifters to separate from germ and broken mixed in the bran. The bran is passed through pneumatic bran separator for better separation efficiency. This equipment separates rice germs and brokens from bran and also conveys the bran pneumatically with the help of air stream at required velocity. It basically consists of a powerful centrifugal fan which aspirates bran from the polisher and bran from the outlet of main cyclone. During this process, fine bran is separated from course materials (germ and brokens).

Glazing

There is huge demand for glazed rice in the country. Process of glazing is carried out only to head rice. This rice appears very shiny and more transparent as the surface is coated with a thin layer of talc and glucose. Glazing can be carried out either continuous or a batch process. Batch process is usually followed as it is cheaper and there is better control on the ingredient fed. During glazing process, white shiny talc powder and glucose are added to rice. Glazing drum is partly loaded with head rice and is slowly rotated. Dry talc powder of about 1-1.2% by weight and glucose is added to the rice in a ratio of 1:1 with water. Subsequently, water is evaporated and talc and glucose remain attached to rice.

Grading

After polishing operation, milled rice contains in addition to head rice, various other fractions like broken, brewers etc. It may also contain other impurities like paddy kernels, damaged and yellow kernels, chalky and immature kernels, foreign matter etc. Grading of milled rice either for foreign or domestic consumption is usually done after milling (Fig. 9). Quality parameters considered in milled rice grading are practically same as in paddy grading with some additional parameters.

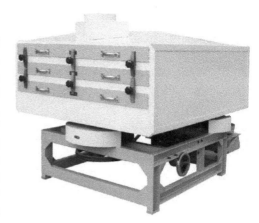

Fig. 9: Rice grader

When rough rice is milled, kernel breakages naturally occur resulting in different kernel lengths, hence, it is necessary to determine the variation. Length differences have been grouped into head rice, broken and brewers with certain length limits.

Head Rice

A kernel or a piece of kernel with its length equal to or greater than $8/10^{th}$ of the average length of the unbroken kernel.

Brokens

Brokens are still subdivided into big, medium and small brokens. Big brokens are pieces of kernels smaller than $8/10^{th}$ but not less than $5/10^{th}$ of the average length of unbroken kernel. Medium brokens are pieces of kernels smaller than $5/10^{th}$ but not less than $2/10^{th}$ of unbroken kernel. Small brokens are smaller than $2/10^{th}$ of unbroken kernel.

Brewers

These are small pieces or particles of kernels that pass through a sieve having round perforations of 1.4 mm in diameter. This is also known as "binlid" or "chips". These brokens and brewers are separated with the help of plansifter, rotating reel graders or a trieur also called as rotating indented separator.

Plansifter

These are either single or double layered sieve which are in oscillated movement provided by the eccentric cam. These basically consists of set of two-three sieves of different perforations to separate polished rice into different fractions viz, head rice, large broken and small broken. In this process, entire broken are not removed as there is a possibility of large brokens getting mixed with the red rice. Further, care should be taken to see that sieves are not clogged for better performance.

Rotating reel graders

This consists of rotating cylinder mounted on a central shaft. Cylinder is made of wire screen or a metallic sheet with opening of required rice, based on the variety to be graded. Milled rice is fed in to reel graders and as milled rice moves through grader, these are divided into three or more fractions based on the size of perforations on rotating cylinder. The different fractions are collected at different outlets.

Trieur

These are also called as rotating indented separator. This is inclined rotating cylinder having indentation similar to small pockets all along its inner surface. This cylinder usually operates at a low speed. The rice is fed into the raised end of cylinder. As rice passes along rotating cylinder, broken gets entrapped into pockets and in the process of movement is picked up by rotating cylinder. As these move to a higher point, they are discharged due to gravity and in the process are collected in the collecting tray placed at the centre of shaft. These brokens are discharged out. Head rice is not collected by the indents and they move along with the surface of the cylinder and will be collected at the head rice outlet. By manipulating the size of the indentation and position of the trough, brokens of different size can be separated. A series of such trieur is used in a commercial paddy milling industry to grade the polished rice into different fractions.

Colour sorter

Presence of discoloured rice in edible rice is not yet studied from the nutritional point of view. It is certain that if they are present in more than accepted levels, consumer will not accept it. Modern rice milling units utilizes colour sorting machines to produce edible rice free of discolored grains (Fig. 10 a&b). These machines are used to sort out discoloured grain from polished rice. These are basically used in basmati rice industry where in a large quantity of rice is exported.

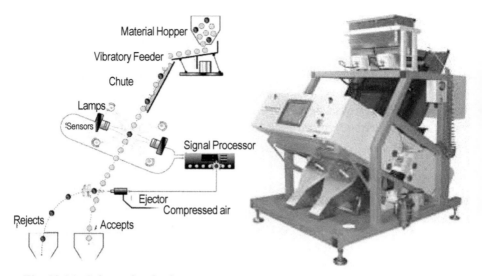

Fig. 10 (a): Schematic of colour sorter   **Fig. 10.(b):** Colour sorter

Factors affecting rice outturn during milling

Outturn of rice after milling differs from sample to sample. This difference arises from three overall factors.

Amount of husk in paddy: Rice outturn depends on the husk content of paddy. More the husk less is the rice and less the husk, more is the rice.

Degree of milling: Rice outturn depends on the degree of milling. The more brown rice is polished; less is the outturn and vice-versa. In India, law fixes the degree of milling as 5 per cent. But, in practice rice is milled as low as 1-2 per cent and as high as 8-10 per cent.

Breakage of rice: Third, and far of greatest importance, is the breakage of rice during milling. Breakage not only reduces the value but also the outturn of rice. Breakage is caused mainly due to cracks in paddy (due to improper drying), immature grain and insect and mould attack during storage.

Ageing of paddy

It is well known fact that raw rice produced from paddy, soon after harvest has a very poor cooking quality. It cooks to a pasty mass, swells less during cooking, gives a thick gruel. The grain also bursts lengthwise during cooking and expands relatively less along the length compared to old rice. These drawbacks of new rice progressively decrease on storage and finally disappear after few months of storage. The old rice, appears short and flat, and becomes round and long after cooking process. In case of stored rice, the loss of solids is minimum in cooking water.

It is reported that cell wall in new rice are relatively more fragile and easily disintegrate during cooking process. Greater dispersal of cell contents in to cooking water further leads to higher degree of pastiness. Changes in physical and chemical properties of starch during the process of ageing might contribute to decrease in pastiness. Decrease in amylase enzyme during storage influences cooking behaviour. Old rice fetches better market price compared to new rice. For higher market price and to reduce time of storage, rice is artificially aged which is called as curing process.

Artificial ageing

Fleshly harvested paddy when steamed and kept hot for some time (30 min to 120 min, depending on the days after harvest), when milled, gives rice similar to stored rice. Milled rice is free from pastiness of fresh rice and is fluffy.

Paddy is steamed in a conical bottom mild steel tank for a period of 15 min till steam vapour appears on top surface of the tank. After steaming is stopped, paddy is kept in the tank for a period of 30 min to 120 min. If condensed steam is allowed to drain off through conical bottom, increase in moisture content of paddy would be to a tune of 5 per cent. Paddy is subsequently dried before milling process.

Mini rice mills and Mobile rice mills

Small rice mills with 1 to 5t/hr capacity are available in Indian markets to mill raw paddy (Fig 11). However, provision of parboiling is not available in such compact mini rice mills. The entrepreneurs may perform the parboiling process in a separate unit and dried paddy after parboiling could feed to such mini rice mills. Nowadays mini mobile rice mills, which can be attached to tractor, are manufactured in India (Fig.12). These mobile mini rice mills can be operated with tractor PTO and hence such machines will be highly useful to places where electricity is not accessible, especially hilly regions. Mobile rice milling units are available with capacity of 1- 1.5 t/hr.

Fig. 11: Mini rice milling unit   **Fig.12:** Mobile rice milling unit

Need of rice mill in villages- a success story from Kerala

It has been observed that the location of rice mills are confined to a few selected production centres. Their development as a village level agro processing unit is yet to take a proper shape. In the absence of village level rice milling unit, farmers have to travel great distances for milling rice (Kachru, *et al.*, 1998). This leads to increased transportation charges and handling losses. Thus, there is a need to develop improved rice mills as a village level agro processing unit for bringing about technical upgradation and development of the sector. Value addition and generation of gainful and sustainable employment opportunities are other possible benefits arising out of this agro processing industry. Though, many small modern rice mills and mobile rice mill units have been developed and are available in the market, they are not popular among potential entrepreneurs. Therefore, a case study of a successful small scale processing centre which is established in a village for processing farmers' produce is briefed below.

M/s. Chandragiri rice mill, Pantharangadi, Malappuram Dist., Kerala state, is a small scale agro processing unit for processing of basmati rice, scented rice, medicinal rice, spices, etc. First of its kind in this region, the unit commenced its production in August 2004. The entrepreneur hails from an agricultural family and has been cultivating basmati paddy in his own fields (Sudheer, 2008$_b$). The seeds were collected from Indian Agricultural Research Institute, Karnal, National Seed Corporation, etc. After successfully experimenting the seeds in his own fields, he distributed seeds to nearby farmers for cultivation. The result was overwhelming and presently nearly 1000 acres of land is indirectly under the control of this processing centre. Since basmati rice is priced more and suitable for higher end, the entrepreneur grew scented rice varieties viz; Thulsi,

Ankur sona, and medicinal rice viz; Rakthasali, Kakkasali, Kumkuma sali, Njavara, etc, which can cater to the lower end market. Besides primary processing of scented and medicinal rice, value addition by producing different grades of roasted rice powder (*Idiyappam* powder, *Puttu* powder, etc), which are used to prepare breakfast items in the State is also produced and marketed under the brand name "Chandragiri". Recently, he started processing parboiled paddy using complete set of equipment for modern rice milling unit.

The entrepreneur ensures quality of raw materials by way of contract farming. He is constantly monitoring the use and application of chemical fertilisers, insecticides and pesticides to ensure that they are within safe limits. He greatly encourages the use of green manure for rice production. The entrepreneur is supplying quality hybrid seeds at competitive price to farmers. Raw paddy for processing was procured from nearby villages for which, the entrepreneur is paying 10-20% higher price than prevailing market price to the farmers. The entrepreneur has promoted intensive and modern farming in nearby villages so that productivity and production of paddy can be improved and more surpluses for processing can be generated. The cumulative effect of the centre in rural sector is enormous due to the creation of labour in the agriculture sector. This is a major intangible advantage of agro-processing and value addition.

Conclusion

Rice milling is the oldest and the largest agro processing industry of the country. Small modern rice mills have been developed and are available in the market but the lack of information is a bottleneck in its adoption by the prospective entrepreneur. The case study showed that processing centre/small scale rice milling unit can play a vital role in primary processing as well as producing value added products. This will prevent the post harvest losses of farm produces and ensures a steady return to paddy growers. It has considerably promoted employment and income generation in rural sector. A strong forward and backward linkage has to be established to procure raw paddy from the farmer and to sell the milled rice to the market. This will not only generate income to the rice milling industry but also will trigger development process in terms of improved agricultural practices for sustainability and food surplus. Rice milling units will also provide opportunity for investment in rural areas and in establishing the infrastructure support for the agricultural produces.

References

Afzalinia, S., M. Shaker & E. Zare. 2004. Comparison of Different Rice Milling Methods, *Canadian Biosystems Engineering*, 46: pp: 3.63-3.66.

Dhankhar, P. 2014. Rice Milling. *International Organization of Scientific Research Journal of Engineering*, 4 (5): pp : 34-42.

Kachru R.P., P.K. Srivastava., B.S. Bisht & T.P. Ojha. 1986. *100 Bankable Post Harvest Equipment Developed in India*. CIAE (ICAR), Bhopal.

Kachru R.P., P.K. Srivastava., S.D. Deshpande & T.P. Ojha. 1998. Using Agro-processing Equipment in Enhancing Rural Industrialization in India – Case Studies. *Agricultural Mechanization in Asia, Africa and Latin America,* 19(2): 55-62.

Sudheer, K.P. 2008[a]. Advances in Rice Milling-Special Emphasis on Basmati. Paper presented during "*Grain tech- 2008*", at Chennai trade centre, Chennai, January 25th to 26th, 2008.

Sudheer K P .2008[b]. Towards Small Scale Basmati Processing, *Indian farming*, Special issue on Global agro-industries forum-2008. 58(1): pp:22-25.

http://www.knowledgebank.irri.org/ericeproduction/PDF_&_Docs/Teaching_Manual_Rice_ Milling.pdf (Accessed on 19.06.2017).

http://royalpunjab.net/process.html (Accessed on 19.06.2017).

http://www.knowledgebank.irri.org/step-by-step-production/post harvest/milling (Accessed on 19.06.2017).

Annexure

Modern rice mill with 1-2 tons/h capacity
Specifications

The details of plant and machinery required for erection of a modern rice milling plant with 12 tons capacity par boiling and dryer per batch and with 1-1.5 tons/hr milling unit is given below.

I. Cleaning section

1.	Item name	Paddy Feed Hopper
	Scope	: One number Paddy Feed Hopper covered with mesh outer frame.
	Type of Construction	: Mild Steel, Fitted at ground level Top Frame made out of 50 x 6mm angle, top covered with Mild steel mesh, 10G MS sheet. Minimum 8ft x 4 ft
2.	Item name.	: Elevator
	Capacity	: Suitable for 4-5 tons/hr pre-cleaner
	Scope	: One number single elevator 24' height, 6" MS Bucket, bottom box 10G & inner boxes made out 14 G M.S Sheet, complete with belts, buckets, motor, motor base and spouting.
	Type of Construction	: Steel, fabricated with rubberized canvas flat black belt, deep drawn mild steel buckets, pulleys, belts. 2HP X 1440 RPM gear motor 3 ö (phase) and with galvanized iron spouting system
3.	Item name	Paddy Pre-Cleaner
	Scope	: Capacity 4-5 tons/hr Complete with Vibro motor, blower, cyclone,Starter & switch arrangements No of screen – TwoScreen perforation cleaning Device – Rubber balls No. of aspiration – TwoBlower Capacity – 7000 cfm at 75 mm WGSPFan Speed - 2880 rpmScreening area - 2+2 = 4 m² With cyclone for dust collection.

II Par boiling section

1.	Item name	: One single elevator for cleaned paddy
	Capacity	: Suitable for 12 T pb unit
	Scope	: One number single Elevator 35- 40 ft' height, Complete with belts, buckets, pulleys,Motor, motor support base and spoutings.
	Type of Construction	: Steel, Fabricated with 8" cup MS bottom 10G &Boxes 14G, with rubberized canvas flat black belt, deep drawn Mild steel buckets, pulleys, belts. 3HP Geared motor 3 ö Motor and with galvanized iron spouting system.
2.	Item Name	: Feed Bin/Surge bin
	Capacity	: 12 tons holding over head tank above the soaking tanks
	Scope	: One number feed bin with top cover and supporting

Contd.

		structures complete with platform, ladders and hand rails.
	Type of Construction	: Mild steel, made of 2.5 mm sheets, reinforced with heavy angles and supporting by beams and channels with sliding door arrangement and chute made out of 3mm S.S. Sheet, paddy bin cover made out of 2 mm HR sheet and 1SA 35 × 6 Angle.
3.	Item Name	: Soaking Tanks.
	Capacity	: 4 Tanks each of 3 tons holding capacity.
	Scope	: 4 number soaking tanks made out of SS 202, 2.5mm thick with suitable inlet, outlet and drain valves for water, 3HP water pump to feed the water to soaking tank with pipelines and hot water circulation arrangements in soaking tank . Insulated Steaming pipelines from boiler to header, slotted pipes for steaming, steam trap, supporting structural complete with MS 8 inch I beam, 50 mm L angles, 100 mm channel, 75 mm channel, foundation bolts, platform, gratings, ladder, hand rails and brazing.
	Type of Construction	: Paddy soaking tank SS 202 grade 2.5 mm thick sheet, Discharge mechanism supported with heavy structural sections, fitted with nosel arrangement which is provided with stainless steel. Tanks are fitted with I.B.R. valves of suitable quality.
4.	Steaming tank / Cooker	: S.S 202(500 kg) × 2 tanks below the soaking tank with steaming slotted pipes, Gravity feeder from 4 soaking tanks for each side, service platform , ladder, handrails arrangements.
5.	One number of boiled paddy elevator	: 2 mm SS.Sheet , suitable height, I.S.A 35× 6 Angle 3HP Motor with Gear box – 8" SS bucket, Rubberized canvas flat black belt, SS 202 deep drawn buckets,_and all connected accessories
6.	Structures	: Standard suitable structures for parboiling unit, Foundation bold, brazing, plat form, steps with handrails, chalf collecting hopper with piping arrangements. Suitable shed so that cleaning, parboiling, drying and paddy conveying to tempering bin should run even in rainy days.

III. Drying section

1.	Bed Cooler	: Paddy cooler made of stainless steel sheet with structural arrangements, 10HP Blower with motor and one set of elevator and its connected accessories Elevator made of 2mm SS 202.
2.	Elevator	: Capable of loading / discharging 12 MT of paddy in 30 min time, purling with truss, with suitable height (35 ft) Two way valve, delivery chute to mill house and to drier (4 mm pipe), blanch connection.
3.	LSU Dryer	: Commodities to be dried – Paddy (Boiled/Raw) Holding Capacity - 12MT (Paddy basis) Material of

Contd.

		Construction – Dryer body, elevator casing made out of SS 2.0 mm thickness, ports made out of 2.5 mm thick SS 202 sheets, elevator bucket (S.S make) Belt , SS feed roller With 1HP gearbox & motor. Hot air duct shall be made out of 3 mm Thick MS sheet.Rates for each version can be given separately. Dust collection blower system from dryer – BB8 Blower with 3HP Motor
4.	Buffer bin (garner) in the dryer :	Provided to accommodate the shrinkage volume of grains only. Remaining boxes (area) should have hot air passage arrangement (i.e. in case of LSU Dryer 'V' ports and hot air ducts should be to the maximum level, leaving empty space only for shrinkage volume) of parboiled paddy containing 30-35% moisture, top covered with hand rails, ladder, dryer to elevator ladder to be provided with hand rails.
5.	Blower :	Sturdy, dynamically balanced 10HP blower, Reputed make motor, volume, and velocity of air may be indicated for 12 MT dryer.
6.	Hot air source :	Steam heat exchanger, hot air temperature at air duct should be up to 100°-125°C with auto temp control device. 2 compartment copper tubes with Aluminum fins, provide steam trap, pressure guage, thermometer, IBR valves, steaming pipe line from header to dryer, duct insulation with glass wool and aluminum foil
7.	Boiled and dried paddy storage tempering bins	Two sets of MS storage bins 12 tons paddy capacity each made of 3.15 mm MS sheets, fabricated and stiffened with suitable structural members and its connected accessories. One set of suitable conveying system from drier to tempering bin

iv. Boiler

1.	Boiler automatic :	Pressure steam boiler 1.5 ton/hr capacity, SS fabricated horizontal pneumatic over feed smoke tube boiler of model with standard fittings and mountings, platform, ladder and fire bars, lining and Insulation of boiler, two assy. of feed water pump with motor, ID fan & FD fan with motor, feed delivery line from feed pump to boiler. Automatic water level controller and one No. of PA fan with motor, husk fired furnace, flue gas ducting between ID fan and chimney and its all connected accessories. Automatic husk feeding blower (bunker), civil ash room, wet bottom, SS chimni 3mm & 5 mm thick, water pump. Steaming pipeline, header & fittings. Insulation of steam piping up to parboiling and heat exchangers Flue Gas Ducting - Boiler to Ash room, Ash room to Chimney with pollution control design.

v. Milling unit 1-1.5 Ton / h

1.	Paddy conveying system	Suitable paddy conveying system from tempering bin to pre cleaner with belt, motor and gear box.

Contd.

2.	Paddy pre cleaner	: 1.5 tons per hour, No. of screen – Two, Screen area – 4 m², Screen inclination, Top – variable, Bottom – fixed; screen perforation cleaning Device – Rubber balls, No. of Aspiration – one ; Blower capacity – 3500 cfm at 75mm WGSP, blower speed – 1440 rpm 5HP
3.	Destoner	: Yanmar type paddy de-stoner – 1.5TPH on paddy fitted with 0.5HP motor & automatic stone discharging arrangement.
4.	Pneumatic sheller	: 10" Pneumatic rubber Sheller with rubber cooling arrangement – Suitable for 1.5TPH on paddy. Power – 10HP.Chute to blow husk to boiler unit with motor
5.	Husk Aspirator	: 1.5TPH on paddy with SS Impeller & 3HP Motor; Chute to blow husk storage tank for boiler unit.
6.	Paddy separator	Gravity paddy separator, 1-1.5 T/ H, tray type with suitable motor
7.	Whitener	: Suitable whiteners(5 stone) – 2 nos with motor & hopper magnet
8.	Silky polisher	Suitable rice silky polisher with motor & standard accessories.
9.	Multi grading sieve	Suitable multi grading sieve with motor and accessories must be minimum 4 sieves.
10.	Bran separator	: Suitable set of bran separator with all driving equipment, bran blower for whitener and silky polisher with motor complete bran separator, cyclone, chute to bran collecting room
11.	Elevators	Suitable number of modified single (11 Nos) or double elevators (21 ft height 5 inchs cup) elevator (6sets) with required motor and driving equipment, bucket nylon, belt – rubberized nylon, geared motor & bottom deck shall have convenient cleaning base
12.	Colour sorter,	: Suitable colour sorter, CCD, air compressor, air conditioned cabin, rice storage hopper and driving equipment complete. cap – 2 TPH 90 channels with resortingSorting accuracy > 99%Air consumption 500-800 L/minPressure at air source > 0.6 MPA Suitable single elevators with required motor and driving equipment, complete set.
13.	Automatic weighing machine	: Range – 5kg-50kg with 10' long conveyor and stitching machine. Accuracy - ± 5 gram

14. One set of complete pipelines, bends, clamps, couplings, pulleys, etc., complete structural works machine hoppers paddy feed hopper, V-pulleys, V-belts, railings, foundation bolts and necessary tools.

15. Suitable centralized electrical push button control panel with connected electrification materials Tool Kit 1 Set

The total cost for the modern rice mill will be approximately 1.5 crores.

6

Entrepreneurial Opportunities in Rice Processing

Durgadevi M, Sinija V R, Hema V & Anandharamakrishnan C

Introduction

Rice is the major staple food for more than 70 percent of the Indian population. India cultivates more than 4,000 varieties and hybrids of rice to cater to varied consumer preferences. A portion of the total cultivated rice is also used for processed products like snacks, savouries and bakery items. India was the largest exporter of rice in 2015-16 followed by Thailand, Vietnam and Pakistan. Rice production forecast is higher at 105 MMT in 2016/17 compared to 2015/16 production of 103 MMT (India Grain and Feed Annual, 2016). Rice milling is the oldest and the leading agro processing industry of the country. Paddy grain is milled either in raw condition or after parboiling, mostly by single hullers of which over 82,000 are registered in the country. Further, over the years, there has been a steady growth of improved rice mills in the country. Most of these have capacities ranging from 2 tons /hr to 10 tons/ hour. At present, it has a turnover of more than 25,500 crore per annum. It processes about 85 million tons of paddy per year and provides staple food grain and other valuable products required by over 60% of the population. Though, rice milling is a wide spread business undertaken in India, commercializing value added products is now gaining customer attraction paving way to merge milling and value addition under a single roof. The desirability of the working population to get foods, particularly rice based products in RTE and RTC forms has thrown a positive approach to new and emerging entrepreneurs to take up a venture in commencing rice based food processing industry.

Health benefits of rice

Rice is used in traditional medicine as a remedy against inflammation, gastrointestinal ailments, hypercholesterolemia, diabetes, and skin diseases (Nam *et al.*, 2008, Burlando and Cornara, 2014). Since, rice is the dominant cereal crop in most Asian countries and is the staple food for more than half of the world's population (including many of those living in poverty), even a small increase in the micronutrient content of rice grains could have a significant impact on human health (Trijatmiko *et al.*,(2016). Rice constituent oryzanol has been intensively investigated for cholesterol regulation and antioxidant/anti-inflammatory activities. Rice carrying the advantage as staple crop can act as vehicle for fortification (Burlando and Cornara, 2014). International Rice Research Institute is developing rice varieties that have more iron, zinc, and β carotene content to help people get more of these important micronutrients. Rice has proved to support biofortification as well as fortification which can greatly minimize hidden hunger as well. Healthier rice varieties have the potential to reach many people because it is widely grown and consumed. These healthier rice varieties can complement current strategies to reduce micronutrient deficiencies. Traditional rice also posses certain phytochemicals which remains hidden to a larger population. For example, Maappillai Samba, one of the traditional rice varieties of Tamil Nadu possess nearly 14 phytochemicals with medicinal value. Consuming the specific variety of rice can act medicine which can heal and prevent ailments (Sulochana & Singaravadivel, 2015)

The byproducts of rice also possess potential functional components. To discuss a few, rice bran contains an array of health enhancing phytochemicals. Rice bran contains high quality oil and protein; cholesterol-lowering waxes; anti-tumor compounds like rice bran saccharide; and antioxidants, including vitamin E and oryzanol; and anthocyanins in pigmented varieties (Ichikawa *et al.*, 2001, Abdel-Aal *et al.*, 2006 and Schramm *et al.*, 2007). Experimental and clinical evidence indicate that brown rice and bran oil reduce hypercholesterolemia and cardiovascular risk. Rice bran is anti-inflammatory and immunostimulatory. The monacolin-rich red yeast rice regulates hypercholesterolemia, and the GABA-rich germinated brown rice has chemopreventive effects. These therapeutic properties of rice reveal the scope of taking up rice as a base product to start an enterprise. With the alarming reports on poor eating habits and diseases, consumers have shown gradual shift from poor to healthy eating habits. This exactly is the right time for entrepreneurs to choose the best out of the available technologies in the market to cater the nutritionally hunger population.

Scope of entrepreneurship development in rice processing sector

Rice growing farmers and rice millers can be involved in value addition of the harvested produce by adopting technologies developed in rice processing by Indian Institute of Food Processing Technology, Thanjavur and NIFTEM, Kundli, Haryana, CFTRI, Mysore etc. People being aware of health changes are now interested in consuming healthy foods with therapeutic properties. Cultivation of traditional rice varieties and safe packaging for marketing will open good scope for the farmers to market the specific rice to the consumers. Farmers when encouraged to become entrepreneurs not only improve their livelihood but also causes tremendous change in the reduction of post harvest losses contributing to food security, long term availability of the produce/processed product and impart variety in the processed products available in the local market. Due to unawareness of proper drying techniques or unavailability of drying and storage areas farmers are forced to sell the harvested paddy to clear their debts. In such cases, rather than selling the harvested paddy for a minimum margin price to FCI/CWC/middle men/agents, if farmers can run/start their own mini modern rice mill, it will greatly help them to get good profit from milled paddy and its by-products through value addition. Gradual growth in the entrepreneurship may support the farmers to process variety of value added products in rice including traditional rice varieties with therapeutic properties enabling them to export high quality products thereby globalizing Indian foods.

Presently, Indian schemes like Start up India, Make in India are supporting new entrepreneurs by providing financial assistance. Under such cases, taking up a rice processing project can be very promising to the entrepreneurs. In addition, IIFPT also extends its support to new entrepreneurs by providing technical guidance in learning the technology of food processing. There are various products that can be processed with rice which can be started as a small enterprise. The following sections will clearly show the different value added products of rice, innovative products and the products that can be formulated with the by-products of rice.

Value added products of rice

Raw rice and parboiled rice are commonly consumed by different methods of cooking. Regular consumption creates monotony and hence value addition of rice provides variety of food products and increases an urge to taste the variety thereby attracting consumers. Value added products can be categorised as processed rice, fermented rice products, RTC rice mix and extruded rice products. The most commonly available value added products of rice are puffed rice, flaked rice, quick cooking rice, RTC puttu flour, idiappam, vermicelli, murukku mix, rice vadam, idli mix, dosa mix, rice starch and rice flour. Plate 1 shows the images of certain value added products of rice.

Ready to eat value added products from rice

Flaked rice/rice flakes/poha: Rice flakes or poha is an important breakfast item in semi-urban and rural areas and middle class families of urban India. Spicy or sweet preparations made from it are not only easy to make but can be made at a short notice as well. Therefore, it is extensively used all over the country round the year. Flaked rice is a very popular rice product in many Asian and rice consuming countries. More than 85% of ûaked rice in India is produced in the traditional tiny-scale production units. The steps involved in processing flaked rice are

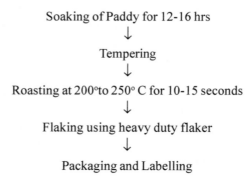

Soaking of Paddy for 12-16 hrs
↓
Tempering
↓
Roasting at 200°to 250° C for 10-15 seconds
↓
Flaking using heavy duty flaker
↓
Packaging and Labelling

Paddy is soaked for 12-16 hrs, tempered and roasted at 200°to 250°C for 10-15 seconds. The roasted paddy is flaked using edge runner (Mujoo & Ali, 1999) or heavy duty flaker after deshelling. The flakes are then packaged for labelling and marketing.

Equipment required

S.No	Name of equipment	Approximate cost (INR)	Purpose
1.	Urli roaster	30,000.00	Roasting soaked paddy
2.	Sand roaster	50,000.00	Roasting bulk quantity of samples
3.	Roller flaker	50,000.00	Flaking
4.	Edge runner	30,000.00	Flaking
5.	Rubber roll sheller	1,00,000.00	Dehusking
6.	Weighing balance	10,000.00	Measuring ingredients
7.	Sealing machine	15,000.00	Sealing packaged product

Puffed rice: Rice is soaked in salt solution and the soaked rice is heated in a pan at 250°C for 30-35 secs.

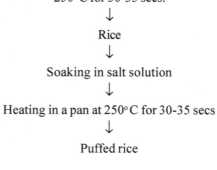

↓
Rice
↓
Soaking in salt solution
↓
Heating in a pan at 250°C for 30-35 secs
↓
Puffed rice

Equipment required

S.No	Name of equipment	Approximate cost (INR)	Purpose
1.	Puffing machine	1.5Lakhs	Puffing
2.	Weighing balance	10,000.00	Measuring ingredients
3.	Sealing machine	15,000.00	Sealing packaged product

Ready to cook value added products from rice

Quick cooking rice: Quick cooking rice is processed by cooking high moisture rice under pressure and drying.

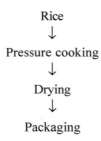

Rice
↓
Pressure cooking
↓
Drying
↓
Packaging

Equipment required

S.No	Name of equipment	Approximate cost (INR)	Purpose
1.	Pressure cooker	10,000.00	Cooking
2.	Tray drier	45,000.00	Drying
3.	Weighing balance	10,000.00	Measuring ingredients
4.	Sealing machine	15,000.00	Sealing packaged product

Rice idiappam: Idiappam is processed by making rice dough from steamed rice flour. The dough is then extruded using hand press and steamed, dried and packaged.

Steamed rice flour
↓
Rice dough
↓
Extruded using hand press
↓
Steaming
↓
Drying
↓
Packaging

Equipment required

S.No	Name of equipment	Approximate cost (INR)	Purpose
1.	Cottage level vermicelli machine	75,000.00	Processing vermicelli
2.	Pasta machine	5,00,000.00	Making vermicelli using dye
3.	Steaming unit	25,000.00	Steaming extruded sample
4.	Tray drier	45,000.00	Drying
5.	Weighing balance	10,000.00	Measuring ingredients
6.	Sealing machine	15,000.00	Sealing packaged product

Rice murukku: Rice flour is made into dough by adding oil, salt and sesame seeds to a required consistency. The dough is extruded through hand press into circular shapes and deep fry in oil. The fried murukku is cooled to room temperature and packaged.

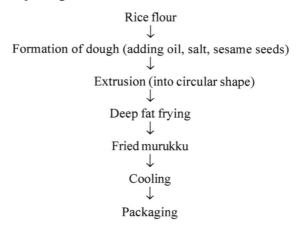

Rice flour
↓
Formation of dough (adding oil, salt, sesame seeds)
↓
Extrusion (into circular shape)
↓
Deep fat frying
↓
Fried murukku
↓
Cooling
↓
Packaging

Equipment required

S.No	Name of equipment	Approximate cost (INR)	Purpose
1.	Urli roaster	30,000.00	Roasting soaked paddy
2.	Mini flour mill	50,000.00	Pulverise rice
3.	Hand press extruder	300.00	Pressing the flour into desired shape
4.	Deep fat fryer	20,000.00	Frying the hand pressed flour
5.	Weighing balance	10,000.00	Measuring the ingredients
6.	Sealing machine	15,000.00	Sealing packaged product

Rice vadam: Rice is cooked to a thick paste. The paste is extruded through hand press and sundried for 6-7 hrs. The dried vadam is packaged and stored.

Rice
↓
Cooking to a thick paste
↓
Extrusion
↓
Sun drying for 6-7 hrs
↓
Dried vadam
↓
Packaged and Stored

Equipment required

S.No	Name of equipment	Approximate cost (INR)	Purpose
1.	Mixie	5,000.00	Grinding rice to make paste
2.	Mini flour mill	50,000.00	Pulverise rice
3.	Hand press extruder	300.00	Press the flour into desired shape
4.	Weighing balance	10,000.00	Measuring ingredients
5.	Sealing machine	15,000.00	Sealing packaged product

Rice starch: Rice starch is used as a thickening agent in food preparation, including infant formula.

Rice flour: Raw rice is soaked, dried and ground in flour mill and used for processing snack items like murukku and adhirasam.

Equipment details

S.No	Name of equipment	Approximate cost (INR)	Purpose
1.	Urli roaster	30,000.00	Roasting soaked paddy
2.	Mini flour mill	50,000.00	Pulverise rice into flour
3.	Weighing balance	10,000.00	Measuring ingredients
4.	Sealing machine	15,000.00	Sealing packaged product

Fermented rice based products

Idli/Dosa: Among the fermented foods of India, idli, a fermented steamed product with a soft and spongy texture is a highly popular and widely consumed breakfast food especially in Southern India (Durgadevi and Shetty, 2014). Idli is prepared by wet grinding soaked rice and black gram dhal in the ratio of 3:1 or 4:1 and allowing natural fermentation overnight. The fermented batter is steam cooked in idli vessel. The shelf life of Idli as well as batter is only one day. Moreover, the quality of idli will vary and it is a tedious process. Hence, standardization including microbial processes for making quality idli/dosa will have high commercial value.

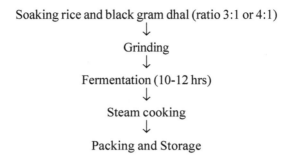

Soaking rice and black gram dhal (ratio 3:1 or 4:1)
↓
Grinding
↓
Fermentation (10-12 hrs)
↓
Steam cooking
↓
Packing and Storage

Equipment required

S.No.	Name of equipment	Appxt. cost (INR)	Purpose
1.	Wet grinder (table top)	4,000.00	Grind the soaked samples
2.	Weighing balance	10,000.00	Measuring the ingredients
3.	Sealing machine	15,000.00	Sealing packaged product

New technologies/innovations in value addition of rice

New technologies/innovations in value addition of rice have enabled consumers to minimise their processing time and thereby the cooking time of rice. The innovation has made even the hard rice which takes longer cooking time into ready to eat item. For example, Rakthashali rice, a traditional Indian rice variety takes more than one hour for cooking to make the rice edible. But, processing of rakthashali variety into flakes has made it into RTE and vermicelli into RTC. Technology has developed in a way that idli which takes nearly 16 hours of processing time, is now available as RTC form like dry idli mix and dehydrated idli which only requires rehydration through steaming. Rice which is generally considered as a high calorie diet is now been available as low glycemic rice which is recommended for diabetics. Micronutrient enriched rice based value added products like idli and vermicelli are also available and can be processed in commercial level. Given below are few of the new technologies /innovative rice based products which can be commercialised to target groups and to the public. Emerging entrepreneurs can involve themselves in processing these innovative products listed below which will gradually gain customer attraction and satisfaction.

Flaked Rice	Puffed rice
Rice idiappam	Rice muruku
Idli	Dosa

Plate 1: Value added products from rice

Fortified/traditional rice vermicelli

Vermicelli is processed from traditional rice, rakthashali rice and also micronutrient enriched/fortified rice flour are used for processing beta carotene enriched rice vermicelli.

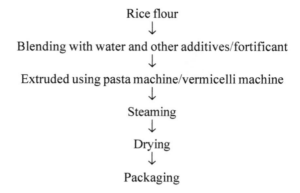

Rice flour
↓
Blending with water and other additives/fortificant
↓
Extruded using pasta machine/vermicelli machine
↓
Steaming
↓
Drying
↓
Packaging

Ready to eat innovative rice based products

- Millets incorporated murukku, idli and dosa mix
- RTE variety rice
- RTE idiappam and puttu
- Flavored rice flakes (sweet/spicy coated)
- Rakthashali rice flakes
- RTE sambar idli

The above listed innovative products consume no time and have a great demand in the market. The people who travel abroad purchase the RTE variety rice and the RTE sambar idli packaged in retort pouches. This makes the people feel at home.

Ready to cook innovative rice based products

- RTC idli batter
- Dehydrated idli
- Instant idli dry mix
- Fortified rice/ rice analogue
- Gamma irradiated rice (Bao *et al.*,2001)
- Low glycemic rice

The ready to cook innovative products reduces the pre processing and cooking time.

Development of Instant idly mix

Instant idly dry mix with liquid culture: Standardized dry form of idly mix with pure identified liquid microbial culture to ferment in 10-12 hours at 28-30°C. Idli prepared from idli dry mix has good texture, smell and taste like idli prepared by wet grinding.

Instant idly dry mix with dry form culture: Instant Idli dry mix technology developed with dry form of patented culture, ferment the dry mix within 2 hours by just mixing the dry mix with two times of water. Shelf life of instant mix is six months. Quality of idly from dry mix is found to be good with soft texture and acceptable quality.

Instant idly dry mix with chemical method: Standardized instant idly dry mix using 1.0% citric acid and 1.5% sodium bicarbonate to prepare idli batter within half an hour.

Ready to serve idli

Ready to serve idly was prepared by steaming the idli batter and packing in HDPE or retort pouches. The shelf life of ready to eat idly is more than three months at atmospheric temperature (28± 2°C)

- Along with ready to eat idly, instant sambar powder mix and coconut chutney powder were also developed as side dishes. Chutney can be served just mixing with water and sambar is prepared by boiling the instant sambar mix with 1:1 ratio of water for five minutes.

- Nutri idli and spicy mini idly with shelf life of more than 3 months was developed from carrot, coriander leaves, curry leaves, ginger, moringa leaves and modakathan leaves. The nutri idly is prepared by adding extracts.

- Sambar idly as ready to eat form is developed with a shelf life of more than 3 months at 15°C.

Initiating a food industry on any of the above said technologies as single or combined will give a new opening to any emerging entrepreneur. The scientific training on learning these technologies are given at Indian Institute of Food Processing Technology through scheduled trainings or as consultancy services to the interested clients on payment basis. The scheduled trainings and its fee structures are available at www.iifpt.edu.in

Success stories in rice processing sector

With the increasing demand in the rice based RTE and RTC, emerging entrepreneurs in rice processing and value addition will have a great scope to extend their market. IIFPT Food Processing Business Incubation Centres under Ministry of Food Processing Industries is involved in offering skill development training and executive trainings equipping the trainees to start their own food processing facilities and aids in technical guidance to run an enterprise. The programmes are designed in a way that would help the people belonging to different strata. With a vision to promote entrepreneurs, programmes offered ranges from the cottage level techniques which can be replicated at home for the budding entrepreneurs to an industrial level for the established entrepreneurs. Some of the successful entrepreneurs from IIFPT Food processing Business Incubation centre are listed in Table 1. Similarly interested entrepreneurs, farmers and rice millers can undertake training and involve in processing value added food products from rice.

Case Study 1: Mrs. Rajeshwari Ravikumar, who had taken up training at IIFPT, established her own range of products through her brand "Suka Foods". She developed innovative products with a good combination of traditional rice varieties and millets as health mixes and variety of instant mixes. Presently, she is running her processing unit with 20 women at Thivaiyaru, Tamil Nadu. She has also won "Srishti Saman" award during the festival of innovation for the year 2015 at Rakshtripathi Bhavan.

Case Study 2: Mrs. R. Rajalakshmi, Proprietor of Ammirdham traditional foods process puttu mix from traditional rice varieties namely mappilai samba, karunguruvai, kattuyanam, poogar rice and thuyamalli. The company was started with the intention to make people consume traditional rice as people are unaware of the ill effects of polished rice consumed regularly. Hard work and perseverance improved the business gradually. Initially, only mappilai samba puttu mix was manufactured. Later, as their product received recognition among the public, the company introduced other rice varieties too. Currently, from each variety of rice one ton of puttu mix is processed every month. The company now possess its own mini flour mill for pulverising rice.

Table 1: Successful entrepreneurs of IIFPT-FPBIC in rice and rice products processing

S.No	Name of the entrepreneur	Company name	Product and quantity
1.	Mrs.Rajeshwari Ravikumar	Suka diet natural foodsjaya sritec@yahoo.com	Red rice puttu mix 0.5 tons/day
2.	Mr.PL.A. Chiambaram	PL. A. hospitality foods	Idli/dosa batter 2500kg/day
3.	Mrs.R.Rajalakshmi	Ammirdham traditional foods	Traditional rice puttu mix 60 kg/month
4.	Mr.T.Chinnappan	Aravind chettinad snacks	Murukku 1.2 tons/day
5.	Mr.Kannan (Chef)	H2H/Health to Happiness	Kauni rice, hand pound rice 600 kg/day
6.	Mrs.Ananthi Elango	Adhisurya foods	Dosa mix/idli mix 0.05 tons/day

Scope of by-product utilization/effluent treatment/ utilization

Pharmacologically relevant compounds could be extracted from rice by-products, providing an economic boost to rice farming and processing. Broken rice, husk, bran and germ are the four major by-products of rice milling industry. By effective utilization of by-products, one can increase the revenue in rice milling industry.

Rice bran as a source of oil

The brown rice grain contains only about 2.5 - 3% fat. Bulk of the grain oil is in a thin layer within the outer parts of brown rice. Pure rice bran contains not less than 15% and even up to 25% oil. The bulk of the oil of rice grains (50-80% of total oil) comes into the bran during its polishing. By effective utilisation of rice bran as a source of oil, the import of edible oil can be reduced considerably.

Rice bran and its utilization

Rice bran in common consists of seed coat, aleurone and sub-aleurone layers of the kernel with part of the germ and a small portion of the starchy endosperm. Considering the proteins, minerals and vitamins in rice bran, the much needed nutrients can be made available for the society by processing or blending and converting it into suitable forms (Premakumari *et al.*, 2013). The nature and composition of bran obtained from rice depends on the system of milling and on any pre-treatment given to the paddy. When paddy is milled by single hullers, a mixture of powdered husk and bran is obtained. Hence, this bran contains very low amounts of oil and protein and high amounts of fibre and ash. It has very little industrial use.

When paddy is milled in sheller mills, it is first dehusked and then polished in a polisher. The main rice bran of industrial importance is the true bran, that is, the outer layer of brown rice separated during its polishing. Bran composition is affected by the degree of milling. As the degree of milling increases beyond 5%, the amount of oil, fibre and ash in it decreases and that of starch increases gradually. Rice bran is the most valuable by-product of rice milling industry. It contains 12-15% protein; 14-20% oil if the paddy is raw and 18-25% if parboiled and is rich in B-Vitamins.

Extraction of rice bran oil: Rice oil is extracted from bran by using solvent hexane in solvent extraction plants. Low FFA oil is used for edible purpose after suitable refining. High FFA oil is used for soap manufacture. The deoiled rice bran is a good source of protein and used as an ingredient in animal feed.

Development of stabilized rice bran: Milled rice bran obtained from a local milling factory is immediately stabilized by heating in an oven at 110°C for 10 minutes. Subsequent to heating, the sample is removed from the oven, cooled to room Temperature (25°C). Stabilized rice bran is milled into flour. The flour is screened through a 30 mesh sieve.

Extraction of protein isolate from rice bran: Rice bran obtained from rice millers are sieved and stabilized using microwave/autoclave and defatted. The protein present in the bran can serve as a high quality human protein supplement.

Broken rice and its utilization

Nearly 5 - 50 per cent of the rice grains may break during milling, the medium and large brokens are mixed with head rice in trade practice. Therefore, only the very small brokens are separated during grading of milled rice. In India, rice brokens are used for preparation of breakfast dishes. Traditionally, broken rice is used in starch manufacturing and brewing. An alcoholic beverage is also prepared from rice brokens in certain parts of Andhra Pradesh and in Nepal. Brokens are processed into flour, semolina and noodles, raw or pre-gelatinized. Extruded ready to cook infant foods and snacks are commercially available. Major constraints to these uses are microorganisms and insects and the tendency of the fat to deteriorate.

Flow diagrams for the processing of rice bran cookies and rice bran muffins are given below with the details of required equipment.

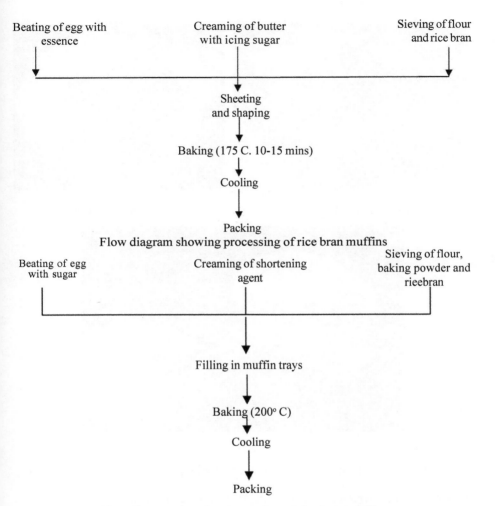

Flow diagram showing processing of rice bran muffins

Flow diagram showing processing of rice bran muffins

Equipment required

S.No	Name of equipment	Approximate cost (INR)	Purpose
1.	Planetary mixer	35,000.00	Mixing/flour dough /beating eggs
2.	Muffin tray per piece	500.00	baking muffins
3.	Cookies cutter set	500.00	Cutting cookies in desired shape
4.	Rotary oven	1,20,000.00	Baking
5.	Weighing balance	10,000.00	Measuring ingredients
6.	Sealing machine	15,000.00	Sealing packaged product

Rice bran muffins   Rice bran Flan

Rice milk

Plate 2. Convenient food products developed from by-products of rice milling industry

Processing of extruded foods from broken rice: Rice bran or brokens are mixed with wheat flour or millet flour and are extruded through a pasta machine using dye to get pasta of required shapes.

Parboiled rice bran flour + wheat flour
↓
Sieving
↓
Addition of water
↓
Feeding in extruder
↓
Extrusion
↓
Drying
↓
Packing
↓
Sealing, Labelling and storing

Flow diagram showing processing of pasta from rice bran/rice broken

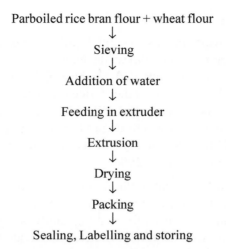

Plate 3. Broken rice and rice bran incorporated pasta products

Rice germ and its utilization

Rice germ is characteristically richer in protein and fat, but lower in fibre than the bran. It is extremely small and is located on the ventral side of the caryopsis. It has 2 - 3% weight of brown rice. About one third of brown rice lipid is in the germ. High quality protein, edible oil and vitamins are the most valuable components of rice germ, but it contains impurities, natural toxicants and allergens besides heavy load of microorganisms exceeding several million per gram (Barber and Benedito de Barber, 1980). Any attempt in utilizing germ for food preparation should involve steps in quick separation, moist heat stabilization, defatting and hygienic handling. Purified germ can be used for the preparation of milk, snacks and sweets. Plate 2 and Plate 3 shows the photos of products developed from rice by-products.

Effluent treatment

Rice mill generates waste mainly in solid, liquid and gaseous forms. Husk and ash constitute the main bulk of solid waste. However, husk is reused in the mill for boiling water. So, residues left after combustion is ash. Gaseous emission is mainly CO, SO_2 and NO_2. Besides these, rice mill generate liquid waste from parboiling process, soaking paddy and boiler blow down. The liquid waste from these operations ultimately passes through a common drain to the outside of the mill boundaries and this is commonly called as combined effluent or combined waste water (Pradhan and Sahu, 2004).

To reduce the rice mill effluent pollution problem, a treatment technique has been developed. The treatment involves three steps viz. aeration, settling and filtration. In aeration, water used for soaking paddy is collected in a tank and aerated using aerator. After aeration, the effluent is treated with alum, to settle all the turbid materials. In the third stage, the clear supernatant effluent is filtered through a battery of black ash column. The filtered effluent is clear without any turbidity and colour. The off-smell is removed and the organic matter and BOD, COD content were reduced to safer limit of 100 mg/L. The acid pH has to be increased to neutral. The treated effluent can be mixed with water and can be used for irrigation. The effluent treatment involves higher initial and running cost and thus increasing the rice production cost. In order to maintain the pollution free environment, these measures should be adopted in rice mills in the national interest (Technology for rice mill effluent treatment-3083/CHE/2008).

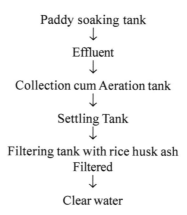

Paddy soaking tank
↓
Effluent
↓
Collection cum Aeration tank
↓
Settling Tank
↓
Filtering tank with rice husk ash
Filtered
↓
Clear water

Flow diagram for effluent treatment

Safety and quality aspects of rice products and future scope

For the safety of any food product, it is necessary to use the Farm to Fork approach. This means that the safety of rice products begins with the cultivation practices involved in paddy and has to maintain until consumption. The moisture content of harvested paddy should be strictly regulated to improve the out turn ration of the paddy. When left unnoticed, the higher moisture content in the paddy leads to increased probability of fungal growth leaving the stored paddy with mycotoxin contamination which ultimately affects the quality of the end product.

Improper storage facility, poor pest management, high fluctuations in storage conditions like temperature and humidity also affects the paddy/rice quality. During processing of food products, HACCP has to be strictly practised to ensure safety and consistency in quality.

In India it is mandatory for the food business owners and food handlers to be registered under FSSAI. Proper certification for the finished products has to be obtained from APEDA and to get the product inspected by Quality Inspection Council in case of export of the products. Besides following the norms of FSSAI, one has to abide by regulations of the country to which the products are exported to.

Packaging of rice products

The main aim of packaging is to keep the product free from contamination (Robertson, 2010) and protecting the quality of the product thereby the shelf life gets extended. The processed foods when exposed to atmospheric air are prone to microorganisms and cross contamination, thereby deteriorating the quality of the product. Packaging helps in extending shelf life and keeping quality of the product. Choosing right packaging material will attract consumers to purchase the product. Packaging materials like LDPE gives transparency to products like murukku, idiappam and vermicelli. Aluminium foils are not transparent but helps in wrapping the foods like idli/dosa. Carton boxes can be used for packaging crispy products like flakes. The product details can be labelled over the packaging materials. The packaging designs, colour and the wrapping can be improvised at regular time period to draw the attention of the consumers.

Conclusion

Food processing sector has a wide range of opportunities to entrepreneurs which not only provides employment but aids in promoting agricultural yield, productivity and the standard of living of the rural population (Negi, 2013). India being the second largest producer of food after China in the world has the potential to become the food bowl for the world. Processing and packaging of RTE and RTC foods is now the emerging field which requires attention of entrepreneurs. This food processing sector plays a major role in value addition, enhancing shelf life of agricultural produce and creating market for export of processed foods. Choosing the right food to be processed based on the availability of raw materials round the year is the first and foremost step in starting the enterprise. Secondly, undertaking suitable training in processing, packaging and labelling to improve the confidence level of the entrepreneurs in commercialization of their products is necessary. Focussing on one particular product round the year may discourage the entrepreneur. In such cases, the entrepreneur can choose a processing line where the same equipment can be used for processing two to three products. Perseverance is the key to a successful entrepreneur. Turning all the challenges into opportunities in this food processing field will enable the entrepreneur to gain good experience in running the business successfully. This chapter has thrown light on the available processing line with rice as a major raw material. The major advantage of developing a food processing business using rice is that it is the staple crop and it is consumed by 60 per cent of the total population. Running an enterprise in rice processing will help to supply the demand of the growing population.

References

Abdel-Aal., J. Young & I. Rabalski. 2006. Anthocyanin Composition in Black, Blue, Pink, Purple, and Red Cereal Grains, *J. Agric. Food Chem.*, 54: 4696-4704.

Bao, J., Q. Shu., Y. Xia., C. Bergman & A. Mcclung. 2001. Effects of Gamma Irradiation on Aspects of Milled Rice (*Oryza sativa*) End-Use Quality, *J. Food Qual.*, 24: 327–336. Doi:10.1111/J.1745-4557.2001.Tb00612.X.

Barber, S. & C. Benedito de Barber. 1980. Rice Bran Chemistry and Technology, *Rice: Production and Utilization*, West Port, AV publishing Co., pp 790-862.

Burlando, B. & L. Cornara. 2014. Therapeutic Properties of Rice Constituents and Derivatives (*Oryza sativa* L.): A Review Update , Trends *Food Sci. Technol.*, 40(1): pp 82–98.

Durgadevi, M. & P.H. Shetty. 2014. Effect of Ingredients on Sensory Profile of Idli. *J. Food Sci. Technol.*, 51(9): 1773–1783.

Ichikawa,H., T. Ichiyanagi., B. Xu., Y. Yoshii., M. Nakajima & T. Konishi. 2001. Antioxidant Activity of Anthocyanin Extract from Purple Black Rice, *J. Med. Food.* 4: 211-218.

India Grain and Feed Annual. 2016. USDA Foreign Agricultural Service, *Grain Report Number*: IN6033. 16.

Mujoo, R. & S. Z. Ali. 1999. Molecular Degradation of Rice Starch During Processing to Flakes, *J. Science Food and Agriculture,* 79(7): 941- 949.

Nam, S., S. Choi., M. Kang., N. Kozukue & M. Friedman. 2008. Antioxidative, Antimutagenic, and Anticarcinogenic Activities of Rice Bran Extracts in Chemical Tests and in Cell Cultures. *J. Agric. Food Chem.*, 53: 516-822.

Negi, S. 2013. Sustainability: Ecology, Economy & Ethics, *Food Processing Entrepreneurship for Rural Development: Drivers And Challenges*, Tata McGraw Hill Education, New Delhi, 186-197.

Pradhan, A. & S.K. Sahu. 2004. Process Details and Effluent Characteristics of a Rice Mill in the Sambalpur District of Orissa, *J. Industrial Pollution Control*. 20 (1): 111-124.

Premakumari, S., R. Balasasirekha., K. Gomathi., S. Supriya., K. Alagusundram & R. Jaganmohan. 2013. Evaluation of Organoleptic Properties and Glycemic Index of Recipes with Rice Bran. *Int. J. Cur. Res. Rev.*, 5 (5): 1-8.

Robertson, G. L. 2010. *Food Packaging and Shelf Life*, Taylor and Francis Group, LLC, Brisbane, Australia.

Schramm, R., A. Abadie., N. Hua., Z. Xu & M. Lima. 2007. Fractionation of the Rice Bran Layer and Quantification of Vitamin E, Oryzanol, Protein, and Rice Bran Saccharide, *J. Biological Engineering*, 1(9): doi:10.1186/1754-1611-1-9.

Sulochana, S. & K. Singaravadivel. 2015. A Study on Phytochemical Evaluation of Traditional Rice Variety of Tamil Nadu-'Maappillai Samba' by GC-MS, *Int. J. Pharm. Bio. Sci.*, 6(3): 606 – 611.

Technology for rice mill effluent treatment – Indian Patent number 3083/CHE/2008.

Trijatmiko K. R., C. Dueñas., N. Tsakirpaloglou., L. Torrizo., F. MaeArines., CherylAdeva., J. Balindong., N. Oliva., M. V. Sapasap., J. Borrero., J. Rey., P. Francisco., A. Nelson., H. Nakanishi., E. Lombi., EladTako., R P. Glahn., J. Stangoulis., P. C. Mohanty., A. A. T. Johnson., JoeTohme., G. Barry & I.H. Slamet-Loedin. 2016. Biofortified Indica Rice Attains Iron and Zinc Nutrition Dietary Targets in the Field. *Scientific Reports 6*, Article number: 19792. doi:10.1038/srep19792.

7

An Insight to Entrepreneurial Opportunities in Millet Processing

Udaykumar Nidoni, Mouneshwari K, Ambrish G, Mathad P F Shruthi V H & Anupama C

Introduction

21st century challenges like climate change, water scarcity, increasing world population, rising food prices, and other socio-economic impacts are expected to generate a great threat to agriculture and food security worldwide, especially for the poorest people who live in arid and sub-arid regions. Millet is one of the oldest foods known to humans and possibly the first cereal grain to be used for domestic purposes. Millet has been used in Africa and India as a staple food for thousands of years. Typical grain texture and hard seed coat of millets increase their keeping quality but make them difficult to process as well as to cook in convenient form. Absence of appropriate primary processing technologies to prepare ready-to-use or ready-to-cook (RTC) products and also the secondary as well as tertiary processing technologies to prepare ready-to-eat value added products from millets are the major limiting factors for their diversified food uses.

Millets have relatively poor digestibility and low bio-availability of minerals due to the presence of inherent anti-nutritional factors. The difficulties in millet grain processing present a challenge but nutritional as well as health benefits and consumer demand for health foods provide opportunities in processing, develop suitable technology for newer products and process mechanization. This change in technology and consumer food preference would help in increasing the area under millets, maintaining ecological balance, ensuring food security, prevent malnutrition and increase the scope for utilization of millet grains on industrial scale.

India ranks first in global consumption and eleventh in per capita consumption of millets. The small millets include six main grain crops namely, finger millet (*Eleusine coracana*), kodo millet (*Paspalum scrobiculatum*), little millet

(*Panicum sumatrense*), foxtail millet (*Setaria italica*), proso millet (*Panicum miliaceum*), and barnyard millet *(Echinochloa frumantacea)* (Singh *et al.*, 2015). The small millets, in addition to nutritional benefits are rich in phytochemicals with health benefits. Dietary fibre protects against hyperglycemia, phytates against oxidation stress and some phenolics and tannins act as antioxidants. Higher antioxidant activity in the phenolic extracts of kodo millet and finger millet than in other millets and cereals has been reported. Small millets have potential benefits to mitigate or delay the onset of complications associated with diabetes. Thus, these grains are considered as nutritious grains (Oelke *et al.*, 1990).

Most of the millets are grown in different regions of the world from east to west. The world total production of millet grain was 7,62,712 metric tons and India tops the ranking with a production of 3,34,500 tons in 2010 (FAO, 2012).

The consumption pattern of minor millets varies from region to region. The people in Himalayan foothills of Uttarakhand use millet as a cereal, in soups, and to make Chapathi. Flat thin cakes called 'Roti' are often made from millet flour and used as the basis for meals. 'Kodo ko jaanr' is the most common fermented alcoholic beverage prepared from dry seeds of finger millet in the Eastern Himalayan regions of the Darjeeling hills and Sikkim. 'Chyang' is also a fermented finger millet beverage popular in Ladakh region. The traditional, naturally fermented finger millet product is called 'Ambali'. Finger millet is the cereal of choice for the preparation of porridges for children and for the sick and old. Germinated finger millet is used to make weaning foods for infants. Finger millet is used for the preparation of popped products. The tribal people in Kumaon hills of Uttarakhand consume weaning food containing malted foxtail millet flour and malted barnyard millet flour. Mostly, minor millets are consumed by the economically weaker sections of the population in India.

Nutritional importance of millets

Millets are unique among the cereals because of their richness in calcium, dietary fibre, polyphenols and protein (Devi *et al.*, 2011). Millets contain high amount of lecithin and is excellent for strengthening the nervous system. Millets are rich in B vitamins, especially niacin, Vitamin B_6 and folic acid. It contains minerals like calcium, iron, potassium, magnesium, phosphorous and zinc. Millet is relished for its nutritional value, being a rich source of carbohydrates. Millets contain no gluten, so they are not suitable for raised bread, but they are good for people who are gluten-intolerant (Hulse *et al.*, 1980). Millets generally contain significant amounts of essential amino acids particularly the sulphur containing amino acids (methionine and cysteine); they are also higher in fat content than maize, rice, and sorghum (Obilana and Manyasa, 2002). Nutritive value of millets in comparison with rice and wheat are given in Table 1.

Table 1: Nutritive composition of millets (per 100 g edible portion)

Food grains	Protein(g)	Fat (g)	Ash (g)	Fibre(g)	Carbo-hydrate (g)	Energy (kcal)	Ca(mg)	Fe(mg)	Thiamin (mg)	Riboflavin (mg)	Niacin (mg)
Rice	7.9	2.7	1.3	1.0	76.0	362	33	1.8	0.41	0.04	4.3
Wheat	11.6	2.0	1.6	2.0	71.0	348	30	3.5	0.41	0.10	5.1
Pearl millet	11.80	4.8	2.2	2.3	67.0	363	42	11.0	0.38	0.21	2.8
Finger millet	7.7	1.5	2.6	3.6	72.6	336	350	3.9	0.42	0.19	1.1
Foxtail millet	11.2	4.0	3.3	6.7	63.2	351	31	2.8	0.59	0.11	3.2
Proso millet	12.5	3.5	3.1	5.2	63.8	364	8	2.9	0.41	0.28	4.5
Little millet	9.7	5.2	5.4	7.6	60.9	329	17	9.3	0.30	0.09	3.2
Barnyard millet	11.0	3.9	4.5	13.6	55.0	300	22	18.6	0.33	0.10	4.2
Kodo millet	9.8	3.6	3.3	5.2	66.6	353	35	1.7	0.15	0.09	2.0

Source: FAO, 1995

Health benefits of millets

Lignans, an essential phytonutrient present in millet, act against different types of hormone-dependent cancers like breast cancer and also help to reduce the risk of heart disease. Regular consumption of millet is very beneficial for postmenopausal women suffering from signs of cardiovascular diseases, like high blood pressure and high cholesterol levels (Shahidi and Chandrasekara, 2013). This form of cereal grain is very high in phosphorus content, which plays a vital role in maintaining the cell structure. Phosphorus helps in the formation of mineral matrix of bone and is also an essential component of ATP (adenosine tri-phosphate), which is the energy currency of the body. A single cup of millet provides around 24% of body's daily phosphorus requirement. It is a very important constituent of nucleic acids, which are the building blocks of genetic code (Lakshmi and Sumathi, 2002).

Millets are associated with reduced risk of type 2 diabetes mellitus mainly due to the high content of magnesium which acts as a co-factor in enzymatic reactions in the body and regulate the secretion of glucose and insulin. Magnesium is also beneficial in reducing the frequency of migraine attack and useful for people who are suffering from atherosclerosis and diabetic heart disease.

Despite their nutritional superiority, the utilization of millets is restricted due to non-availability of refined and processed millets in ready to eat form. Hence, millets are confined to traditional consumers and to the people of lower economic strata. The reasons for limited utilization of millets are poor grain quality characteristics, such as rough texture, high fibre content, lack of gluten and typical flavour (Shankaran, 1994).

Processing of millets

The rural people are still dependent on manual methods of harvesting millet crops, bullock treading, storage in mud bins and gunny bags and milling by manual chakkis (Dandsena and Banik, 2016). The common household practises of processing millets include hand pounding, germination or sprouting, malting, fermentation and cooking. Each of these processes qualitatively modifies the nutritive value of the food (Singh and Raghuvanshi, 2012).

Pre-milling treatments for millets

Acid treatment

Decorticated grains are treated with mild organic acids, such as acetic, fumaric or tartaric and also with the extracts of natural acidic materials like tamarind. Acid treatment reduces polyphenols, phytic acid and other anti-nutritional factors (Hadimani and Malleshi, 1993).

Dry heat treatment

Lipase activity is the major cause of spoilage of millet meal, so its inactivation by dry heat treatment before milling improves the meal quality. Dry heat treatment is carried out in a hot air oven at 100±2°C for 60 minutes and then cooling to room temperature (Palande *et al.*, 1996).

Parboiling

Parboiling is a hydrothermal treatment which helps in enhancing the grain hardness and improves milling characteristics. Parboiled grains decorticate more efficiently in removing the germ and the pericarp. Parboiled decorticated grains have slightly lower protein digestibility than the raw grains decorticated to the same extent. The parboiled grains are used in the preparation of various snack items. Parboiled grains are also cooked to produce rice-like products.

Milling of millets

Traditional method of pounding is usually applied to decorticate millet grains partially or completely before further processing and consumption. Millets are milled either by using a hammer mill or a roller mill. The flour produced using a hammer mill has large particle size and is not uniform. Whole grains are directly dry-milled to give a range of products *viz.*, broken or cracked grains, grits, coarse meal and fine flour. The flour thus obtained is used in the preparation of variety of simple to complex food products. They can also be mixed with other flours to form composite flour for soft and stiff porridges. Flow chart for processing of millets is given in Fig. 1 (Bangu *et al.*, 2000).

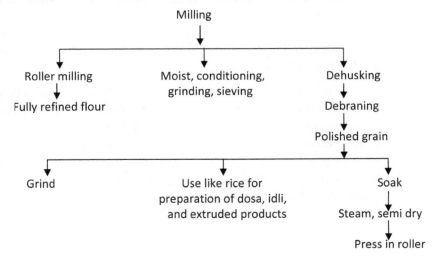

Fig. 1: Flow chart for processing of millets
(*Source:* Verma and Mishra, 2010)

Traditional milling equipment

The simplest type of processing is to grind the whole grain in a stone hand mill, a village stone mill or a hammer mill driven by diesel engine. Many consumers decorticate (dehull or dehusk) the kernel before grinding it into various particle sizes for use in different products. Millets are usually decorticated by washing the grain in clean water. The water is strained and the grains are crushed using a stone mortar and wooden pestle. The bran is removed by washing or winnowing the sun-dried crushed material (Kajuna, 2001). It is usually milled daily in quantities of 2-3 kg. Water is added to moisten the pericarp, and facilitate bran removal. The moisture often promotes fermentation and growth of microorganisms which affect the keeping quality of the products (Rooney and McDonough, 1987).

In India, women use mortars made of stone and pestle made from wood to decorticate the millets. The stone mortar is sometimes fixed in the ground and appears as a pit or hole in the ground. Usually, an amount of grain less than 1kg is washed and placed in mortar, and moistened to soften the bran. Millet grains are pounded vigorously for 5 minutes and the bran obtained after pounding is washed, and the clean endosperm recovered is left for sun drying (Kajuna, 2001).

Improved milling equipment

Due to the small grain size, milling is often more complex and tedious with millets. Decortication is sometimes accomplished by using rice dehullers or other abrasive dehullers. A hammer mill is used for milling of millet into flour. Some machines equipped with carborundum stones are used to polish the grains. The key raw material characters for grain milling quality are the size, form and structure of the seed, including the development of its outer (bran) layers and the endosperm hardness (Kajuna, 2001).

Recently, improved millet deshusking machines namely, rubber roll sheller type, single stage and double stage centrifugal type dehuskers (Plate 1) having capacity of 100 kg to 1000 kg per hour have been introduced. The dehusking efficiency of these machines varies from 80-95%. The dehusking of the remaining grains is accomplished by pearling. The supporting equipment namely, cleaner cum grader, destoner and pearler (Plate 2) of matching capacities are used to produce high quality dehusked millets.

Plate 1: Rubber roll sheller type, single stage and double stage centrifugal type millet dehusker (from left to right)

Plate 2: Cleaner cum grader, destoner and pearler (from left to right) suitable for millet processing

Production of millet flour

When millet grains are ground by using hammer mill or pestle and mortar into flour (without addition of water) and winnowed, sifted and reground and sieved through a 100 mesh sieve, the yield of meal is 70 per cent and husk & bran is 28-30 per cent (Hulse *et al.*, 1980). Fig. 2 shows a process flow chart for milling millets into flour.

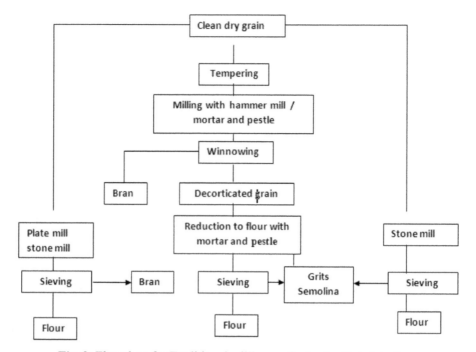

Fig. 2: Flowchart for Traditional milling methods of millets into flour

Packaging and storage of processed millets

Scientific attention has not been given to the storage of millets despite their relative importance as a staple food in many countries. The other notable reason is that farmers in the arid and semi-arid countries where millets are grown achieve quite impressive performance in grain storage by employing relatively simple traditional methods such as sealed storage drum, mud straw bins and earthenware pot and jar. Underground storage of millet has been reported in different countries like Somalia and Sudan (Kajuna, 2001).

Most of the millets have excellent storage properties as the grains are harvested and stored in dry weather conditions. The millets are kept up to 4-5 years in simple granaries as the grains are protected from insect attack by the hard hull covering the endosperm.

Millets are milled to remove the inedible husk or outer bran, prior to cooking. Simple milling process might trigger oxidative rancidity and development of off flavour during storage of both grains and flour, so it is best to grind the flour right before it is to be used. The grain should be stored in tightly closed containers, in a cool dry place with a temperature of less than 40°C in ambient or in the refrigerated condition. The packaging of dehusked millets in polyethylene teraphthalate (PET) or gunny bags with polyethylene lining are recommended for safe storage.

Millet based value added products

Millet grains are customarily milled before being cooked. Dry milling embraces a wide range of technologies from simple grinding of the whole seed between stones or in a pestle and mortar to the complex continuous system of precision rollers. Once millet has been processed (milled) into flour, the latter can further be processed into various secondary products. The details of value added millet based products are given below.

Thick & thin porridges

The most common and simple food prepared from millets is porridge. Thick (stiff) porridges are consumed in almost all countries where millets are cultivated. Thin (soft) porridges are also a simple food product from millets. The basic difference between thick and thin porridges is the concentration of flour. Generally, thick porridges are solid, and can be eaten with the hand, while thin porridges are fluid and can be drunk from a cup, or eaten by a spoon. The preparation of thick porridge entails adding flour to boiling water in increments accompanied by vigorous stirring. The flour is cooked until it forms a thick, homogeneous and well gelatinized mass devoid of lumps. The flour is soured and fermented for at least 18 hours before cooking, resulting in a fermented porridge. Flow chart of preparing thick porridge from millets is illustrated in Fig. 3.

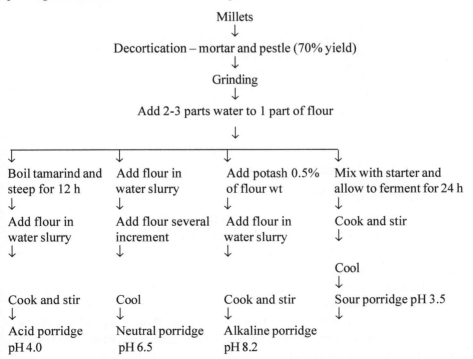

Fig. 3: Flow chart for producing thick porridge from millets (*Source:* Malleshi, 1993)

The millets are also used for the preparation of various value added products (Plate. 3) *viz.,* malted porridge, popped millet, millet bread, muffins, biscuits, extruded snacks, beer flakes and grits.

Malted porridge	Popped millet	Millet bread
Millet muffins	Extruded millet	Millet biscuit
Millet flakes	Millet grits	Beer

Plate 3: Millet based value added products

Value addition in small millets

Traditionally, dry, moistened or wet grain is normally pounded with a wooden pestle in a wooden or stone mortar. Moistening the grain by adding about 10% water facilitates not only the removal of fibrous bran, but also the separation of germ and endosperm, if desired. Although this practice produces slightly moist flour, parboiling increases the dehusking efficiency of kodo millet and eliminate the stickiness in cooked finger millet porridge. The polished grain called 'millet rice' is either used directly or further milled in plate or hammer mill to semolina or flour. These processing add value to these millets three to four-fold and make them acceptable to the elite urban consumers as niche food or health food (Dandsena and Banik, 2016).

Kodo millet

Kodo millet is a nutritious grain and a good substitute to rice or wheat. The fibre content of the whole grain is very high.

Traditionally, it is cooked like rice and a fermented beverage called 'Landa'. Kodo millet is ground into flour and used to make pudding. Two commercial products such as noodles, rusk and two homemade recipes such as dosa and chapathi can also be prepared by using composite flour comprising major portion of kodo millet flour. Process flow chart for value addition of kodo millet is given in Fig.4.

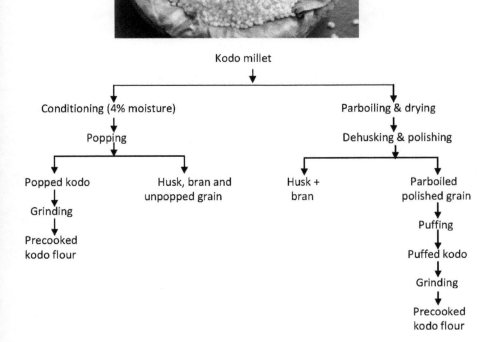

Fig. 4: Process flow chart for value addition of kodo millet
(*Source*: Verma and Mishra, 2010)

Finger millet

Finger millet is grown extensively in southern part of Karnataka. It is also called as nutritious millet. This is used in the form of whole meal for preparation of traditional items like *mudde* (dumpling), *roti* (pancake) and *ambli* (porridge). The grain is nutritious with balanced protein, higher calcium, iron and dietary fibre (Chetan and Malleshi, 2007).

Finger millet flour is used for convenience food preparations *i.e.*, spaghetti, macaroni, vermicelli and noodles through cold extrusion system. Finger millet malt, laddu, chakkuli, hurihittu, papad etc can also be prepared. Ready-to-eat (RTE) puffed finger millet mix which requires no further cooking can be given to children and pregnant and lactating women. Bakery products like bread and biscuits can also be made out of finger millet (Verma and Patel, 2012). The process flow chart of various finger millet based value added products are given in Fig. 5, 6, 7 and 8.

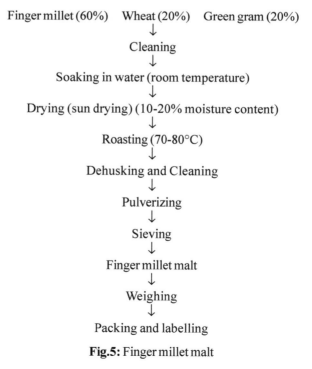

Finger millet (60%)   Wheat (20%)   Green gram (20%)
↓
Cleaning
↓
Soaking in water (room temperature)
↓
Drying (sun drying) (10-20% moisture content)
↓
Roasting (70-80°C)
↓
Dehusking and Cleaning
↓
Pulverizing
↓
Sieving
↓
Finger millet malt
↓
Weighing
↓
Packing and labelling

Fig.5: Finger millet malt

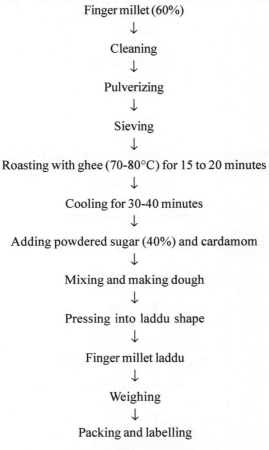

Fig. 6: Finger millet laddu

Finger millet (60%) black gram dal (20%) Roasted bengal gram dal (20%)
↓
Cleaning
↓
Mixing of grains
↓
Pulverizing
↓
Sieving
↓
Adding sesame and cumin seeds, salt and heated oil
↓
Making dough
↓
Pressing in chakkuli press
↓
Frying
↓
Finger millet chakkuli
↓
Weighing
↓
Packing and labelling

Fig. 7: Finger millet chakkuli

Finger millet (50%)
↓
Cleaning
↓
Puffing
↓
Pulverizing
↓
Sieving
↓
Adding powdered sugar (30%), Deffated soy flour (10%)
Milk powder (10%) and cardamom
↓
Mix well
↓
Finger millet hurihittu
↓
Weighing
↓
Packing and labelling

Fig. 8: Finger millet hurihittu (*Source:* Yankanchi and Majula, 2016)

Little millet

It is called Kutki in Hindi and locally called 'Chikma'. The seeds of little millet are smaller than those of common millet. Little millet has the highest fat content of all the millets with good level of protein.

Little millet is processed into rice, semolina and flour. The flour can be used for the preparation of local food items like *bread, roti* and *dosa.*

Foxtail millet

Foxtail millet is ovoid in shape, pale yellow to orange, red, brown or black in colour. The kernel enclosed in thin hulls can be separated by de-husking before being processed for food. It is a good source of dietary fiber and â carotene. The de-husked grain can be cooked and consumed in the same manner as rice.

Foxtail millet is processed into rice, semolina and flour. It has highest mineral content and can be used in the preparation of many foods such as porridge, pudding, bread, cake, flour, chips, rolls and noodles. Foxtail millet is fermented to make vinegar and wine in China and to make beer in Russia and Myanmar. Sprouted grains are eaten as vegetable. The kernels are also directly cooked to make rice and the flour can be used for making *roti, dosa* and *idli.* It could also be exploited for the nutritional benefits by developing the recipe for nutritive health foods.

Banyard millet

Barnyard millet is a good source of slowly digestible carbohydrate, fair source of protein, excellent source of dietary fibre and minerals (Dandsena and Banik, 2016).

It has been reported that, the barnyard millet could be successfully used for the preparation of traditional foods *viz.,* rice, *roti, dosa, idli* and *chakli* (Veena *et al.,* 2004). Barnyard and finger millet based *khichadi, laddu* and *baati* can also be prepared by fortifying with legumes and fenugreek seeds (Arora and Srivastava, 2002).

Proso millet

Proso millet is also one of the important millets as it has the highest protein content and very high in carbohydrates as well.

Proso millet based convenience mix for infants and children can be prepared by malting and popping techniques (Srivastava et al., 2001). Sweet and salt gruels, *halwa, burfi* and biscuits based on the convenience mix can also be prepared with acceptable sensory attributes (Dandsena and Banik, 2016).

Some traditional recipes from millets

Idli

Mix, wet milled sorghum-millet batter with wet milled black gram batter, ferment overnight and pour into small cakes or moulds and cook by steaming to give a product called *idli* which is soft, moist and spongy with a slightly sour taste (Murty and Kumar, 1995).

Dosa

In Southern India and Sri Lanka, fermented batter is mixed with legume batter and cook into a thin pancake called *dosa*. Rice is preferable to other cereals, but millets can also be used. De-hulled millet grains are soaked, wet milled and fermented overnight. Pour the batter in small quantities onto a hot metal plate with some oil, and cook for about one minute. The resulting crispy but flexible 2-4 mm thick and 20-25 mm diameter pancake is called *dosa*, and is consumed with curd, vegetables, chillies or other sauces (Kajuna, 2001).

Roti

In India, an unfermented pancake, called *roti*, is produced from pearl millet, small millets, sorghum or maize flour. *Roti* is also known as chapati in other parts of India and East Africa, such as Tanzania. It can be consumed with vegetables, dal, meat, milk, curd, sour milk, pickles and other sauces. To prepare *roti*, take about 50 g of flour, mix with about 45 ml of warm water. Knead the flour-water mixture on a wooden board to obtain cohesive dough. Make balls with the dough, press by a wooden rod into a thin (about 1.3 to 3 mm thick) circular sheet, and then cook for about one minute on an earthen or iron pan (Kajuna, 2001).

Snack foods

There is numerous number of snack foods which can be derived from millet flour. Different types of snack foods can be prepared by deep frying the dough. Thick porridges of fermented dough can be extruded or molded or sundried and then deep fried. Sugar, salt and other spices may be added to improve flavour of the snack.

Millet malt : Flow chart to prepare millet malt is given in Fig. 9.

Millet
↓
Cleaning
↓
Soaking for 12h
↓
Germination for 16 h
↓
Removal of vegetative portion
↓
Roasting
↓
Grinding and sieving
↓
Storing in air tight container

Fig. 9: Flow chart to prepare millet malt

Different traditional items like bisbelebath, kadambam, little millet pudina rice, diabetic friendly methi pulao, pumpkin rice etc. can also be prepared using different types of millets. Techno economic feasibility of millet processing unit is given in Annexure.

Methi pulao   Bisbelebath

Kadambam   Pumpkin rice

Little millet pudina rice

Entrepreneurship development through millet processing

Since the women's groups are stepping towards the economic development of their family through income generating activity, there needs to be wide range of opportunities for undertaking entrepreneurial activity. Processing of millets provide an opportunity for income generation and also ensures food security.

To increase the value of millets and to make them widely available, ICAR-KVK Raichur under UAS, Raichur (Karnataka) had undertaken demonstrations on entrepreneurship development through millet processing for self help groups of Raichur District. Major activities included promotion of high yielding variety of foxtail millet, establishment of processing facility, training members of SHGs on value addition and establishing market linkages for the processed products. Foxtail millet varieties namely, HMT-100-1, SIA-2644, DHFt-103-9 released by University of Agricultural Sciences, Dharwad was found to be promising compared to the local varieties. The millets brought by SHGs were subjected to cleaning, dehusking and packaging in PET pouches and branded in the name of UAS Raichur. Groups were facilitated with sealing machine and printed labels having the information about the nutritional composition of millets. Trainings on value addition of millets were one of the important components of the entrepreneurship programme. Under the banner of vocational training, six weeks training programme on value added millet products was organized for the prospective entrepreneurs. Marketing linkages were established through organic markets at Bengaluru., Sahaja Samriddhi, Bangalore., KVK, Raichur., hospitals in local area and melas conducted in the nearby areas. Marketing of products was done either directly by the members of the same SHG which produced the products or by another SHG which undertook specialized task in packaging and marketing the products. The economic analysis is given in Table 2.

Table 2: Details of economic analysis per 1000 kg

Cost of raw millets @ Rs. 18/- per kg (unprocessed millets)	18000
Cost of processing @ Rs. 6/- per kg	6000
Cost of packaging and labeling @ Rs. 2.5/- per kg	1700
Total cost (Rs.)	Rs. 25,700
Recovery rate	68%
Quantity of processed millet	680 kg
Price of processed millets (Rs.per kg)	50
Gross returns (Rs.)	34000
Net returns per 1 ton	8300
B:C Ratio	1.32

Thus, it may be inferred that, small millets are important for the food security and income generation of farmers in certain agro-climatic regions. These crops need to receive adequate attention of the policy makers for the purpose of investment on their research and development, mainstreaming of these grains in developmental programmes, in public procurement and distribution system.

References

Arora & Srivastava, 2002. Suitability of Millet Based Food Products for Diabetics. *J. Food Sci. Tech.,* 39(4): 423-428.

Bangu, N.T.A., S.T.A.R. Kajuna & G.S. Mittal, 2000. Storage and Loss Moduli for Various Stiff Porridges. *Int. J. Food Properties,* 3(2): 275-282.

Chetan, S. & N. G. Malleshi. 2007. Finger Millet Polyphenols: Characterization and their Nutraceutical Potential. *Amer. J. Food Tech.,* 2(7): 582-589.

Dandsena, N. & A. Banik, 2016. Processing and Value Addition of the Underutilized Agriculture Crops and Indigenous Fruits of Bastar Region of Chhattisgarh. *Int. J. Multidisciplinary Res. Develop.,* 3(3): 214-223.

Devi, P. B., R. Vijayabharathi., S. Sathyabama., N.G. Malleshi & V. B. Priyadarisini, 2011. Health Benefits of Finger millet (*Eleusine coracana* L.) Polyphenols and Dietary Fiber: A Review. *J. Food Sci. Technol.,* DOI: 10.1007/s13197-011-0584-9.

FAO. 1995. Sorghum and Millets in Human Nutrition. *FAO Food and Nutrition Series No. 27.* Rome: FAO.

FAO. 2012. *Economic and Social Department*: The Statistical Division. Available from FAO [http://faostat.fao.org/site/567/ DesktopDefault.aspx? PageID=567]. Posted September 29, 2012.

Hadimani, N. A. & N.G. Malleshi, 1993. Studies on Milling, Physico-Chemical Properties, Nutrient Composition and Dietary Fiber Content of Millets. *J. Food Sci. Technol.,* 30: 17–20.

Hulse, J. H., E. M. Laing & O. E. Pearson, 1980. *Sorghum and the Millets: their Composition and Nutritive Value.* New York, NY: Academic Press. pp. 187–193.

Kajuna, S.T.A.R. 2001. Millet: Post-Harvest Operations. *Information Network on Post-Harvest Operations (FAO).* pp. 25-34.

Lakshmi, K. P. & S. Sumathi, 2002. Effect of Consumption of Finger millet on Hyperglycemia in Non-Insulin Dependent Diabetes Mellitus (NIDDM) Subjects. *Plant Foods Hum. Nutr.,* 57: 205–213.

Malleshi, N. G. 1993. Processing of Coarse and Minor Millets for Food and Industrial Uses. *Proceedings of International Food Convention,* Mysore, pp. 349-359.

Murty, D. S. & K. A. Kumar, 1995. Traditional Uses of Sorghum and Millets. In: D.A.V. Dendy, (ed.), *Sorghum and Millets: Chemistry and Technology.* St. Paul, Minn. *Am. Assoc. Cereal Chem.,* pp. 185-221.

Obilana, A. B. & E. Manyasa, 2002. Millets. In: P. S. Belton and J. R. N. Taylor (eds.), *Pseudo Cereals and Less Common Cereals: Grain Properties and Utilization Potential.* Springer-Verlag: New York. pp. 177–217.

Oelke, E. A., E. S. Oplinger., D. H. Putnam., B. R. Durgan., J. D. Doll & D. J. Undersander, 1990. Millets. *Alternative Field Crops Manual. (Purdue University: West Lafayette, IN, USA) Available at: www. hort. purdue. edu/newcrop/afcm/millet. html (accessed 25 March 2017).*

Palande, K. B., R. Y. Kadlag., D. P. Kachare & J. K. Chavan, 1996. Effect of Blanching of Pearl Millet Seeds on Nutritional Composition and Shelf life of its Meal. *J. Food Sci. Technol.,* 33(2): 153–155.

Rooney, L. W. & C. M. McDonough, 1987. Food Quality and Consumer Acceptance of Pearl Millet. In: Witcombe J. R. & S. R. Beckerman editors. *Proceedings of the International Pearl Millet Workshop.* Andhra Pradesh, India: ICRISAT. pp. 43-61.

Shahidi, F. & A. Chandrasekara, 2013. Millet Grain Phenolics and their Role in Disease Risk Reduction and Health Promotion: A Review. *J. Funct. Foods.,* 5: 570-581.

Shankaran, S. 1994. Prospects for Coarse Grains in India. *Agricultural Situation in India,* 49(5): 319-323.

Singh, B., A. Bahunga & A. Bhatt, 2015. Small Millets of Uttarakhand for Sustainable Nutritional Security and Biodiversity Conservation. *Int. J. Management Social Sci. Res. (IJMSSR)*, 4(8): 26-30.

Singh, P. & R. S. Raghuvanshi, 2012. Finger millet for Food and Nutritional Security. *African J. Food Sci.*, 6(4): 77-84.

Srivastava, S., A. Thathola, & A. Batra, 2001. Development and Nutritional Evaluation of Proso millet-Based Convenience Mix for Infants and Children. *J. Food Sci. Technol.*, 38(5): 480-483.

Veena, B., B. V. Chimmad., R. K. Naik & G. Shantakumar, 2004. Development of Barnyard millet Based Traditional Foods. *Karnataka J. Agril. Sci.*, 17(3): 522-527.

Verma, P. K. & N. Mishra, 2010. Traditional Techniques of Processing on Minor Millets. *Res. J. Agr. Sci.*, 1(4): 465-467.

Annexure
Techno economic feasibility of millet processing unit
Assumptions

Capacity of the Unit (Processed produce per day)	-	500 kg
Equipment to be installed	:	
Cleaner cum grader	-	1No. – Rs, 1,00,000/-
Destoner	-	1No. – Rs, 1,00,000/-
Specific gravity separation	-	1No. – Rs, 1,00,000/-
Dehusker	-	1No. – Rs, 1,00,000/-
Pearler	-	1No. – Rs, 1,00,000/-
Total Cost - Rs, 5,00,000/-		
Shed Area	:	0 m x 5 m
Cost of shed & land	:	15 lakhs
Life of machinery & shed	:	8-10 years
Total fixed cost	:	15 + 5 = Rs. 20.0 lakhs
Working days per year	:	250 days
Depreciation		
Depreciation on fixed cost	=	(20,00,000-2,00,000)/8 = Rs.2,25,000/annum
	=	Rs 75/ day

Labour

Labour cost	:	1 skilled @ Rs.350 / day
	:	2 unskilled @ Rs. 250 / day
Per day labour charges	:	Rs. 850/-
Per month	:	Rs. 25,500 / month
Labour cost per year	:	Rs. 3.06 lakh
Raw material cost		
Cost of raw material for production of 500 kg processed millet with recovery @ 65% recovery	:	770 kg
Average price of raw material (millet)	:	Rs. 25/- per kg
	:	770 x 25 = Rs. 19,250/-
Annual requirement of raw material	:	Rs. 48.12 lakh

Total cost of processing	=	**48.12+2.25+3.06=53.43 lakh**
Returns by selling processed millets	=	500 @ Rs.60/- = 30,000 per day
For 250 working days in a year	=	Rs. 75.0 lakh per year
Returns by selling husk	=	(770-500) @ Rs. 3.0/kg = Rs.810/- day
	=	2.02 lakhs per year
Total returns per year	=	75.0+2.02 = Rs. 77.02 lakh
Net profit per year	=	**77.02 – 53.43 = 23.59 lakh**
Benefit cost ratio	=	**77.02 / 53.43 = 1.44 : 1**
Payback period	=	**20/23.59 = 0.85 years**

8

Entrepreneurship Developments in Pulse Processing

Visvanathan R, Ravindra Naik, Vennila P & Borkar P A

Introduction

Pulses are major and cheap sources of protein in Indian vegetarian diet, providing most of the essential amino acids. Pulses are mainly consumed in the form of dehusked split pulses (Dal); and have 15-24% proteins. Various types of pulses grown are pigeon pea, chickpea, black gram, green gram, lentil, etc. Pulses are consumed because of body building properties due to the presence of various amino acids. By-products of pulses like leaves, pod coats and bran are fed to animals in the form of dry fodder. Some pulse crops like gram, lobia, urd bean & moong bean are fed to animals as green fodder.

Milling

Milling is an overall process and it includes size reduction, hulling, scarification, polishing, sorting, mixing and in some instances, also refers to certain chemical reactions. Through milling, outer husks are removed and grain is split into two equal halves. The kernel tightly holds the husk. Therefore, de-husking poses a problem. Method of alternate wetting and drying is used to facilitate de-husking and splitting of pulses. In India, dehusked split pulses are produced by traditional methods of milling in which loosening of husk by conditioning is insufficient. Therefore, a large amount of abrasive force is applied for complete de-husking of grains which results in the loss of broken and powdered grains. Yield of split pulses in traditional mills is only 65-70 per cent in comparison to 82-85 per cent potential yield. De-husking improves appearance, texture, cooking quality, palatability and digestibility of pulses (Chakravarthy, 1988; Dhal, 2014). In India, there are two conventional pulse milling methods; wet milling (Fig.1) and dry milling (Fig.2). The latter is more popular and used in commercial mills. Pre-milling treatment plays an important role in improving dal recovery. During pre-milling treatment, loosening of husk from cotyledons takes place. But, removal of husk and splitting of grains is achieved by means of various machines, which work on principles of compression, shear, abrasion and impact (Kulkarni, 1993; Chakarvarthy *et. Al.*, 2003). Different milling machines used are detailed below.

Chakki or Disc-sheller

Hand-operated chakkies are used for de-husking and splitting grains since olden times in domestic and traditional milling. It consists of two cylindrical stones – one stationary and the other rotated by means of a wooden handle. Unhusked or full grains are fed from the center and de-husked grain and split dal is recovered at the periphery of the cylinder. Improved power operated chakkies or emery-coated roller machines have been used for dehusking operations.

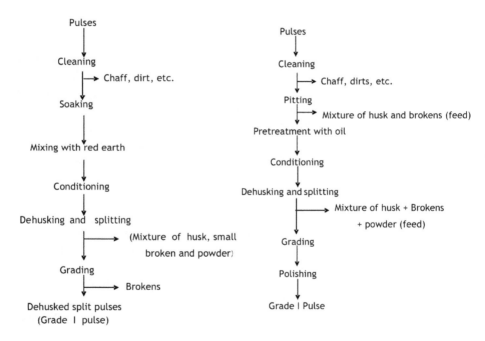

Fig. 1: Flow diagram of wet milling   **Fig. 2:** Flow diagram of dry milling

Huller

Commonly used Engelburg rice huller can also be used for dal milling. It consists of ribbed iron cylinder on rotating shaft in a concentric cylindrical housing. Bottom of the housing is provided with slots for removing the husk. It is used for milling black gram and green gram in some South Indian dal milling industries.

Cylindrical concave de-husker

It consists of a tapered carborundum roller. The diameter increases from feeding to discharge end, thus reducing the annular space between the roller and cylindrical screen casing. Reduction in the annular space increases the pressure on the grains and thus gradually increases the de-husking rate as grains move forward. Metal screen casing has a 2-mm diameter circular perforation through which husk and powder are discharged.

Rubber roller sheller

It consists of two rollers rotating in opposite directions with different speed. When grains are passed through the rollers, they are subjected to shear and compression leading to husk removal. As husk is more tightly attached to cotyledons in case of pulses, rubber roller sheller can be used only to a limited extent. However, the machine causes minimum scouring and can be used to polish split pulses.

Some of the designs of the equipment used for dehusking and polishing pulses are given in Table 1.

Table 1: Some Improved mini dal mills and allied machines

Sr.No.	Name of prototype	Capacity	Electric motor
1.	IARI dehusking cum splitting machine	Small	1 hp, three phase
2.	CIAE dehusking and splitting machine	100 kg/h	2 hp, three phase
3.	CFTRI mini dal mill	125 -150 kg/h	Two motors of 1 hp single phase
4.	Pantnagar dal mill	50 kg/h	2 hp, single phase
5.	IIPR dal mill	75 -125 kg/h	2 hp, single phase
6.	TNAU mini dal mill	30 kg/h	1 hp, single phase
7.	PKV mini dal mill	100 -125 Kg/h	3 hp, three /single phase
8.	Higher capacity PKV mini dal mill	250 – 350 Kg/h	5 hp, three phase
9.	Integrated PKV mini dal mill	100 – 125 Kg/h	3 hp three phase
10.	Small capacity PKV mini dal mill	30 -35 Kg/h	1 hp, single phase
11.	Laboratory working model of PKV mini dal mill	6 -7 Kg/h	0.5 hp, single phase
12.	PKV cleaner-cum-grader-cum-polisher	500 Kg/h	1 hp, single phase
13.	PKV cleaner-grader	200 -250 Kg/h	1 hp, single phase
14.	PKV screw polisher	150 – 200 Kg	1 hp, single phase
15.	PKV waste fired dryer	6 q grains (Fuel requirement is nearly 12 kg/h)	3 hp, three phase for blower

Details of the designs developed by ICAR, CFTRI and TNAU for dehusking pulses are given below

ICAR – CIAE design

An abrasive emery roller cylinder mill to dehusk and split pulses to make dal was developed at ICAR- Central Institute of Agricultural Engineering (ICAR-CIAE), Bhopal (Fig.3). It consists of 250 mm diameter cylinder coated with emery paste and an outer layer of Carborundum. The clearance between the outer screen cage and the inner abrasive roller is maintained throughout at 10 mm.

CSIR- CFTRI design

A mini dehuller of capacity of 150-200 kg/h has been developed by CSIR-CFTRI, Mysore, India (Fig.4). Mill consists of emery coated metal cone fixed to a vertical shaft and rotating under a fixed conical wire mesh screen. The screen and the cone are concentric, and their clearance is about the diameter of the grain.

TNAU mini dal mill

Mini dal mill is working on the principles of attrition or rubbing in size reduction (Fig.5). Dal mill consists of a milling chamber, feed hopper, power drive system, frame, etc. Two attrition discs are provided in the milling chamber. One of the discs made of rubber is stationary and fixed on the inner side on the milling chamber. Other disc made of cast iron is mounted on a shaft with screw and made to rotate through a suitable drive mechanism. An auger to feed the pulse to the dehusking chamber provided in the milling chamber direct the pulses into the milling zone. Depending on size and type of pulses, clearance between the rotating disc and the rubber disc can be adjusted with the help of a hand wheel provided outside the dehusking chamber. This dal mill has a capacity of about 20 kg/hour and dal recovery is 80%. Improved model of this dal mill with pitting roller and cleaning arrangements are also available.

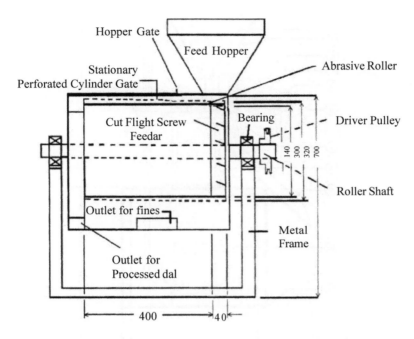

Fig. 3: Horizontal abrasive roller for pulse dehusking (ICAR- CIAE design).

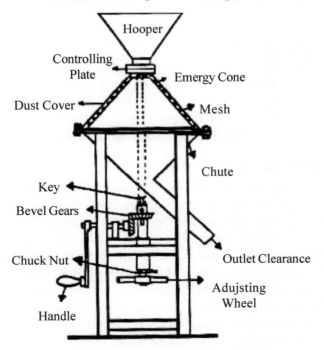

Fig. 4: CSIR- CFTRI mini mill for dehusking pulses.

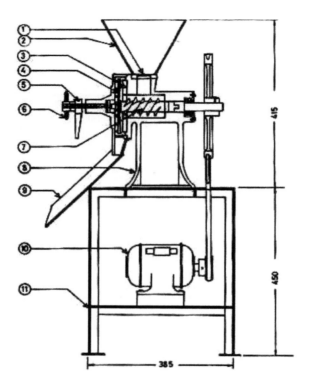

1. Feed Control Shutter
2. Feed Hopper
3. Rubber Pad
4. Grinding Disc
5. Lock Nut
6. Hand Wheel
7. Screw Auger
8. Main Body
9. Dhal Outlet
10. Motor
11. Stand (Dimensions in MM)

Fig. 5: TNAU model mini dal mill

Health benefits of pulses

Pulses form an important part of a healthy, balanced diet and have an important role in providing food and nutritional security. In addition, the pulses have health benefitting properties and help in preventing illnesses. Typical nutritional composition of common pulses is given in Table 2. Pulses are a source of carbohydrates (60-65%) of which starch is the primary one. Slowly digesting nature of starch in pulses makes it diabetic friendly with low glycaemic index (GI). Pulses are very high in fibre, containing both soluble and insoluble fibres. Soluble fibre helps to decrease blood cholesterol levels and control blood sugar levels, and insoluble fibre helps with digestion and gut health. Pulses are also a source of vitamins and minerals. Some of the key minerals in pulses include iron, potassium, magnesium and zinc. Pulses are also abundant in B vitamins including folate, thiamin and niacin. Pulses typically contain about twice the amount of protein found in whole grain cereals like wheat, oats, barley and rice. They contain high amounts of lysine, leucine, aspartic acid, glutamic acid and arginine and provide well balanced essential amino acid profile when consumed with cereals (Asif *et.al.*, 2013; Agrawal, 2016).

Pulses are not consumed raw and are processed to improve the eating quality, digestibility, nutritional and health significance. Dehusking, soaking, germination, cooking, roasting, and fermentation are commonly used processing techniques to make them edible. Soaking and germination help to eliminate the trypsin inhibitor activity, proteolytic enzyme inhibitors, phytates and tannins and increases the protein digestibility and bioavailability of minerals.

Germination of pulses is a simple and popular technique to enhance the palatability, digestibility and nutritive value. Increase in vitamin B and C; utilization of available proteins and carbohydrates and decrease in anti-nutritional and flatulence factors are useful outcomes of germination. Malting, roasting and fermentation are other preferred methods for developing many local food products in different parts of the world.

Entrepreneurship through value addition in pulses

Pulses have been used as ingredients for several years in order to develop products with unique functional properties as well as enhanced nutritional profiles. Apart from carrying out the entrepreneurship development through primary processing, there is scope for entrepreneurship development in pulse processing by value addition through product development.

Pulses being a cheaper source of plant protein than nuts, milk, cheese, meat and fish can be used in bakery products like pasta, bread, snacks etc. Pulses provide ample opportunities to be used in processed foods, as an ingredient in designer foods for snacks, baby foods and sports foods.

Value added products of pulses

Malted / germinated pulse flour

Different steps involved to prepare malted/germinated pulses flour are

Pulses (green gram, bengal gram, peas) → Cleaning → Soaking (12 h) → Germinating (48 h) → Drying (shade drying - 24 h) → De-vegetating → Kilning (70° - 75°C) → Milling (local flour mill) and sieving (BS 80 mesh) → Malt / Germinated flour.

Weaning food mixes

Weaning food mixes are prepared by blending malted cereal and pulse flour. This could be further fortified or enriched with skim milk powder and vitamins if desired. Standard procedure for pulse based weaning food mix is malted millet flour, malted pulse flour, ground nut flour and powdered jaggery mixed in the ratio of 3:2:1:2 (cereals: pulses: oilseed: jaggery).

Instant adai mix

To prepare instant *adai* mix, clean parboiled rice, raw rice, black gram dal, red gram dal, green gram dal and bengal gram dal to remove immature, weeviled, broken and damaged grains and dry in the sun for two to three hours. Steps to prepare instant *adai* mix is given in Fig. 6

Instant vada mix

For the instant vada mix, powder bengal gram dal, black gram dal and raw rice in a grinder/mixie separately. Sieve, bengal gram flour using BS 14 sieve and rest of the items in BS 36 sieve. Then, dry the flour separately in a cabinet drier at 80°C for two hours, cool and store in airtight containers.

Pakoda and murukku mixes

Pakoda and murukku are deep fat fried snack items prepared using pulse and cereal combinations. For the instant pakoda mix, bengal gram flour (100g), rice flour (25g), salt (3g), chilli powder (1g), sodium bicarbonate (0.1%), aniseed (1g) and vanaspathi (1g) are required. Mix, all flours thoroughly and sieve twice. Mix aniseed and make dough by adding water and extrude through a die and fry in oil. For murukku instant mix, powder raw rice and black gram in a grinder mill separately and sieve through 80 BS sieve. Murukku can be prepared after making the dough with this instant mix by adding salt and white sesame seeds.

Table 2: Nutritional composition of common pulses (per 100g)*

Pulses	Energy (kcal)	Proteins (g)	Fat (g)	Mineral (g)	Carbohydrate (g)	Fibre (g)	Calcium (mg)	Phosphorous (mg)	Iron (mg)
Bengal gram whole	360	17	5	3	61	4	202	312	5
Bengal gram dal	372	21	6	3	60	1	56	331	5
Bengal gram roasted	369	22	5	2	58	1	58	340	9
Black gram dal	347	24	1	3	60	1	154	385	4
Cow pea	323	24	1	3	54	3	77	414	9
Field bean, dry	347	25	1	3	60	1	60	433	3
Green gram whole	334	24	1	3	57	4	124	326	4
Green gram dal	348	24	1	3	60	1	75	405	4
Horse gram whole	321	22	0	3	57	5	287	311	7
Khesari, dal	345	28	1	2	57	2	90	317	6
Lentil	343	25	1	2	59	1	69	293	7
Moth bean	330	24	1	3	56	4	202	230	9
Peas green	93	7	0	1	16	4	20	139	1
Peas dry	315	20	1	2	56	4	75	298	7
Peas roasted	340	23	1	2	57	4	81	345	6
Rajmah	346	23	1	3	61	5	260	410	5
Red gram dal	335	22	2	3	58	1	73	304	2
Red gram tender	116	10	1	1	17	6	57	164	1

Source: Nutritive value of Indian Foods (Gopalan *et al.*, 2004)

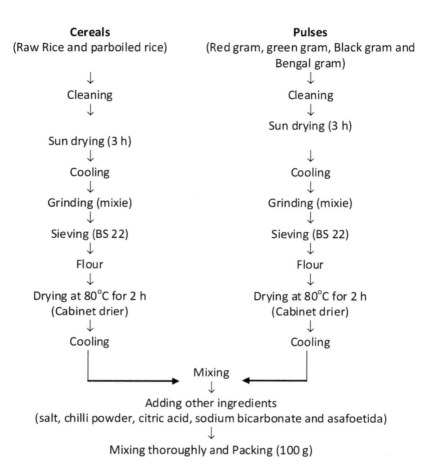

Fig. 6: Flow chart for the preparation of instant *adai* mix

Instant dhokla mix

Dhokla is a fermented food prepared from rice, bengal gram dal and black gram dal. It is also known as khaman in Gujarat, Rajasthan and Maharashtra. Optimum levels of various ingredients and the process flow diagram are given in Fig.7.

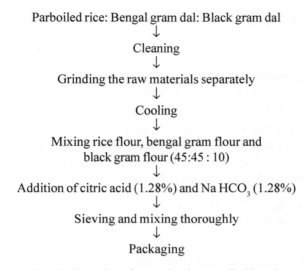

Parboiled rice: Bengal gram dal: Black gram dal
↓
Cleaning
↓
Grinding the raw materials separately
↓
Cooling
↓
Mixing rice flour, bengal gram flour and
black gram flour (45:45 : 10)
↓
Addition of citric acid (1.28%) and Na HCO$_3$ (1.28%)
↓
Sieving and mixing thoroughly
↓
Packaging

Fig. 7: Flow chart for making instant dhokla mix

Ready - to - cook foods

Extruded foods: Different ingredients and steps to prepare high protein pulse based noodles are:

Maida (60%) + Composite pulse flour (40%) → Sieving (BS 60 sieve) → Steaming (10-15 min.) → Sieving (BS 60 sieve) → Adding into the extruder → Wetting with water (350 - 400ml/kg) and salt (2g/kg) → Kneading (15-20min.) and extruding → Gelatinizing (steaming for 12 min.) → Tempering overnight (10 h.) → Drying (50°C for 2 - 4 h) → Cooling and Packaging.

Puffing of pulses: Soya bean, pea and bengal gram are commonly puffed. Moist conditioning before roasting helps in good puffing. Continuous gram roasters are in commercial use for puffing. The roasted grains get dehusked, puffed and split as they are subjected to mild impact between a knurled roller and a hot plate. Expansion during puffing varies with variety and process conditions and range from 1.2 to 2 times. Steps involved in puffing are

Whole pulses → Cleaning and washing → Soaking in 5% salt and 3% NaHCO$_3$ for 3h. → Draining and puffing (250°C for 5 min.) → Sieving and dehulling → Winnowing → Puffed pulses.

Flaking of pulses: Bengal gram, black gram, soya bean and red gram dal flakes are prepared by soaking in sodium chloride and carbonate solutions, cooking under 1 kg/cm^2 steam pressure, conditioning to reduce the moisture, flaking and drying at 100°C. Flaked dals reconstitutes within eight minutes when mixed with hot water. The steps include

Cleaning → Dehulling → Blanching in boiling water for 60 min. → Drying (25 to 30% moisture content) → Flaking → Drying to 8% moisture content → Storing.

Quick cooking pulses (*Instant pulses/precooked pulses*): Many pulses are to be cooked to become soft before consumption. Cooking time needed for softening will vary from 15 - 45 minutes. This can be reduced by cooking in a pressure cooker. Instant or quick cooking pulses are a necessity for modern urban consumers and for special defence needs for which time saving is important. Precooked pulses are dried under controlled conditions to make it porous enabling quick cooking. The steps include

Pulses → Soaking in water (1:2) for 6 - 10 h → Pressure cooking for 20 min. → Freezing at -10 to -18°C → Drying (5% moisture content) → Cooling and packaging.

Instant red gram dal: Cook red gram and dry in fluidized bed drier at 60°C. Cooking time of the instant dal can be reduced to 80 to 84% when cooked in boiling water.

Instant mixes based on precooked dehydrated products: Important among this group are instant bisibelebhat, curried dal, peas, curried chholay and sambhar. These products are prepared by precooking the ingredients preferably under pressure, followed by dehydration and mixing of the dehydrated ingredients, salting and spicing. Depending on the conditions employed for dehydration and nature of the product mix, these can be reconstituted in boiling water within 6-20 min.

Instant kitchdi mix: Steps involved are soaking of rice and whole green gram in water containing 1% salt and autoclaving at 1 kg/cm^2 for 15 min. for rice and 20 min. for green gram. Then, conditioning is done to reduce the moisture level of rice to 20% and green gram to 30%. Next step is tumbling the two ingredients in roller flaker and drying in a fluidised bed drier at 60°C to bring down the moisture level to 5 – 6%. Finally, blend the precooked and dried rice and green gram with spices (cumin, black pepper, cardamom, clove, ginger, chilli, turmeric and garlic powders) and salt uniformly by dispersing them in vanaspathi in a planetary mixer.

Papads : These are thin wafer like product prepared from a variety of base ingredients like black gram flour or black gram flour and other pulse flour combinations with and without spices. These are mostly prepared at household level or as cottage scale industry. Prepare stiff dough with the flour and make small balls and roll into thin round discs by means of rolling pins and dry to a level in which it remains pliable. Before consumption, papads are either deep-fat-fried or toasted on a hot plate or on open flame. Recently, there has been considerable growth in papad production because of their demand from overseas market.

Ready - to- eat /ready - to - use foods

Canned foods: Process of sealing food stuffs hermetically in containers and sterilizing them by heat for long storage is known as canning. Canned peas, field beans and bengal gram are commercially available in the market.

Retort pouch foods: Retort pouch processed products are beans in sauce (Rajma). Retort pouches are flexible packages made from multilayer plastic films with or without aluminium foil as one of the layers. The 3-ply laminate consisting of PET/Al foil / PP is the most common material used in retort pouches and is the only one used in India at present. Unlike usual flexible pouches, they are made of heat-resistant plastics, thus making them suitable for processing in retort at temperatures of around 121°C normally encountered in thermal sterilization of foods. Besides, retort pouches should possess toughness and puncture resistance normally required for any flexible packaging. Apart from being heat sealable, the pouch material should possess good barrier properties to give the desired shelf-life to the product and be suitable for food contact applications.

Traditional sweets: These include candies (chikki) which are prepared from puffed bengal gram and roasted peanuts. Moisture content in these products varies from 4-8% and the products are highly hygroscopic and tend to become soggy and sticky. At present, these are mostly prepared manually in small batches by preparing jaggery syrup, mixing with roasted grains and nuts and rolling and cutting into small slabs. Sweets like Laddu, Boondi, Jilebi, and Jhangiri are also prepared using pulses.

Traditional fried products

Fried dals and whole pulses are prepared by soaking in water along with salt and sodium bicarbonate solution and frying in vegetable oil and mixing with spices. Bengal gram, green gram, horse gram and black gram dals are most frequently used. Fried dals are highly crisp. Chakli, (murukku), tengolal, boondi and sev are some of the popular fried products prepared from pulse flours. Stiff dough is extruded through a hand operated press and fried. Crispness and soft texture are the most desirable characteristics, and these are mostly determined by dough composition.

Bakery products

Pulse flour upto 20 per cent are incorporated with main ingredient to prepare high protein bakery products like bread, cake, biscuits and soup sticks.

New technologies/innovations in value addition of pulses

Pulses are central to many culinary traditions. With rapid increase in global food needs on the horizon, role of pulses will become even more significant, especially with regard to dietary protein and micronutrients. Local and regional food industries can offer new niche markets for pulse crops, especially where commercially viable uses can be found for all pulse fractions (i.e. protein, starch). Focuses have to be on consistent, high-quality production of specific pulse varieties with tailored properties (e.g. protein content, ease of cooking and processing). 'Consumers' for small and medium sized food products (e.g. baby food, breweries) have to be targeted. Some of the processing areas which need more attention are puffing, steaming, germination and canning.

Some pulses require considerable time for their preparation into palatable form or for subsequent use in traditional dishes. Owing to rapid urbanization, novel systems of processing pulses are required. Some of them are listed below.

Quick-cooking and instant food products based on pulses

Many quick cooking pulse based products were developed by Defence Food Research Laboratory, Mysore to meet varied operational requirements of Armed Forces. Products are light in weight with instant rehydration and easy cooking characteristics. Products were developed by precooking and dehydration and freeze thaw dehydration, which can be instantly reconstituted within 8-10 minutes by adding the contents in boiled water. The shelf-life of these products is 12 to 18 months.

Instantised pulse based food products

Besides quick cooking, a good number of food items have been developed which could be reconstituted by just adding hot water. Products get reconstituted within 5 minutes and will be ready for consumption. This category of instantised food items includes instant curried dals, instant kichidi mix and instant whole pulses.

Pulse powders

Pulse powders that can be used in familiar foods increase the consumption of pulses of all economic categories. Pulse powder can be prepared by three methods of processing i.e soaking the whole beans, cooking them to slurry, and drying the slurry using a drum dryer. In a second process, powdered pulse is immediately blended with hydrochloric acid solution to inactivate the enzyme responsible for causing bitter flavour followed by cooking the slurry for five minutes and addition of sodium hydroxide for neutralization. Cooking is continued to make flour into palatable form. Slurry is drum dried to make flour for the preparation of baked foods and other food mixtures. In the third method, precooked (blanched, soaked) whole pulses is dried and milled into powder. Bean powders produced by these three methods possess good characteristics of rapid hydration.

Pulse protein concentrates

Development of protein concentrates and isolates have received wide attention in recent years. Until recently, researchers had overlooked simpler ways of preparing protein concentrates and isolates. There are two methods to make protein concentrates i.e the age-old method of agglomerating flour by water absorption and a new method for preparing protein concentrate by air classifying finely milled flour. Both these methods can be adopted for medium and large scale production.

Shelf –life extension of pulses and their milled products by infra Red (IR) and microwave processing

Shelf life of many pulses and their milled products are affected by pest infestation. Infestation by pulse beetle, red flour beetle and other storage pests make pulse and their milled products inedible. Microwave treatments rely on the absorption of electromagnetic energy by water molecules in the pests infesting the foodstuff under treatment, which is transformed into heat, leading to the pests' death within a few minutes. In fact, most infesting biological agents do not survive over a certain temperature called lethal temperature, which is between 58°C and 70°C. Infrared treatment particularly at 300-350°C provides an effective and environmental friendly, chemical free disinfestation technique without affecting the product quality significantly. Besides disinfestations, IR heating also helps to inactivate lipoxygenase and lipase enzymes in food products and prevent microbial spoilage

Success stories in pulse processing sector

Agro processing centre at Kokarda, Tq. Anjangaon Surji, Dist. Amravati

Shri. Nivrutti Barabde, native of village Kokarda, Amravati District, under the technical guidance from, AICRP on PHET, Dr. PDKV, Akola, purchased equipment for pulse milling with 50% subsidy under Central Sector Scheme and started pulse processing business. In a period of four months, he processed

235q of pigeon pea and 80q of chick pea. He also purchased sheller, pulveriser and cleaner grader and established an Agro processing centre. Shri Nivrutti Barabde received Rs. 1,32,560/- as profit through this centre during the year 2013-14.

S.No.	Operation	Quantity Processed	Rate (Rs)	Amount (Rs)
1.	Cleaning, grading	8 q	@Rs 100/q	800
2.	Dal milling (pigeon pea, green gram, black gram)	315 q	@Rs 500/q	1,57,500
3.	Flour mill	72 q	@Rs 200/q	14,400
4.	Grinding of spices/pulverizing	80 kg	@Rs 10/kg	800
	Total			1,73,500
	Annual profit			1,32,560

This innovative activity of Shri Nivrutti Barabde encouraged the rural youths, farmers and budding entrepreneurs in the region.

Manufacturer of pulse milling equipment

The address of the manufacturer of pulse milling equipment is given below.

Er. UdayKumar, M/s Perfura Technologies (India) Private limited

7, Maruthamalai Gounder Lay out, Ramakrishnapuram, Ganapathy Coimbatore – 641006

Tamil Nadu, India

Phone : 91-9894800009; 91-98948500009

Email : Sales@perfuratech.com

Website : www.perfuratech.com

Cost economics

The cost economics of pulse processing unit of 1ton/h (8t/day) capacity is given in Table 3. The details of the equipment are given in Fig. 8 and Table 4.

Scope of by-product utilization/effluent treatment/ utilization

Various by-products which come out from the milling of pulses include different fractions of pulses and it should be separated from the whole/split pulses. These by-products of dal milling like husk, powder and small brokens are usually sold as cattle feed. Husk has been traditionally used as cattle feed because of their low bulk density and it forms about 10% of the raw material and sold at a lower price. Dal powder and small brokens which are richer in nutrients sold at a higher price are also used as cattle feed. Recent reports indicated that activated carbon can be prepared from the waste of pulse processing industries by carbonization and activation with sulfuric acid which has potential to adsorb metal. Activated carbon produced from pulse processing waste can be utilized for designing adsorption process to control water pollution due to heavy metals

Table 3: Cost economics of pulse processing unit

A. Capital investment	Rs. in Lakhs
Land: 600 m² @ Rs.800/ m²	4.80
Building: 500 m² @ Rs.6000/ m²	30.00
Plant and equipment, erection, contingency, etc.	27.00
Other assets, expenses, margin, etc	2.50
B. Working capital (for 1 month or 25 days)	70.00
Total capital investment	134.30
Profitability (Rs. in lakhs)	
a) * 2000 tons of pulse @ Rs.35,000/ton	700.00
b) Packaging (gunny bags @ Rs.20/-)	1.60
c) Utilities electricity 1,41,000 units @ Rs.6/unit + water 600 cu.m	9.00
d) Labour 4 nos. + 1 no. mechanic @Rs.250/ day + Manager @ Rs.10,000/ month	2.80
e) Overheads	1.60
f) Administrative expenses	2.50
g) Oil purchase/selling and other expenses	9.00
h) Depreciation (building @ 5%; plant @ 15%)	4.30
i) Interest on capital @ 14%	19.00
Total manufacturing cost	**749.80**
Sales returns	
Dal (77%) - 1540 tons @ Rs.50,000/ton	770.00
Husk & powder (20%) - 400 tons @Rs.8000/ton	32.00
Total returns	**802.00**
Profit before taxation	52.5
Commission for marketing (20% of profit)	10.4
Net profit before taxation	42.1
Pay back period	**3.3 years**

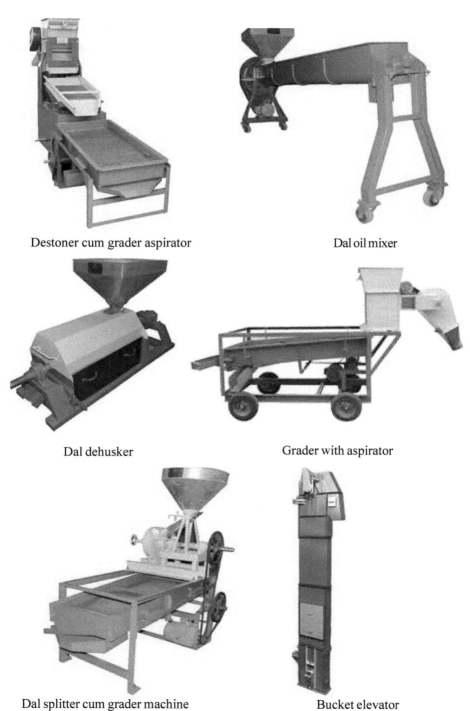

Destoner cum grader aspirator   Dal oil mixer

Dal dehusker   Grader with aspirator

Dal splitter cum grader machine   Bucket elevator
Fig. 8: Various dal milling equipment and accessories

Table 4: Details of commercial dal milling equipment and accessories

Destoner cum grader cum aspirator	Description	Details
	Type	Continuous type
	Purpose	Removal of impurities like stones and foreign materials.
	Capacity	1000 kg/h
	Feed hopper Capacity	15kg
	Motor	Destoner: 5 hpGrader: 1 hp
	Sieves	Grades provided with two set sieve
	Approximate cost (Rs)	1.40 lakh
Dal oil mixer	Type	Continuous type
	Purpose	Mixing oil with dal (preparation for dehusking)
	Capacity	350 kg/h
	Motor	1 hp
	Approximate cost (Rs)	50,000
Dal dehusker	Type	Continuous type
	Purpose	Dehusking various Dal
	Capacity	350 kg/h X 3
	Motor	15 hp
	Approximate cost (Rs)	1.33 X 3 lakh = 4 lakhs
Grader with aspirator	Type	Continuous type
	Purpose	Grading of grains
	Capacity	1000 kg/h
	Feed hopper Capacity	15 kg
	Motor	Grader: 1 hpAspirator: 0.5 hp
	Approximate cost (Rs)	0.85 lakh
Dal splitter cum grader machine	Type	Continuous type
	Purpose	Splitting various dals into two halves
	Capacity	350 kg/h
	Feed hopper Capacity	15 kg
	Motor	Destoner: 3 hpGrader: 1 hp
	Approximate cost (Rs)	0.90 lakh
Bucket elevator	Type	Continuous type
	Purpose	Transferring grains from outlet of one machine to the inlet of the other machine.
	Capacity	350 kg/h X 6 nos.
	Feed hopper Capacity	15 kg
	Motor	1 hp 3phase
	Approximate cost (Rs)	0.55 X 6 lakh = 3.3 lakhs

Future scope of research

There exists market potential for the production and placing of pulse-based health foods with the aim of providing high protein foods with significant nutraceutical and functional benefits for specific population groups. Primary processing and value addition technologies in pulse-based foods have to be developed with enhanced quality, nutrient bio-availability, functional benefits, and sensory appeal. Pulses are source of constant supply of nutrition round the year and nutritional value of pulses per 100 gram is much higher than any other vegetarian food. Pulses have tremendous scope to be popularized as 'Health Food' or 'Nutri-Rich Food'. Presence of anti nutritional factors limits the utilization of pulses but at the same time these act as bioactive substances exhibiting significant favourable effects on health in reducing the risk of coronary heart disease, diabetes and obesity.

Functional properties such as solubility, water and fat binding capacity and foaming are influenced by genetic makeup of pulses and amino acid type, water/ oil absorption and protein solubility which determine their utility in the development of bakery products, soups, and extruded products and ready to eat snacks. Ratio of amylase to amylopectin also determines the functional properties of pulse starch such as texture, rheological and swelling properties that are important in food application. High amylase starch retrogrades to greater extent than high amylopectin starch, resulting in higher degree of crystalline, syneresis and gel firmness. Technological processing may evoke positive effects like protein coagulation, starch swelling and gelatinization, texture softening and formation of aroma components. Biofortification can improve nutrition composition, antioxidant capacity, chlorophyll content and soluble sugars by several folds which can support developing attractive, convenient, ready-to-eat and tasty pulse-based food formulations as per consumer demand. Thermal treatment (pressure cooker) can increase water absorption capacity in certain pulses, with reduction in fat absorption and foaming capacities

Safety and quality aspects of pulses

The Codex Alimentarius is a food code established in 1963 by FAO and WHO which sets International standards. For over 50 years, Codex Alimentarius has helped to ensure food safety, quality and fairness in International trade. The Codex Alimentarius of certain pulses, CODEX STAN 171-1989, applies to whole, shelled or split pulses including beans, lentils, peas, chick peas, field beans and cow pea, and it outlines fundamental characteristics pulses should have in order to ensure the quality and safety of the products.

Essential composition and quality factors

Pulses should be safe and suitable for human consumption and must be free from abnormal flavours, odours, living insects and filth (impurities of animal origin, including dead insects). The specific quality factors are given below.

Moisture content: Two maximum moisture levels are provided to meet different climatic conditions and marketing practices. Lower values in the first column of Table 5 are suggested for countries with tropical climate or when long-term storage is a normal commercial practice. The values in the second column are suggested for more moderate climate or when other short-term storage is the normal commercial practice. Various moisture ranges for marketing of pulses are given in Table 5.

Extraneous matter: Pulses shall have not more than 1% extraneous matter of which not more than 0.25% shall be mineral matter and not more than 0.10% shall be dead insects, fragments or remains of insects, and/or other impurities of animal origin.

Table 5: Moisture range for pulses to meet climatic and marketing practice

Pulse	Moisture content (%)	
Beans	15	19
Lentils	15	16
Peas	15	18
Chickpeas	14	16
Cowpeas	15	18
Field beans	15	19

Toxic or noxious seeds: Products covered by the provisions of this standard shall be free from the toxic or noxious seeds like *Crotolaria*, Corn cockle (*Agrostemma githago* L.), Castor bean (*Ricinus communis* L.), Jimson weed (*Datura* spp.) and other seeds that are commonly recognized as harmful to health.

Contaminants: General Standard for Contaminants and Toxins in Food and Feed (CODEX STAN 193-1995) sets the maximum levels for heavy metals in pulses. For example, the maximum level for lead in pulses is 0.2ppm. Pulses should comply with the maximum pesticide residue level, established by the Codex Alimentarius Commission.

Hygiene

Pulses should be prepared and handled in accordance with appropriate sections of the General Principles of Food Hygiene (CAC/RCP 1-1969), Codes of Hygienic Practice as well as other relevant Codes of Practice. They should be free from parasites and from micro-organisms in amounts, which may represent a hazard to health.

Methods of analysis & sampling

Visual examination can help identify various defects of pulses. This include seeds with serious defects which, for example, have been affected by pests, seeds with slight defects, e.g. those which have not yet reached normal development or broken pulses. Other sampling criteria include seed discoloration and presentation. A percentage limit for defects and seed discoloration serves as a guideline to help ensure that a certain amount of defective or discoloured pulses does not exceed the accepted limit. Details of method of analysis for pulses are given in Table 6.

Table 6: Method of analysis for pulses

Defects	Limit
Seeds with serious defects. Seeds in which cotyledons have been affected or attacked by pests; seeds with very slight traces of mould or decay; or slight cotyledon staining	MAX: 1.0%
Seeds with slight defects. Seeds which have not reached normal development; seeds with extensive seed coat staining, without cotyledon being affected; seeds in which seed coat is wrinkled, with pronounced folding, or broken pulses	MAX: 7.0% of which broken pulses must not exceed 3.0%
Broken pulses. Broken whole pulses are pulses in which cotyledons are separated or one cotyledon has been broken. Broken split pulses are pulses in which cotyledon has been broken	
Seed discolouration	
Seeds of a similar colour but a different commercial type (except in beans with white seeds)	MAX: 3.0%
Seeds of different colour (other than discoloured seeds)	MAX: 6.0%
Discoloured seeds	MAX: 3.0%
Discoloured seeds of same commercial type	MAX: 10.0%
Beans with green seeds and peas with green seeds with slight discolouration of seed	MAX: 20.0%

Adulteration in pulses

Many times pulses purchased from market are either adulterated with khesari dal or some low-priced dals or with some undesirable colours to improve its appearance which are harmful for health. A common consumer may not have sufficient knowledge about the impurity and quality of pulses. Awareness about necessary means to verify or test its quality is also lacking. Mere visual inspection does not serve purpose. Consumer awareness is the remedy for eliminating this evil of adulteration and sale of sub-standard product.

Test for adulteration

Some simple and quick tests are available which can be easily performed to ascertain the purity of dal or besan if it is adulterated with khesari dal or the presence of some harmful chemicals in the form of colouring agents. Available tests are given in Table 7.

Packaging of pulse products

Packaging is an important function in the marketing of pulses. It is a practice to protect the produce from any damage during storage, transportation and other marketing operations. Packaging also has a great potential for value addition in pulse marketing. When the product is packaged in sacks, these must be clean, sturdy and strongly sewn or sealed. Containers, including, packaging material should be made of substances which are safe and suitable for their intended use. They should not impart any toxic substance or undesirable odour or flavour to the product. Properly graded and packed grains not only facilitate convenience in transportation and storage but also attract consumers to pay more and hence fetch more income from farm produce.

Pulses shall be packaged in containers which will safeguard the hygienic, technological, nutritional and organoleptic (taste, colour, odour, and feel) qualities.

Plant fibre bags

A number of plant fibres like jute, cotton, and sisal are used for making bags for packaging of pulse grains. The choice of packaging material depends upon packaging requirement.

Plastic fibre bags

PP woven bags: These bags are made of plastic (polythene) woven fabrics, or of mixed fabrics (plant fiber and plastic fiber). Polythene bags are widely used for packaging grains and they seriously rival the traditional jute bag. These bags have the advantage of being resistant to moisture transfer, rot-proof, and impermeable to insect pests. However, they should be suitably treated in order to resist sunlight, since polythene deteriorates when exposed to light. With good treatment, a polythene bag can be reused for 6-12 times. They cost less than jute bags and are harder to handle. Their surface is very slippery, and so they cannot be stacked very high.

Polyethylene bags: The most commonly used material is polyethylene and polyester. These bags are not moisture-proof but are moisture resistant. They deteriorate easily if exposed to sunlight, therefore, should be protected from direct sunlight. They are difficult to stack as bags may slide and fall down.

Table 7: Screening tests for detecting adulteration in pulses

Sl. No	Food articles	Adulteration	Test
1.	Pulses/Besan	Khesari dal (Lathyrus sativus)	Add 50 ml of HCl to small quantity dal and keep on simmering water for about 15 minutes. The pink colour, if developed, indicate the presence of Khesari dal.
2.	Pulses	Metanil yellow (Dye)	Add concentrate HCl to small quantity of dal in little amount of water. Immediate development of pink colour indicate the presence of metanil yellow and similar colour dyes.
3.	Pulses	Lead chromate	Shake 5 g of dal with 5 ml water and add a few drops of HCl. Development of pink colour indicate the presence of lead chromate.

Labelling

In addition to the requirements of the *Codex General Standard for the Labelling of Pre_Packaged Foods*, the following applies: The product name, which is shown on the label, should refer to the commercial type of the pulse. Information for non-retail containers shall either be given on the container or in accompanying documents, except that the name of the product, lot identification and the name and address of the manufacturer or packer shall appear on the container.

Conclusion

The post harvest technology in pulses in recent times has attracted attention as a powerful tool for entrepreneurship. With the onset of industrialization, most of the processing activities have been shifted to urban areas where infrastructural facilities are available. Easy availability of raw materials, opening up of exports, better packaging and marketing for finished products are some of the needs of the hour for encouraging entrepreneurship in pulse processing. However, most crucial step that needs to be taken is to maximize the use of improved dal mills to increase yield and better output. It will take a conscious effort by all dal milling associations and dal mills across India to implement suggestions that will best work towards uplifting dal milling industry aid entrepreneurship. Some of the factors which will aid entrepreneurship in pulse processing to improve turnover and profitability are improved protocol for value addition of by-products, up gradation in pulse processing machinery etc.

References

Agrawal, A. 2016. Value Addition of Pulse Products in India. *J. Nutr. Health Food Eng.* 5(2): 00166. DOI: 10.15406/jnhfe.2016.05.00166.

Asif, M., L. Rooney., R. Ali, & M. N. Riaza, 2013. Application and Opportunities of Pulses in Food System: A Review. *Crit Rev Food Sci Nutr.* 53(11): 1168-1179.

Chakravarty, A. 1988. *Milling of Pulses in Post Harvest Technology of Cereals, Pulses and Oilseeds.* Oxford and IBH. New Delhi.

Chakravarty, A., A. S. Mujumdar., G.S.V. Ragahavan & H.S. Ramaswamy. 2003. Handbook of Post Harvest Technology for Cereals, Fruits, Vegetables, Tea and Spices. Published by Marcel Dekkar inc, New York. ISBN No 0-8247- 0514-9., Pages 884.

Dhal, L. 2014. Overview of Pulse Industry in India and Importance of Minor and Imported Pulses. *In Hand Book of Minor and Imported Pulses in India*, Bangalore 8-12. Foretell solutions. Commodity India. com/pulse/hand Ebook.

Gopalan, C., B.V R Sastri & S.C. Balasubramanian. 2004. *Nutritive Value of Indian Foods*, National Institute of Nutrition, ICMR, Hyderabad.

Kulkarni, S. D. 1993. Development of Food Legumes Machinery in India, *National Seminar on Dhal Milling Industry in India – Its Future Needs*, p. 118-138, National Productivity Council (NPC).

9

Prospects in Value Addition of Soybean

Ajesh Kumar V & Sumeda S Deshpande

Introduction

Soybean (*Glycine max*) is also known as the wonder bean or miracle bean or "golden grain" due to its diversified applicability. Soybean has an important place in world's oilseed cultivation scenario, due to its high productivity, profitability and vital contribution towards maintaining soil fertility. The crop also has a prominent place as the world's most important seed legume, which contributes 25% to the global vegetable oil production, about two thirds of the world's protein concentrate for livestock feeding and is a valuable ingredient in formulated feeds for poultry and fish. The major soybean producing nations are United States, Brazil and Argentina. These three countries dominate in global production, accounting for 80% of the world's soybean supply. Global production of soybean has grown at a CAGR of 2.78% from 215.69 million metric tons in 2004-05 to 283.79 million metric tons in 2013-14.

India stands 5th in global production of soybean. In India, soybean cultivation started in early 1970s, but it has emerged as number one among oilseed crops covering around 11 m ha area with total production of around 11.5 m tons. At present, the average productivity of soybean is about 1.3 tons/ha and it can be easily doubled in near future. Production of soybean in India is dominated by Maharashtra and Madhya Pradesh which contribute 89% of the total production. Rajasthan, Andhra Pradesh, Karnataka, Chhattisgarh and Gujarat contribute the remaining 11% production.

About 85% of world's soybeans are processed annually into soybean meal and oil. Approximately 98% of soybean meal is crushed and further processed into animal feed with the balance used to make soy flour and proteins. Of the oil fraction, 95% is consumed as edible oil; the rest is used for industrial products such as fatty acids, soaps and biodiesel.

Health benefits of soybean

Soybean grain legume has around 40% good quality proteins and 20% oil. Protein content in soybean is almost double in comparison with all other pulses. In addition to its high protein content, soybean has a wealth of other nutrients such as dietary fibre, a host of vitamins, vitamin C and K, riboflavin, folate and thiamine. Mineral elements abundant in soybeans include iron, magnesium, zinc, selenium and calcium. Soybean is rich in antioxidants also. Nutritive value of soybean is given in Table 1. Soy products can easily meet the protein requirements of a vegetarian diet. Unfortunately, soybean in India is used mainly as oil crop, whereas, it is used as food crop in most of the East and South East Asian countries including Japan and China for over 5000 years.

Besides nutritional benefits, soybeans provide several therapeutic benefits too. Soybean is one of the very few plants that provide high quality protein with minimum saturated fat. They have been reported to contain certain special biochemical constituents that have been implicated for positive effect on human health. DASH- (Dietary Approach to Stop Hypertension) is one important aspect where it has been evidenced that consumption of vegetarian diets that are high in dietary fibre are, in general, extremely useful in controlling hypertension. In this context, it may be mentioned that soybean fibre plays an important role in controlling hyperlipidaemia and hypercholesterolemia. Regulating the serum cholesterol levels and thereby hypertension, gastro-intestinal functions and coronary heart diseases has thus been evidenced and correlated with soybean consumption.

Table 1: Nutritive value of soybean per 100g.

Components	Quantity	Components	Quantity
Moisture (g)	8.5	zinc (mg)	4.9
Energy (Kcal)	446	Copper(mg)	1.7
Protein (g)	36.5	Manganese(mg)	2.52
Fat (g)	19.9	Selenium (mg)	17.8
Carbohydrates (g)	30	Vitamin C (mg)	6.0
Fibre (g)	9.3	Riboflavin(mg)	0.874
Ash (g)	4.9	Thiamine (mg)	0.874
Isoflavones (mg)	200	Niacin (mg)	1.62
Calcium (mg)	277	Panthotenic acid (mg)	0.79
Iron (mg)	15.7	Vitamin B_6 (mg)	0.38
Magnesium (mg)	280	Folate (µg)	375
Phosphorus (mg)	704	Vitamin B_{12} (µg)	0.0
Potassium (mg)	1797	Vitamin A (µg)	2.0
Sodium (mg)	2.0	Vitamin E (mg)	1.95

Source: USDA Nutrient Data base for Standard Reference Release 28 (2015)

Foods derived from soybean have a potential role in the prevention and treatment of chronic diseases; notably, cancer, osteoporosis and heart diseases. By fitting into estrogen receptors, soy's isoflavones play a special role in helping women get through menopause and in reducing hot flashes. Soy foods rich in isoflavones also help to prevent the bone loss that often accelerates after menopause, which can progress to osteoporosis.

Soybean extract containing chymotrypsin has been indicated to inhibit cancer in colon, lungs and mouth. Soy polysaccharides are known to reduce fasting blood glucose levels and significantly reduce insulin response thus delaying glucose absorption in both insulin dependent and non-insulin dependent diabetes mellitus.

Though, soybean production has increased significantly in the country, consumption did not increase in the matching order. This can be attributed mainly to lack of awareness and familiarity with proper processing methods. Thus, creation of awareness of utility of soy foods in providing nutrition through simple methods of processing at domestic scale significantly improves the nutritional status of population at low cost. With increase in acceptability and demand of soy based foods in Indian population, there exists a vast scope for establishing soy based entrepreneurship in large, small and cottage scales.

Scope of entrepreneurship development in soybean processing sector

Economic development of a country is supported by entrepreneurship in several ways. It is a key contributor to innovativeness and product improvement and plays a pivotal role in creating employment. At Centre of Excellence on Soybean Processing and Utilisation (CESPU), CIAE, Bhopal R & D efforts were made to develop process technologies for soy based food products matching to conventional food habits of the country. The centre has also developed set of equipment suitable for processing soybean at domestic and cottage levels. The dissemination of research output to the end user is achieved through awareness programmes, training to the home makers and upcoming entrepreneurs.

Value added products of soybean

Soy products find wide application as a versatile ingredient virtually in every food system, including bakery, breakfast cereals, beverages, infant formulas, dairy analogue and meat analogue. CESPU, CIAE, Bhopal has developed technology for more than 20 soy products and some of the important soybean based products are discussed below.

Soybean oil

Soybean oil is a natural extract from whole soybeans. This clear, odourless and flavourless oil is excellent for stir-frying as it brings out the flavour of foods. Soybean oil has a high smoking point, facilitating the cooking process. The oil content in the seed is about 20% on dry weight basis. The crude soybean oil of good quality has a light amber colour, which upon alkali refining turns to yellow. Soybean oil is low in saturated fatty acids without any cholesterol and hence considered as heart-healthy oil. Two fat components essential for health and wellbeing namely linoleic and linolenic acids are also present in right proportion in soybean oil. It is also a good source of vitamin E. Like fish oils, soybean oil contains omega-3 fatty acids which are known to be protective against heart diseases and cancer. Soybean oil is obtained from the seeds either by pressing or solvent extraction method.

Soy nuts

Soy nut is a ready to eat product made by either deep fat frying or roasting. The process of preparation of deep fat fried soy nut starts from soaking of whole soybeans in water containing 5% table salt and 3% sodium bicarbonate for three hours followed by dewatering and removal of surface moisture and then frying in oil for 10 minutes at 180°C. The roasted soy nut can be prepared by roasting the soaked and dewatered beans for 10 minutes at 200°C. Chick pea flour can also be coated on these nuts for better taste. Soy nuts have less fat and more protein when compared to conventional nuts. These are similar in texture and flavour to peanuts and comparatively cheaper. Soy nuts have 50% more protein and 50% less fat than peanuts.

Soy flour

Soy flour can be prepared by processing soybean to get full fat soy flour, medium fat soy flour and defatted soy flour. It is a good source of quality protein. These soy flours can be used with cereal (wheat, maize, millets) flour at 10% level to get high protein products.

Full fat soy flour (FFSF): It is one of the simplest products that can be made from soybean (Fig.1). The process for the production of full fat soy flour developed at CIAE which is recommended for commercial

Fig. 1: Full fat soy flour

use is shown in Flow chart No.1. This flour contains 40% protein and 20% fat. To avoid the problems in milling due to high oil content in soybean, it is suggested to mix the processed dal with wheat or other cereals before grinding.

Medium fat soy flour (MFSF): It is a by-product of soy oil extraction in which the cake obtained is milled to produce medium fat soy flour. It contains 45-46% protein and 6-9% oil. The equipment required to process 2 tons MFSF per day may cost Rs. 25-30 lakh.

Defatted soy flour (DFSF): It contains more than 50% protein and less than 1% fat. It is produced by milling of desolventised soybean. DFSF unit is normally associated with solvent extraction unit and it demands heavy investment for modern technique of desolventisation to get DFSF with good functional quality. Enterprise to produce 1 ton of DFSF supplemented with cereal flour per day can be established at a cost of Rs.5-7 lakhs.

Soybean
(100 kg)
↓
Cleaning and Grading
(Cleaned soybean, 95 kg)
↓
Dehusking and Splitting
(Split cotyledon, 80.8 kg)
↓
Blanching in hot water
(For 25 min, dal: water, 1:3)
↓
Blanched splits
(Dried, 80 kg)
↓
Drying
(Dried splits, 79.6 kg)
↓
Milling
(78.8 kg)
↓
Sifting
↓
Full fat soy flour
↓
Weighing and Packaging

Flow chart No 1. Production of full fat soy flour

Soy semolina (soy suji)

Soy suji can be used like wheat suji to prepare *upma, halwa, laddu*, etc. and is useful in nutritional enhancement of conventional products prepared using wheat suji. It contains about 40% proteins and 20% oil and can be used up to 50% with wheat suji. It is prepared by the method of extrusion cooking of soy grits followed by milling. Protein content of soy suji is almost four times of wheat suji.

Soy milk

Soy milk is one of the simplest dairy analogues that can be prepared from soybean. Protein content is same as that of cow's milk (around 3.5%) with 2.9% fat. Soy milk can be made at home with simple traditional tools or with a soy milk machine or juice maker. Large scale commercial processing is also possible. Soy milk can be used as a substitute for cow milk but it is having different aroma and flavour. Soy-based infant formula (SBIF) is used for infants who are allergic to cow milk. Common method to extract soy milk is described below.

Properly cleaned soybeans are soaked overnight (10-12 h) in 4-volumes of water until soft. If dehulled soybeans are used, the soaking time may be reduced to 4-6 h or until soft. They are then thoroughly washed in tap water and ground into pulp using about 2-3 volumes of hot water (> 80°C). The easy way is to grind small batches with twice the volume in hot water in the blender or food processor. The resulting mash is then transferred into container preferably stainless steel and boil for at least 20 min. Cooking inactivates Trypsin Inhibitor (TI) and some non-desirable enzymes such as lipoxygenase present in soy. To the boiled mash, add sufficient boiled water so as to make its volume up to about 10- 12 litre (including soy slurry volume). Mash is then strained through a double layer cotton cloth and squeezed to get the liquid. The liquid thus obtained is soy milk (Fig.2). Unfortunately, this oriental beverage has a distinct beany flavour. This flavour is quickly generated by the endogenous lipoxygenase enzyme whenever the soybeans are ground and exposed to water. To make soy milk palatable, different flavours could be added to attract the consumers. This method will yield about 6-8 litres of soy milk from one kg of raw soybean. The residue left behind is

Fig. 2: Soy milk

called okara, which contains about 20-22% protein (db). It is also a good quality edible by-product and can be utilized to prepare various other dishes. It may be dried and preserved for future use because its shelf life, in fresh form may not be more than 5-6 hour.

Soy paneer (Tofu)

Soy milk is the base material for making soy paneer (Tofu), yogurt and other dairy analogues. Tofu is a product somewhat similar to dairy paneer and is made by curdling soy milk (Fig.3). Calcium sulphate (dihydrate) or magnesium chloride at the rate of 2 g per litre of soy milk or their combination is traditionally used for curdling. Citric acid may also be used. So far as yield of tofu is concerned, calcium sulphate gives best yield of tofu (1.75 kg per kg of raw soybean). The coagulant is dissolved or suspended in warm water and spilled over the soy milk and mixed gently. Curd thus get separated from water (whey) is then set in perforated wooden boxes or stainless steel trays and pressed under a weighted lid for half an hour or more. One kilogram of dry soybean yields 6-8 litres of milk which gives about 1.5-1.75 kg tofu. Tofu comes in variety of forms. Silken tofu has a creamy consistency and can be used with salad dressings and as a replacement for sour cream; firm tofu is solid and can be cubed or cut; and soft tofu is softer than firm tofu but is not a liquid like silken tofu. Tofu is also called soybean curd, and it absorbs the flavour of the food with which it is cooked. Nutritionally, tofu is high in protein and B vitamins, with firm tofu containing the most nutrients. At 72% moisture, it contains 14% protein and 9% fat. Tofu is a popular ingredient for stir-fry dishes. It may be used instead of meat, fish or chicken. It can also be used in numerous dishes as a replacement for eggs and cheese. Though, tofu is sold as already cooked, additional cooking makes it more digestible. The process for production of soy paneer (Tofu) developed at CIAE which is recommended for commercial use is shown in Flow chart No 2. Soybean tofu plant is illustrated in Fig. 4.

Textured soy protein

Textured soy protein also known as soy chunks or soy nuggets is an extruded product prepared from de-oiled cake (Fig.5). The chunks are dried after extrusion. They have become very popular and are used along with other vegetables for making curries. They are low cost protein rich substitutes for cheese, paneer, meat and fish. The process flow chart to prepare textured soy protein is given in flow chart No 3. Chunks can also be converted into flakes or granules.

Whole Soybean
(Cleaned and graded, 1kg)
↓

Dehulling and splitting → Hulls
↓ (1kg)
Soaking in water (1:3)
With 1% NaHCO$_3$ Soaked soybean with 45% m.c
↓

Wet Grinding with hot water (1:8)
(Soy puree)
↓

Boiling for 15-20 min
↓

Filtration through muslin cloth → Soy residue (Okara)
↓   (Soy milk, 6-8 litres)
Coagulation using CaSO$_4$, or citric acid
(Proteinate complex)
↓

Filtration and Pressing → Whey
(Soy paneer)
↓

Washing and Storage
(Soy paneer, 1.5-1.75 kg)

Flow chart No.2: Production of soy paneer (Tofu)

Fig. 3: Soy paneer (Tofu)

Fig. 4: Soybean tofu plant

Mixing of ingredients
↓
Extrusion
↓
Drying
↓
Packing

Flow chart No. 3: Production of texturised soy protein

Fig. 5: Textured soy protein (Soy chunks)

Soy based bakery products

Centre of Excellence on Soybean Processing and Utilization (CESPU), CIAE, Bhopal has developed process technology for soy fortified biscuits and cakes. The cost of equipment for an enterprise to produce one ton of product per day costs about Rs. 20 Lakh. Small units of 50, 200 or 500 kg biscuits/day can also be established. By incorporating 10-12% soy flour, the protein content of the bakery products can be enhanced to 50% from 7 to 8% without soy incorporation. Soy based bakery products are good in taste and appearance.

Soy yogurt

Yogurt is a tasty and nutritional rich product made by fermenting cow milk to form an acidic gel. A yogurt analogue can be prepared from soybean beverage base. The procedure for the preparation of yogurt from soy milk is given in flowchart No.4.

3 parts soybean beverage base (10% total solids) and 1 part water
↓
Neutralize to about pH 7
↓
Add 3% Sucrose and 3% cerelose (dextrose)
↓
Heat to 88°C (190°F) to pasteurize
↓
Homogenize at 3500 psi/500 psi
↓
Cool to 43°C by cool water and stirring
↓
Add 3% of culture containing 1 to 1 ratio of *S. thermophilus* and *L. bulgaricus* in low fat (2%) cow's milk, mix thoroughly
↓
For the "Sundae style" put 15-20% fruit flavour (preserves) at bottom of the cups in advance.
↓
Fill in the yogurt cups (or in bucket for batch production)
↓
Incubate at 40°C until pH reaches 4.4-4.3 (approximately 5-6 hrs)
↓
Cool to 20°C with cold water – (optional)
↓
Immediately mix with 15-20% fruit flavour and stir until uniform
↓
Pack, label, and store under refrigeration

Flow chart No.4. Production of soy milk based yogurt

The expenditure to be incurred to establish soy based food enterprises is given in Table 2. Entrepreneurs who are interested in establishing any of these soy based enterprise should get technical knowhow of the process through proper training so as to ensure the production of good quality products.

Table 2: Details of soy based food enterprises

S. No	Soy product	Production capacity Kg/day	Equipment cost Rs in Lakh	Employment potential Man -years
1.	Soy flour			
	(a) Full fat	100	4-5	4-6
		500	15	12-14
		1000	23	16-20
	(b) Medium fat	2000	28	10-12
2.	Soy fortified wheat flour	1000	7	8-10
3.	Soy suji	2000	22-25	10
4.	Soy nuggets	1500-2000	25	10
5.	Soy milk	200 Litre	4	4
6.	Soy paneer	50	4	4-8
7.	(a) Soy sattu	50		
	(b) Ready to eat mixes	2000	1.530	310
8.	Soy based bakery products	2001000	5-618-20	615
9.	Soy snacks	50-100	3-4	2-3
10.	Soybean oil			
	(a) Solvent extraction	As per requirement		
	(b) Extrusion - Expelling	1800	30	10-12

Scope of by-product utilisation

The common industrial practice of discarding food processing by-products leads to economic loss and socio-environmental problems. Hence, the search for their alternative uses and value addition has gained much global attention in recent years. Okara, also known as soy curd residue, is the major food processing by-product derived from soybeans. It is the ground soybean residue remaining after filtering the water-soluble fraction during soy milk or Tofu (soybean curd) production. For every 1 kg of soybean used in manufacturing soybean curd, about 1.1-1.2 kg of okara is obtained (Khare *et al.*, 1995). The direct incorporation of okara into animal feed or human food is possible. Dried okara can also be used as an ingredient in various foodstuffs, especially in bakery items. Biovalorisation of okara is also possible through fermentation to produce a variety of functional ingredients and foodstuffs. The health benefits and nutritional quality of okara are often enhanced by fermentation, and the fermented okara is an inexpensive substrate for extraction of bioactive substances (Vong and Liu, 2016). The best method of utilisation of okara is to use it as a food additive. The general composition of okara is shown in Table 3.

Table 3: Composition of okara (per100g dry matter)

Macro components	Quantity (g)	Micro components	Quantity (mg)	Phytochemicals	Quantity (mg)
Carbohydrate	3.8-5.3	Thiamine	0.48-0.59	Isoflavone aglycones	5.41
Protein	15.2-33.4	Riboflavin	0.03-0.04	Isoflavone glucosides	10.3
Dietary fibre	42.4-58.1	Niacin	0.82-1.04	Malonyl glucosides	19.7
Fat	8.3-10.9	K	936-1350	Acetyl glucosides	0.32
Insoluble dietary fibre	40.2-50.8	Na	16-96	Phytic acid	0.5-1.2
Soluble dietary fibre	4.2-14.6	Ca	260-428	Saponins	0.10
Ash	3.0-4.5	Mg	130 -165		
		Fe	0.6-11		
		Cu	0.1-1.2		
		Mn	0.2-3.1		
		Zn	0.3- 3.5		

Source: Vong & Liu (2016)

Soy whey water is another by-product generated during the curdling of soy milk for the preparation of tofu. Soy isoflavones are mostly dissolved in the whey. They have significant pharmacological actions because of their similar structures to that of estrogen.

New technologies/innovations in value addition of soybean

In recent times researchers have come up with some innovative value addition methods and new technologies in soybean processing. Use of sprouted or germinated soybean is one of such new innovations. During germination, qualitative and quantitative changes occur due to the activation of growth enzymes. The optimized process for sprouting of soybeans includes cleaning of whole soybean seeds to remove dust, dirt and foreign matter, sterilization with 0.1% (w/v) potassium permanganate, rinsing with distilled water to remove traces of potassium permanganate, soaking in 4 times the amount of water for 4 h and draining excess water. After this, keep the seeds in seed germinator in a single layer on filter paper in sterile containers in the dark at 25°C and 90% Rh for 48 h. This will give good quality sprouted seeds which are further dried at 60°C overnight in an oven to arrest the enzymatic activity and stop the germination process (Agrahar and Jha, 2009). Sprouting enhances protein, mineral and vitamin (ascorbic acid, riboflavin, choline, thiamine, tocopherols and pantothenic acid) content in soybean (Sangronis and Machado, 2007). The dried sprouts can either be used as such or made into flour and grits to be used as start materials in various products.

Many researchers have reported the use of membrane technology in soybean processing. Membrane techniques represent a type of unit operation which is having the potential to replace conventional separation and concentration processes to an extent. Membrane filtration/separation processes have gained popularity in the food processing industry. The main attractions of membrane technology in biological and food applications are mild operating conditions, simplicity, potentially high selectivity and, low energy requirements (in the case of low pressure processes such as ultra filtration). Main applications of membrane separation techniques in soy processing are in soy protein concentration, purification and fractionation. Two important soybean protein products manufactured are soy protein concentrate (SPC) and soy protein isolates (SPI). SPC is an edible protein product with a protein content of at least 65% on dry weight basis, whereas SPI is a product with at least 90% protein on dry weight basis (Wang and Jhonson, 2001). These are valuable ingredients in foods due to their high nutritional value and possible health benefits. The versatile uses of SPI can be attributed to the wide range of functional properties that SPI can confer to a food product. Critical functional properties necessary in protein ingredients include solubility, water and fat absorption, emulsion stabilisation, whippability, gelation, foaming and good organoleptic properties (Kinsella, 1979). Many physical, chemical and enzymatic modifications have been used to expand the range of functional properties in soy proteins. Extensive research has been done on enzymatic, mechanical and thermal modifications of soy proteins to improve their functional properties (Bernard *et al.*, 2007). Their use has been limited due to undesirable flavours, reduced bioavailability due to the presence of high levels of phytic and nucleic acids and functional problems such as insolubility.

Traditional processing techniques for producing soy protein concentrates and isolates partially overcome these problems. These methods involve extraction, heat treatment, precipitation by addition of acid or alcohol, and centrifugation to separate protein from other components. These conventional methods are time consuming and sometimes result in products with poor functional properties, and can generate a whey like waste stream which contains some of the proteins and nutraceuticals. An increasing concern these days is that large quantities of water are required, especially to "wash" the curd (the precipitated protein) to remove as much of the adhering non protein components.

A good alternate process which is found for purifying vegetable proteins and removing many objectionable flavour compounds is by using membrane technology. Since, the undesirable oligosaccharides, phytic acid and some of the trypsin inhibitors are smaller in molecular size than proteins and fat components, it should be possible, by careful selection of the membrane and operating parameters, to selectively remove these undesirable components and produce a purified protein isolate. Production of soy-protein isolate using membrane separation technology in the laboratory scale has been studied by many researchers and marked its advantages over the conventional separation processes.

Safety and quality aspects of soybean products

The first step in any food processing is to get good quality raw materials to produce quality product. Raw soybean when procured in the season should be cleaned thoroughly to remove the broken and shrivelled grain, dirt, dust, infested grain etc. Then, grains are stored in air tight metallic bins on raised platform so as to minimise moisture migration in the bin. Damage to soybean grain and exposure of grain to oxygen and moisture will develop a beany flavour in soy products during processing. Therefore, undamaged and sound soybean should be selected for processing. Different processing approaches like dry or moist heat treatments are effective to inactivate the antinutritional factors in soybean and unfolding of protein structure for improved protein digestibility.

Packaging of soybean products

Soybean products being nutritive are more susceptible to spoilage during storage. Development of off-flavour, often described as beany, grassy or bitter during processing and storage seriously limits full utilization of these products. The deterioration in quality has been attributed to the action of lipoxygenases. However, oxidative rancidity in soy products during storage cannot be ruled-out which might occur due to non-enzymatic decomposition of lipid hydro peroxides. Deterioration in quality may be more rapid if the environment of storage is adverse either with higher temperature or with humidity. General packaging details of soy products are discussed here.

Soy flour, soy nuts, soy sattu, soy biscuits and other soy based dry products can be safely packed in LDPE bags of 125 to 200 microns. During packaging, the moisture content of any of the dry soy products should never be more than 10%. To ensure this, the packing should be done immediately after production. Special care should be taken during monsoon season and in areas adjacent to any water bodies (river, ocean, cannels, etc. where the environment is humid). The LDPE packaging materials can provide safe storage up to 3 months. However, after opening, these products must be consumed within 10-15 days. The shelf-life of FFSF can also be enhanced by modifying the atmosphere in the head space of the package. For this purpose, before sealing the package, the oxygen present in its head space is replaced by an inert gas (nitrogen) to retard the rate of oxidation inside the package and thereby checking the deterioration in product quality (Bargale & Griffin, 1991).

For soy based dairy analogues such as soy milk and soy paneer, the most common and easily available packaging's are reusable glass bottles and low density polyethylene (LDPE) bags. However, glass bottle may be a better option for soy milk. These can be stored for more than 24 hours under refrigeration. The main constraint in the popularization of tofu is its poor keeping quality of 3-4 hr at ambient storage conditions and 2-3 days under refrigeration. This is mainly because; this product is highly susceptible to microbial spoilage due to high water activity. Unpacked tofu can be stored for 2-3 days by immersing them in water under refrigerated condition (5°C). Pasteurization can extend the shelf life of both tofu and soy milk. Pasteurization is effective for packed tofu to enhance its shelf-life to more than 15 days under refrigerated storage in glass bottles or LDPE bags. The expected shelf life of tofu and milk is 7-9 days.

Shelf life of soy milk can be extended up to 90 days through sterilization process. Batch type sterilization is performed in autoclave. In sterilization, all the microbes which are not destroyed by pasteurization are killed because of high temperature treatment of 121°C for 30 min. This temperature can be attained in autoclave at 1.1 kg/cm². For this, special grade bottles (autoclavable bottles) and autoclavable LDPE bags are used. Aseptic packaging of soy milk in tetra packs extends the shelf life up to 6 months without refrigeration.

Success stories in soybean processing sector

In northern and western parts of India, mainly in Punjab, Haryana, Utter Pradesh, Gujarat, and Maharashtra; many entrepreneurs are successfully running soybean processing based industries. Centre of Excellence in Soybean Processing and Utilization (CESPU), formerly known as Soybean Processing and Utilization Centre (SPU), was established in 1985 at CIAE, Bhopal to develop and disseminate technologies to extend the nutrition and health benefits of soybean to population. To achieve this objective, processing technologies and suitable equipment were developed. To transfer these technologies effectively through enterprise development, the SPU centre started the training of upcoming entrepreneurs since 1995. The training is organized for preparation of soy milk and soy paneer, full fat soy flour, and soy based bakery products and snacks. So far, more than 165 training programmes have been organized for soy milk and soy paneer preparation for upcoming entrepreneurs of different states and 2300 trainees have been trained. Out of these, over 500 cottage scale soy food based enterprises have been established in different parts of India. These enterprises process soybean mainly for production of soy milk, soy paneer, soy flour, soy based bakery products, soy nuts, etc. These enterprises have succeeded in making available nutritious soybean products to the population of different states.

Cost of equipment required in soybean processing units

Interested entrepreneurs can select any of the soybean product for establishing new enterprise subjected to the market potential and availability of raw materials. Size of the plant varies according to the targeted production and financial capacity of the entrepreneurs. The information on equipment with approximate cost for soybean processing plant is given below. Soy milk/soy paneer plant, textured soy protein plant and soybean oil plant are categorised into small, medium and large scale units respectively.

Soy milk / Paneer plant: Soy milk and soy paneer can be prepared from the same processing plant. The details of equipment required in soy milk/ soy paneer plant is given in Table 4.

Table 4: Machinery/equipment required for soy milk/ soy paneer plant

Equipment description	Capacity	Approximate cost (Lakh)
Commercial soy milk/paneer machine	200 litre soy milk/hr	2.9
Grinder		
Milk sterilizer		
Tank		
Pump		
Pneumatic tofu press		
Soya milk and paneer making machine	100 litre soy milk/hr	1.5

The details of the suppliers of machines are listed below:

1. Pushpanjali Agro Industries, Plot No. 29, Gaurav Park, Near Tangri Bridge Ambala - 133001 , Haryana

2. Khalsa Engineers, Machhiwara Road, Kohara, Ludhiana-141112, Punjab India

3. M/s SSP Pvt. Ltd. 19 DLF, Industrial AreII,13/4, Mathura Road Faridabad-121003

4. M/s Sanjay Gupta, DGS Group, Gandhi Nagar, Quarter No.5 Pathenkheda, Dist. Betul, Madhya Pradesh-460449

5. M/s Hilal Ahmed Jalib, 3rd floor, 38A, Amritpuri (Garihi) East of Kilash NewDelhi-110065

6. M/s Gujrat Engineering Enterprises, 31/1, Paiga Estate, Jehangarabad Bhopal- 460008

7. central Institute of Agricultural Enginering , Nabibagh, Berasia Road Bhopal-462038

8. M/s Nul Life Consultants & Distributors Pvt. Ltd, 31 C, Second Floor Mohammadpur, Bhikaji Cama Place, New Delhi-110066

9. M/s Gaur Engineering Pvt . Ltd,217 , Nagarjuna Hills, Panjagutta Hyderabad-500482

10. M/s Raylon Metal works P.O Box No. 17426, J.B Nagar, Anderi (E) Mumbai- 400059

11. M/s Monica Processing Plant, 44-D2, Sanver Road, Industrial Area Indore-452015

Texturised soy protein plant: The details of equipment required in a plant to produce 2 tons of texturised soy protein per day are given in Table 5.

Table 5: Machinery/equipment required for texturised soy protein plant

Equipment description	Qty	Approximate cost (Lakh)
Mixing-cum-grinding machine with 5 HP motor and other accessories	1	2.00
Soya Nugget Extrusion Plant with 30 HP motor and complete set of screws, barrel and suitable dies	1	14.00
Vibrating sieve with 3 HP motor	1	0.60
Platform type weighing scale; Capacity 200 kg.	1	0.20
Bag sewing machine	1	0.20
Total		17.00

Suppliers of equipment for textured soy protein

1. Brimco Engg. Works, M-24/1, Street No 9, Anand Parbat Inds Area New Delhi 110 005.

    Tel. No. : 25726347- 25761786, Fax: 22145040

2. Flavourite Foods & Services Pvt Ltd, 208 Manas Bhavan, 11 RNT Marg Indore 452 008, Tel. No. 2527644, 5046509, Fax: 5040953

Soybean oil extraction plant: The entire process to extract oil from soybean using Hexane as the solvent is divided into main extraction, desolventisation and distillation. The extraction involves complex machineries and processes. Table 6 shows the details of equipment required in a soybean oil solvent extraction plant.

Suppliers of solvent plant machineries

1. M/s. M. M. Tekno Engineers, A – 65, MIDC, Taloja , Navi Mumbai – 410208

2. M/s. Muez Hest India Pvt. Ltd., 231, Blue Rose Industrial Estate, Near Cable Corporation Western Express Highway, Borivali (E), Mumbai – 400 066.

Suppliers of Boilers

1. M/s. Ross Boilers, 33, Burhani Industrial Estate, Kondhwa Bhudruk, Pune – 411 037. Tel. No. 020 - 24269393, 24272293

2. M/s. Micro Dynamics Pvt. Ltd., T – 181 – 1/A, MIDC Bhosari, Pune – 411 026. Tel. No.020-27120839 / 30685454

Table 6: Machinery/equipment for soybean oil solvent extraction plant

Equipment description	Capacity	Approximate cost (Lakh)
Soybean oil solvent extraction plant Elevators, Seed Cleaner, Aspiration System, Cracker, Cooker, Flaker, Roll Grinding Attachment, Hydraulic System, Conveyors, Rotary Air Lock, Feed Bin, Micro Level Indicators, Extractor, Rising Hoppers, Discharge Bin, Bulk Flow Conveyor, Rotar Air Lock, Rotary Air Lock, Toaster, Dust Catcher New Design, Horizontal Tubular Condenser, Sealing Device, Vapour Cooler, Miscella Holding Tank, Water Solvent Separator, Spent Water Desolventiser, Evaporator, Separator, Pre Heater, Condensers, Oil Stripping Column, Heater, Drier, Oil Holding Tank, Vacuum Equipment, Final Vapour Absorber, Heat Exchangers, Final Vertical Stripper, Accessories and Misc. including installation charges.	50 ton/ day	150.00
Boiler Coal Fired with Chimney, motor & all accessories	1.5 ton/hr	4.00
Total		154.00

Conclusion

Entrepreneurship in soybean processing can fulfil both nutritional security and employment opportunity among rural as well as urban population of India. Government agencies like Centre of Excellence in Soybean Processing and Utilization, CIAE, Bhopal has been working on soybean related research for the last four decades. The future of soybean production and soybean utilization is bright because of the growing demand for protein from non-animal sources. There are ample opportunities for soybean based food units in India which will transform India in the coming future. For such transformation, support of both at the governmental and societal level is needed. Entrepreneurship development is the key factor to fight against unemployment, poverty and to prepare ourselves for globalization to achieve economic progress. This could also provide some affordable nutritious foods to the population to achieve food and nutrition security.

References

Agrahar, M. D. & K. Jha, 2009. Effect of Sprouting on Nutritional and Functional Characteristics of Soybean (*Glycine max* L). *J. Food. Sci. Technol.* 46(3): 240-243.

Bargale, P.C. & R. C. Griffin, 1991. Modified Atmosphere Packaging of Full Fat Soyflour. *The Indian J. Nutri. and Dietetics*, 28: 118-121.

Bernard, E. C., S. G. Alistair & J. L. Michael, 2007. Some Functional Properties of Fractionated Soy Protein Isolates Obtained by Microfiltration. *Food Hydrocolloids*. 2: 1379–1388.

Khare, S. K., K. Jha & A. P. Gandhi, 1995. Citric acid Production from Okara (soy residue) by solid-state fermentation. *Bioresource Tech.* 54: 323-325.

Kinsella, J. E. 1979. Functional Properties of Soy Proteins. *J. of American Oil Chemical Society*, 56: 242–258.

Sangronis, E. & C. J. Machado, 2007. Influence of Germination on the Nutritional Quality of *Phaseolus vulgaris* and *Cajanus cajan*. *LWT J. Sci. Technol.* 40: 116-120.

USDA Nutrient Data Base for Standard Reference Release 28, 2015.

Vong, W.C. & S. Q. Liu, 2016. Biovalorisation of Okara (soybean residue) for Food and Nutrition. *Trends in Food Sci. & Technol.* 52: 139-147.

Wang, C. & L. A. Jhonson, 2001. Functional Properties of Hydrothermally Cooked Soy Protein Products. *J. Am. Oil Chem. Soc.* 78: 189–195.

10

Scope of Entrepreneurship Developments in Groundnut Processing

Vishnu Vardhan S

Introduction

Groundnut (*Arachis hypogaea* L.) commonly called as Poor man's cashew, is the 6th most important oil seed crop in the world. It contains 48-50% oil, 26-28% protein and 11-27% carbohydrates. Groundnut is grown on 26.4 million hectare worldwide, with a total production of 37.1 million metric tons and an average productivity of 1.4 metric tons/ha. Developing countries constitute 97% of the global area and 94% of the global production of this crop (FAO, 2011). The production of groundnut is concentrated in Asia and Africa, where the crop is grown mostly by smallholder farmers under rain-fed conditions with limited inputs. Groundnuts (peanuts) are popularly used as seed oil in India. Since, they are obtained from the ground, they are called as groundnuts and also known as 'Mungfali' in Hindi, 'Palleelu or Veru sanaga' in Telugu, 'Kadalai' in Tamil, 'Nilakkadala' in Malayalam, 'Kadale kaayi' in Kannada, 'Singdana' in Gujarati and 'Shengdaane' in Marathi. Groundnuts are easily available round the year and almost everywhere in India. Groundnuts are also taken as snacks in many households especially in India. In reality, groundnuts are actually legumes. Since, they have all the properties of nuts like almonds, cashew nuts, etc., these are also included in the family of nuts.

Groundnut is the major oil seed of India. It accounts for 25% of total oil produced in India. Annual production of groundnut is around 9690 thousand tones. As the crop is grown in rainfed conditions, production and productivity are highly vulnerable to rainfall and exhibit huge fluctuation between the years. In India, groundnuts are produced both during Kharif (75%) and Rabi seasons (25%). Major production catchments of groundnut are located in Gujarat holding a share of 50.77% of total groundnut production, Andhra Pradesh holding a second place (12.69%) followed by Tamil Nadu (9.91%). Following are the top groundnut producing states of India (Table 1).

Table 1: Production and per cent share of total production of groundnuts in India (2013-14)

S.No.	State	Production (thousand tons)	Share of total production (%)
1.	Gujarat	4920.00	50.77
2.	Andhra Pradesh	1260.00	12.69
3.	Tamil Nadu	960.00	9.91
4.	Rajasthan	910.00	9.39
5.	Karnataka	660.00	3.41
6.	Maharashtra	330.00	2.06
7.	Madhya Pradesh	200.00	2.06
8.	Uttar Pradesh	90.00	0.93
9.	Odisha	80.00	0.83
10.	Others	310.00	3.20

(*Source:* www.indiastat.com)

Groundnut produced in different countries is consumed in various ways depending on consumer preference and food habits. The seed can be consumed raw, boiled or roasted. The oil extracted from it is a very important cooking medium in India and many Asian countries. Bulk of groundnut is used for extraction of oil. The quality characters and uses of groundnut vary among the developed and developing countries. In USA, Canada and Australia, groundnut is grown to make peanut butter rather than to extract oil. Groundnut is also used to make confectionery and its flour to make baked products. In developing countries, it is mainly used for oil extraction and its by-product is utilized for feed and food purposes (Jambunathan, 1991).

Health benefits of groundnut

Groundnut is a rich source of energy due to its high oil and protein content. It supplies about 5.6 calories grain^{-1} when consumed raw and 5.8 calories grain^{-1} when consumed after roasting. It is a rich source of essential amino acids, minerals, and vitamins. Groundnut has good digestibility in both raw and roasted forms (Nagaraj, 1988). The composition of groundnut with its essential amino acid content is given in Table 2.

Table 2: The chemical composition and essential amino acids of groundnut

Content	Percentage	Content	g100gProtein[1]
Protein	25.2	Lysine	4.0
Oil	48.2	Threonine	3.12
Starch	11.5	Valine	4.59
Soluble sugar	4.5	Methionine+ Cystein	2.56
Crude fiber	2.1	Isoleucine	3.69
Moisture	6.0	Leucine	6.95
		Phenylalanine+ Tyrosine	10.12

Jambunathan (1991)

Groundnuts are considered as healthy snacks and offers amazing health benefits as detailed below.

- Rich source of minerals: Potassium, manganese, copper, calcium, magnesium, iron, selenium and zinc are some of the minerals present in groundnuts which are important for many body functions.

- Helps promote fertility: Groundnuts contain a good amount of folic acid. Studies have shown that women who had a daily intake of 400 micrograms of folic acid before and during early pregnancy had reduced risk of having a baby born with neural tube defect.

- Antioxidants: Groundnuts contain antioxidants in high concentrations. These antioxidants become more active when groundnuts are boiled. There is a 2-fold increase in Biochanin-A and 4-fold increase in Genistein content. These reduce the damage done by free radicals produced in the body.

- Helps fight depression: Groundnuts are good sources of tryptophan, an essential amino acid which is important for the production of serotonin, one of the key brain chemicals involved in mood regulation.

- Reduces chances of stroke: The antioxidant, resveratrol in groundnuts prevents heart strokes by increasing the production of nitric oxide.

- Boosting of memory power: Groundnut contains niacin essential for normal brain function and boost memory power.

- Aids in blood sugar regulation: One fourth cup of groundnuts can supply the body with 35% of the Daily Value of manganese, a mineral which plays a role in fat and carbohydrate metabolism, calcium absorption, and blood sugar regulation.

- Cancer protection: A form of phytosterol called beta-sitosterol (SIT) found in groundnuts protect against cardiovascular diseases by interfering with the absorption of cholesterol and protect against cancer by inhibiting tumor growth.

- Helps prevent gallstones: Research studies have shown that groundnut lowers the risk of developing gallstones by 25%.

- Fight against heart diseases, nervous diseases, alzheimer's disease, and infections: The poly-phenolic antioxidant, resveratrol present in groundnut prevents heart diseases, cancers, nervous diseases or fungal infections efficiently. Groundnuts are also rich in heart-friendly monounsaturated fats such as oleic acid.

- Helps lower cholesterol: Groundnut contains copper which helps to lower and control cholesterol levels. This is also essential to reduce bad cholesterol and to increase good cholesterol levels.

- Consumption of groundnut also makes the skin smooth, supple and younger looking. Groundnut kernels also contain several hair friendly nutrients that are beneficial for maintaining healthy hair.

Scope of entrepreneurship development in groundnut processing sector

Groundnut processing sector offers greater scope for entrepreneurship development through various unit operations in primary processing and/or through value addition and manifests economic progress through.

- Identifying, assessing and exploiting business opportunities in primary, secondary and tertiary processing of groundnut.

- Creating new smaller capacity groundnut processing plants and/or renewing existing ones by making them more dynamic catering to local demands.

- Driving the economy forward through innovation, competence, job creation and by generally improving the well being of society.

Groundnut processing

The initial step in processing is harvesting, which typically begins with the mowing of mature groundnut plants. Then, the groundnut plants are inverted by specialized machine called groundnut inverters which dig, shake, and place the groundnut plants with the groundnut pods on top into windrows for field curing. After open-air drying, mature groundnuts are picked up from the windrow with combines that separate the groundnut pods from the plant using various threshing operations. The groundnut plants are deposited back onto the fields and the pods are accumulated in hoppers. Some combines dig and separate the vines and stems from the groundnut pods in one step, and groundnuts harvested by this method are cured during storage. Some small producers still use traditional harvesting method of ploughing the plants from the ground and manually stacking them for field curing. Harvesting is normally followed by mechanical drying. Moisture in groundnut is usually kept below 12 per cent, to prevent aflatoxin contamination. This low moisture content is difficult to achieve under field conditions without over drying vines and stems, which reduces combine efficiency (less foreign material is separated from the pods). On-farm dryers usually consist of either storage trailers with air channels along the floor or storage bins with air vents. Fans blow heated air (approximately 35°C) through the air channels and up through the groundnuts and are dried to moisture of content of 7 to 10 per cent. Local groundnut mills take groundnuts from the farm and cure it if necessary, clean, store or process for oil extraction, roasting, production of butter, milk or other value added products.

Next unit operation is shelling which begins with separating the foreign material with a series of screens, blowers, and magnets. The cleaned groundnuts are then sized with screens (size graders). Grading of groundnut pods is required to minimize damage to the kernel during decortication of pods. During decortication, shells of the sized groundnuts are crushed, typically by passing the groundnuts between rollers that have been adjusted for groundnut size. The gap between rollers must be narrow enough to crack the groundnut hulls, but wide enough to prevent damage to the kernels. A horizontal drum, with a perforated and ridged bottom and a rotating beater is also used to hull groundnuts. The rotating beater crushes the groundnuts against the bottom ridges, pushing both the shells and groundnuts through the perforations. The beater can be adjusted for different sizes of groundnuts, to avoid damaging the groundnut kernels. Shells are aspirated from the groundnut kernels as they fall from the drum. The crushed shells and groundnut kernels are then separated with oscillating shaker screens and air separators. The separation process also removes undersized kernels and split kernels.

Following crushing and hull/kernel separation, groundnut kernels are sized and graded. Sizing and grading can be done by hand, but most mills use screens to size the kernels and electric eye sorters for grading. Electric eye sorters can detect discolouration and can separate groundnuts by colour grades. The sized and graded groundnuts are bagged in 45.4 kg bags for shipment to peanut butter plants and nut roasters. Mechanization of various unit operations in value chain in groundnut processing are listed in Table 3.

Table 3: Unit operations in groundnut processing

S.N.	Unit operation	Available Technology	Technical specifications
1	Groundnut stripping	TNAU Model	Capacity : 100 kg/h Power requirement:1.5 Hp Labour : 4 No Stripping efficiency : 93% Pod breakage : Nil
		UAS Raichur Model	Capacity :70 kg/h Power requirement:3 Hp Labour : 2 No Stripping efficiency : 90% Pod breakage : Nil
2	Groundnut Pod drying	Bin dryer	Capacity : 80 kg/batch Power : 2 HP Time of drying :15 h
		PAU Waste Fired Dryer	Power : 5 Hp

Contd.

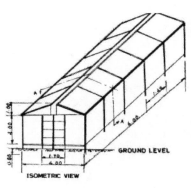

Drying time will be half the drying time of sun drying

Solar polyhouse dryer

3	Groundnut pod graders	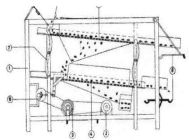	Type : Slotted oscilating sieves Capacity : 600 kg/h

Two deck Groundnut grader

4.	Groundnut Decorticator		Capacity: 260 kg/h (kernel) Shelling efficiency : 95.8% Power : 5 HP Labour : 2 No

TNAU Decorticator

Contd.

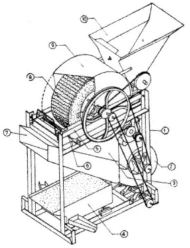

Capacity : Power : 2 HP
Efficiency : 94-96%

Motorized Rubber Tire decorticator
(CIAE Model)

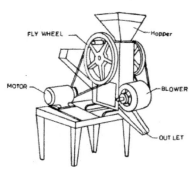

Capacity : 150 kg/h
Power : 1 HP
Efficiency :94-96%

Motorized Mini Decorticator
(JNTU Anantapur Model)

5   Oil expression

Crushing capacity :
30-40 kg/h
Power : 7.5 HP

4 bolt oil expeller

Contd.

6   Oil extraction

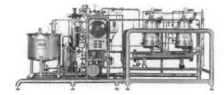

7   Briquetting of   Capacity : 500 kg/h to 2 ton/h
     groundnut shells

8   Groundnut roaster   Capacity : 100-200 kg/h

9   Groundnut blancher   Capacity : 200-400 kg/h

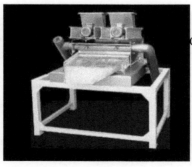

Contd.

10   Grinding mill  Capacity :100-200 kg/h

Value added products of groundnut

A brief outline of various value added products that can be prepared from groundnut is depicted in Fig.1.

Roasted groundnuts

Roasting imparts the typical flavour to groundnut kernels. During roasting, amino acids and carbohydrates react to produce tetrahydrofuran derivatives. Roasting also dries the groundnuts further and causes them to turn brown as groundnut oil stains the groundnut cell walls. Roasted groundnuts are further processed into candies or butter. Typical groundnut roasting process is shown in Figure 2. There are 2 primary methods for roasting groundnuts ie. dry roasting and oil roasting (Woodroof, 1983).

Dry roasting

Dry roasting is either a batch or continuous process. Batch roasters are typically natural gas-fired revolving ovens which are drum-shaped. The rotation of the oven continuously stirs the groundnuts to produce an even roast. Oven temperature is set to 430°C so that kernel temperature is raised approximately to 160°C and roasting generally completes in 40 to 60 min. Actual roasting temperatures and time vary with the condition of the groundnut batch and the desired end characteristics. Continuous dry roasters vary considerably in type. Continuous roasting reduces labour, ensures a steady flow of groundnuts for other processes (packaging, candy, butter production, etc.), and decreases spillage. Continuous roasters may move groundnuts through an oven on a conveyor or by gravity feed. In one type of roaster, groundnuts are fed by a conveyor into a stream of countercurrent hot air that roast the groundnut. In this system, the groundnuts are agitated to ensure that air passes around the individual kernels to promote an even roast. Dry roasted groundnuts are cooled and blanched. Cooling occurs in cooling boxes or on conveyors where large quantities of air are blown over the groundnuts immediately following roasting. Cooling is necessary to stop the roasting process and to maintain a uniform quality. Blanching removes the skin of the groundnut as well as dust, moulds and other foreign materials. There are several blanching methods including dry, water, spin, and air impact blanching.

Dry blanching is used primarily in peanut butter production, because it removes the kernel hearts which affect the flavour of butter. During dry blanching, the groundnuts are heated to approximately 138°C for 25 minutes to crack and loosen the skin. The heated groundnuts are then cooled and passed through either brushes or ribbed rubber belting to rub off the skin. Screening is used to separate the hearts from the cotyledons (groundnut halves). In water blanching, the groundnuts are passed on conveyors through stationary blades that slit the groundnut skin. The skin is then loosened by hot water spraying and removed by passing the groundnuts under oscillating canvas-covered pads on knobbed conveyor belts. Water blanching requires drying the groundnuts back to a moisture content of 6 to 12 per cent. In spin blanching, steam is used to loosen the skin of the groundnuts. Steaming is followed by spinning the groundnuts on revolving spindles. As the groundnuts move down a grooved conveyor, the spinning unwrap the groundnut skin. Air impact blanching uses a horizontal drum (cylinder) in which the groundnuts are placed and rotated. The inner surface of the drum has an abrasive surface that helps to remove the skin as the drum rotates. Air jets inside the drum blow the groundnuts counter to the rotation of the drum and create air impact which loosen the skin. The combination of air impact and the abrasive surface of the drum remove the skin. Either batch or continuous air impact blanching can be conducted.

Oil roasting

Oil roasting is also done on a batch or continuous basis. Before roasting, the groundnuts are blanched to remove the skin. Continuous roasters move the groundnuts on a conveyor through a long tank of heated oil. In both batch and continuous roasters, oil is heated to 138 to 143°C, and roasted for 3 to 10 minutes. Roasting time varies on desired characteristics and groundnut quality. Oil roaster tanks have heating elements on the sides to prevent charring the groundnuts present on the bottom. Oil is constantly monitored for quality. Frequent filtration, neutralization, and replacement are necessary to maintain quality. Coconut oil is preferred, but groundnut and cottonseed oils are frequently used. Next step is cooling, so as to get a uniform roast. Cooling is achieved by blowing large quantities of air over the groundnuts either on conveyors or in cooling boxes.

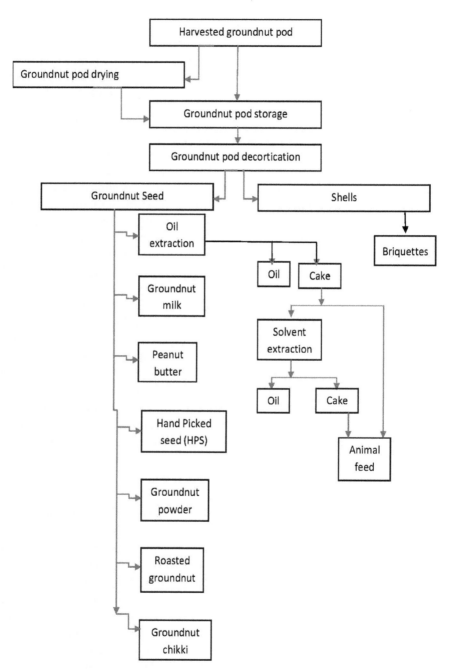

Fig. 1: Value chain in groundnut processing

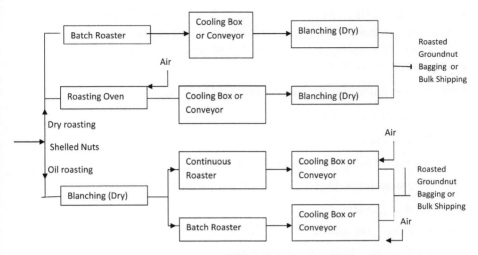

Fig. 2: Typical process flow diagram of roasted groundnuts

Peanut butter

Peanut butter is an important value added product consumed in large quantities especially in western countries. It is not very popular in India and the domestic market is dominated by milk butter. Hence, the promoters must target the export market. Peanut butter is an ideal substitute for milk butter. It is a low calorie, high protein product. Compared to milk butter, its price is very competitive. Gujarat and Maharashtra are the preferred locations in view of good quality groundnuts cultivated in Gujarat and two well developed ports being available for export.

To manufacture peanut butter, good quality groundnut pods are sorted out and destoned before shelling. Shelled groundnuts are graded according to size to ensure the selection of only big or bold groundnuts for the manufacturing process. Roasting is done at around 160°C for 40-60 minutes depending upon the moisture content. Roasting reduces water content to around 1% and helps to develop flavour. After roasting, groundnuts are cooled and then blanched to remove outer red skin. After blanching, each groundnut is inspected to remove the discoloured (grey or black) nuts. Groundnuts are then ground in peanut butter mill in two stages to produce fine and creamy butter. Outlet temperature is maintained at 65-75°C. All ingredients like salt, sugar and stabilisers are added during this process.

Air is incorporated into peanut butter during milling and subsequently removed in a vacuum. Next step is cooling and a scraped surface heat exchanger is used for cooling. Outlet temperature depends upon the type of stabiliser used.

Peanut butter is filled in pet jars or metal drums as per the instructions of the buyer. Immediately after filling, jars are vibrated to remove any remaining air bubbles. After keeping jars or drums for around 35-40 hours at around 20°C, peanut butter sets completely.

Machinery required for production of peanut butter:

- Hoppers and elevators
- Seed cleaner
- Vibrating sieve with dust aspiration system
- Decorticator with pneumatic husk separator
- Kernel grader
- Picking/sorting table
- Radiant ray roaster for dry roasting
- Cooling sieve
- Ammonia chilling plant for refrigeration
- Whole/broken nut blancher
- Piston feeler for peanut butter processing line
- Ingredient feeder
- Chamberless electronic vacuum packing machine
- Pellet truck for loading

Groundnut milk

Milk obtained from groundnut can be used as a supplement in the diet of children or as a substitute for ordinary milk in allergic conditions. Groundnut milk provides nutritional benefits which are lacking in cow's milk such as vitamin E, magnesium and vitamin B_6, and is packed with heart-healthy unsaturated fats (Giyarto *et al.*, 2012). To process groundnut milk (Fig.3), good quality groundnut kernels are selected. Roast the kernels lightly and remove the red skin. Separate the germ by sieving and handpick the spoiled seeds. Soak the kernels in one per cent sodium bicarbonate ($NaHCO_3$) solution for 16 to 18 hours. Kernels are ground in aqueous medium. Wet mass is steeped in water for 4 to 5 hours and filtered through cheesecloth to remove the product. The paste is mixed with 7 times the weight of water in a blender. Calcium hydroxide solution is added till the pH of the milk is adjusted to 6.8. A mixture of disodium phosphate and acid potassium phosphate having pH 7.0 is added to stabilise the milk. Milk is filtered through a fine cloth and fortified with vitamins A, D, B_2, folic acid, B_{12}, calcium and iron. Cane sugar is added at 7% level to add taste. The milk is homogenised, steamed, bottled and kept in a refrigerator till distributed.

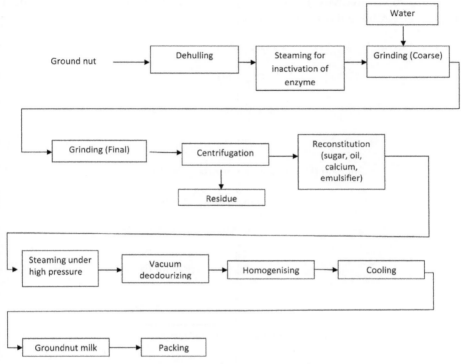

Fig. 3: Production of groundnut milk

Groundnut milk is usually fermented with *Lactobacillus acidophilus* and used as probiotic drink. Groundnut milk is one option to try for people who follow a casein-free diet. Sweeteners or seasonings such as cinnamon can also be added.

Groundnut oil

Oilseeds and edible oils are two of the most sensitive agricultural commodities in the country. Out of all oilseeds, groundnut is the major source for obtaining edible grade oil. Apart from oilseeds, the byproducts obtained during the processing operation viz. deoiled cakes; oil meal and other minor oil products are also of high economic value. Groundnut oil is used for cooking food and as a shortening or as a base for confectioneries and they can be used to make peanut butter. Groundnut oil ranks at the top among edible oils exported from India. It is premium oil and its cost is also high. Groundnut oil is available in the market in refined and filtered forms. It is estimated that one metric ton of groundnut seed produces an average of 420 L of groundnut oil, 420 kg of groundnut cake, and 40 kg of groundnut sludge. In order to get high quality edible oil, various processing techniques are used. The process of obtaining oil from oil seeds involves the separation of oil by mechanical means, chemical means, etc. A typical oil extraction process is given in Fig.4.

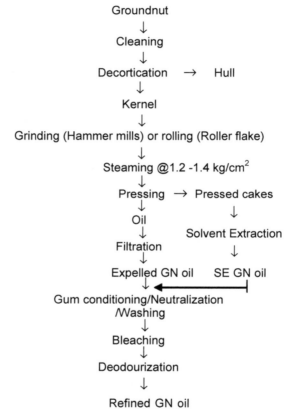

Fig. 4: Groundnut oil expression process

Cleaning is the first step in preparing groundnut for oil extraction. Inspection of the seeds is carefully done to remove stones, sand, dirt and spoilt seeds. Dry screening technique is often used to remove all materials that is over or under sized. During decortication, the outer hull of groundnut pod is removed. A power operated decorticator is generally used for the operation. Removal of outer groundnut hull is necessary as it does not contain oil and inclusion of it in the unit operations makes oil extraction process less efficient. Grinding is the process for reducing the particle size. Small motor powered hammer mills are used for this unit operation. Another alternate process used for reduction of particle size is rolling oilseeds to produce flakes for oil extraction. Many large scale commercial plants find this as the most effective approach. Steaming is the final step for preparing the raw material for oil extraction. Steaming at 1.2-1.4 kg/cm^2 for 5- 10 min leads to increased oil yield and also helps in killing those enzymes present in the plant tissue which have a deteriorating effect on oil quality. Cooking process coagulates the proteins present in the seed causing coalescence of oil droplets and making the seed permeable to the flow of oil. Then, oil seeds/oil bearing material is pressed using a lever press, hydraulic press or a mechanical expeller to remove oil. Refining helps to remove undesirable cloudiness, colour and flavour from extracted oil.

Machineries required for groundnut oil production are listed below.

- Pre-cleaner for cleaning the oil seeds
- Decorticator cum groundnut kernel grader
- Cans and trays for handling oil seeds
- Batch type solar dryer
- Tapering screw type mechanical oil expeller
- Filter press
- Steel drums for storing edible oil and sedimentation of impurities
- Weighing balance
- Semi-automatic bottle filling machine
- Molded polycarbonate bottle capping cum sealing machine.

In establishing a groundnut oil production unit, following license and registration from different Government authorities are required.

- Company Registration with ROC
- Trade License
- SSI Registration
- Factory License
- Food Operator License from FSSAI
- BIS Certification
- 'No Objection' from Pollution Control Board
- AGMARK Certification
- Fire License.

The Food Standards and Safety Rules, 2011, and rules made under IS 544: 1968, permit additions of antioxidants in groundnut oil for a higher shelf life of oil. However, under the scheme of labelling of environment-friendly products, the presence of antioxidants within a prescribed limit is a requirement as per notification of the Ministry of Environment and Forests.

Fermented products from groundnut

Oncom is a popular dish of Indonesia and can be prepared from groundnut cake obtained after taking oil. It is usually done by soaking the cake in water for 24 h and then draining it. High starch material such as cassava is also added to it. After steaming this, incubate with *Neurospora intermedia* or *Rhizopus oligosporus* and ferment for 1 to 2 days at 25 to 30°C after wrapping in banana leaves. It can be fried in oil or in margarine. Oncum is readily digestible, tasty and nutritious.

Tofu from groundnut is a famous product in China and Japan. Soak the groundnut kernels overnight and grind into an emulsion. Boil the emulsion and filter. Precipitate the curd by adding calcium or magnesium sulphate. Leave the product to settle and transfer to boxes lined with cloth filters or spread on trays. It can be sold as slices or slabs, served in soups or deep fried in oil (Giyarto *et al.*, 2012).

New technologies/ innovations in value addition of groundnut

Protein isolates

Groundnut protein isolates are akin to soy protein isolates. The protein isolate extraction requires pre soaking of split groundnut seeds in sodium chloride solution (brine) overnight. This is washed with hot water to wash out the water-soluble components and flavour components. The protein can be easily extracted from the cake with dilute alkali and is later precipitated by lowering the P^H. The other way is to use water, 70% alcohol, 10% NaCl, 0.25% KOH. The isolated protein contains 16.2% N, 0.74% ash, and 0.06% lipid. Groundnut protein consists of two globulins namely, arachin (93% of defatted seed protein) and conarachin. When a salt extract (0.1% NaCl buffered with 0.01 M sodium sulfate at pH 7.9) is treated with saturated $(NH4)_2 SO_4$, arachin gets precipitated first at 40% saturation followed by conarachin between 40 and 85% saturation. Since, globulins contain 18.3% N, the conversion factor for protein used is 5.46 (Nagaraj, 1988). The protein isolates of groundnut have high solubility; they are white and are free from nutty flavour. These protein isolates and oil can be used in the manufacture of cheese analogues. Replacement levels of 40% and 50% were found to be optimal in producing cheese analogues. The groundnut protein isolates are used in various food products. These isolates can replace about 80% of milk solids without changing the texture, and about 60% without loss of flavour, colour, or overall acceptability in the preparation of frozen desserts.

Groundnut cake composite flour

Groundnut cake flour is used to improve protein content and quality of several cereal- based food products in India, Kenya, Malawi, Nigeria, Senegal and Zimbabwe (Desai *et al.*, 1999). In India alone, there have been several agriculture-products with groundnut as the protein-enriching medium. The partially defatted flour is used to improve the nutritional quality of various cereal and millet based products. The addition of defatted groundnut flour results in an improvement of colour and texture of baked products. Fortification with groundnut and subsequent fermentation improves the *in vitro* digestibility of the sorghum flour.

Scope of by-product utilisation/effluent treatment/ utilization

Briquetting of groundnut shells

Groundnut shells, a by-product of processing, being less dense are difficult to transport, store and handle. Direct burning of loose biomass in conventional grates is associated with very low thermal efficiency and widespread air pollution. The conversion efficiencies are as low as 40% with particulate emissions in the flue gases in excess of 3000 mg/Nm³. In addition, a large percentage of unburnt carbonaceous ash has to be disposed of. At present, two main high pressure technologies: ram or piston press and screw extrusion machine, are used for briquetting. The briquettes produced by a piston press are completely solid, while the screw press briquettes on the other hand have a concentric hole which gives better combustion characteristics due to a larger specific area. The screw press briquettes are also homogeneous and do not disintegrate easily. Having a high combustion rate, this will substitute for coal in most applications and in boilers. Briquettes can be produced with a density of 1.2 g/cm³ from loose biomass of bulk density of 0.1 to 0.2 g/cm³. These briquettes can be burnt clean and therefore are eco-friendly. The success in briquetting technology and the growing number of entrepreneurs in this sector indicate that biomass briquetting will emerge as a promising option for the new entrepreneurs and other users of biomass.

Safety and quality aspects of groundnut products

According to the Indian Oilseeds and Produce Export Promotion Council (IOPEPC), India has exported 5, 36,929 tons groundnut in 2015-16, against 7, 88,307 tons in 2014-15. "Vietnam banned the groundnut import from April 2015 to January 2016 due to quality issues. It is the biggest drawback for the country which has translated into lower export by 32 per cent. Hence, knowledge on CODEX standards on safety and quality aspects of groundnut is mandatory for farmers, traders and exporters. The norms and procedures for export of groundnut and groundnut products shall be elucidated vide APEDA/PPP/Q/ 2015 Dated 12.03.2015. The details pertaining to quality standards of groundnut are given in Table 6.

Table 6: Quality standards of groundnut (CODEX STAN 1-1985)

S.No.	Quality aspect	Norms
1.	Moisture content	Groundnuts in-pod 10% Groundnut kernels 9%
2.	Mouldy, rancid or decayed kernels	0.2% m/m max*Note : Rancid kernels are defined as those which have undergone oxidation of lipids (should not exceed 5 meq active oxygen/kg) or the production of free fatty acids (should not exceed 1.0%) resulting in the production of disagreeable flavours*
3.	Filth	Impurities of animal origin (including dead insects) 0.1% m/m max
4.	Other organic and inorganic extraneous matter	Groundnuts in-pod 0.5% m/m max Groundnut kernels 0.5% m/m max
5.	Aflatoxins	As detailed below

Levels of aflatoxins shall not exceed the following in their respective categories. The authorized laboratories shall analyze groundnuts and groundnut products for determination of aflatoxin levels for the following:

Sl.No.	Product categories	Maximum aflatoxin levels j.	
		B_1	Sum of B_1 +B_2+GU
(i)	Groundnuts (peanuts) and processed products thereof, intended for direct human consumption or as an ingredient in foodstuffs, with the exception of crude vegetable oils destined for refining and refined vegetable oils for exports to EU (maximum levels of aflatoxins in μg/kg related to a product with maximum moisture content of 7%).	2	4
(ii)	Groundnuts (peanuts) to be subjected to sorting or other physical treatment or further processing, before human consumption or use as an ingredient in foodstuffs with the exception of groundnuts (peanuts) for crushing for refined vegetable oil production for exports to EU (maximum levels of aflatoxins in μg/kg related to a product with maximum moisture content of 7%).	8	15
(iii)	Groundnuts (peanuts) as bird feed for exports to EU (maximum levels of aflatoxins in μg/kg related to a product with maximum moisture content of 7%)	20	20
(iv)	Groundnuts (peanuts) for exports to Japan and Korea (maximum levels of aflatoxins hi μg/kg related to a product with maximum moisture content of 7%).	10	10
(v)	Groundnuts (peanuts) for exports to countries other than EU (maximum levels of aflatoxins in μg/kg related to a product with maximum moisture content of 7%)	15	15

S.No.	Factor or Defects	Limit
1	In-Pod Defects	
1.1.	Empty pods	3% m/m
1.2.	Damaged pods	10%m/m
1.2.1.	Shrivelled pods	
1.2.2.	Broken pods	
1.3.	Discoloured pods	2% m/m
2.	Kernel Defects	
2.1.	Damaged kernels	
2.1.1.	Freezing injury causing discoloured kernels	2% m/m
2.1.2.	Shrivelled kernels	5% m/m
2.1.3.	Damaged due to insects	2% m/m
2.1.4.	Mechanical damage of kernels	2% m/m
2.1.5.	Germinated kernels	2% m/m
2.2.	Discoloured kernels	
2.2.1.	Flesh discolouration	3% m/m
2.3.	Broken and split kernels	3% m/m
3.	Groundnuts other than designated type	5% m/m

Packaging of groundnut products

The term shelf life, with respect to groundnuts or groundnut oil, can be described as "the number of days before the onset of oxidative rancidity, a process which is generally induced in either the whole groundnut or groundnut oil by exposure to heat and air" (Mercer *et al.*, 1990). Stability of groundnut oil is quite good when compared to other vegetable oils, partly due to the fatty acid composition. Traditionally, runner-type groundnut contain approximately 50% fat or oil, which consists of 41-67% oleic acid (18:1n-9) and 14-42% linoleic acid (18:2 n-6). Because of the high amount of oil in groundnut, the quality can deteriorate quickly due to lipid oxidation, depending on number of factors, such as the presence of oxygen, light, moisture, and high temperature. Low level of linolenic acid (18:3 n-6) in groundnut oil is thought to be partially responsible for oxidative stability, where the rate of oxidation is approximately 1:10:100:200 for 18:0, 18:1, 18:2, and 18:3, respectively (O'Keefe *et al.*, 1993). Lipid oxidation has been found to be a major source of the off flavour and decreased quality in groundnuts. Painty and cardboardy off flavours are formed during lipid oxidation, making the product rancid and unacceptable to the consumer.

One of the most important parts of oil processing is packaging that protects the product from the point of manufacture till usage by consumer. Packaging is an important unit operation that directly affects shelf life of oil. Carefully processed oil may be damaged by improper selection of packaging materials.

Packaging system for groundnut oil in India

Traditionally, oil and fats have been packed in 15 kg square tinplate containers. The other types of packages like plastic containers, lined cartons and flexible pouches have been recently introduced. Even though, packaging has witnessed many changes, till today about 52% of oil and fats continue to be traded in loose/unpacked form. This includes retail selling of loose oil from 15 kg tins as well. This allows a lot of scope to pursue the dangerous practice of adulterating the oil with less expensive and unhygienic varieties. Plastic packaging provides safe hygienically packed oil at competitive cost to consumers. It is extremely important that, whatever be the packaging material used, it should be food grade and non-toxic. Packaged oil and fats offer various advantages such as:

- Ease in quick disposal at retail points
- Ease of identification
- Tamper evident and therefore minimum chances of mixing or adulteration
- Quality is guaranteed
- No need for consumer to carry own container
- Convenience in storage and use by the consumer
- No wastage due to spillage at retail shops/containers
- Brand identification can be established.

The array and availability of packaging materials, sizes and shapes of package construction are unlimited. In the present day, consumer is willing to try and use new materials. Modern packaging technology provides many opportunities to maintain product protection while reducing the cost. The main requirements for a packaging system for groundnut oil or any value added products of groundnut should be:

- Non – toxic and compatible
- Protect against environmental factors
- Machineable
- Leak-proof and transport-worthy
- Easy to store, use and handle
- Printable.

The various options of packaging materials with advantages and limitations are listed in Table 8.

Table 8: Various packaging materials for groundnut oil

S.No.	Types of packing materials	Specific features and advantages	Remarks
1	Tinplate containers 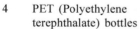	Widely used for packaging of groundnut oil in sizes of 15 kg. Suitable for high filling and packaging operations.	High cost and uncertainty about availability. Used containers should not be used for packaging (banned under GSR 575 (E) dated 4/8/95).
2	Glass bottles	Glass bottles provide excellent protection and can also be used for high-speed operation.	They are not commonly used for edible oil packaging because of their fragility and high tare weight.
3	High density polyethylene containers (HDPE) 	Blow molded HDPE containers in the form of bottles (200 g, 400 g), jars (1 kg and 2 kg) and jerry cans (2 kg, 5 kg and 15 kg) are widely used for packaging. They provide a moderately long shelf life, are light in weight and are transport-worthy.	Although they do not provide long shelf-life as the tinplate container, they are economical as compared to tinplate container, and therefore, suitable for long shelf-life.
4	PET (Polyethylene terephthalate) bottles	PET bottles have excellent clarity, are odour–free and have good gas barrier properties. PET bottles are also accepted internationally for edible oil packaging.	

5.	PVC (Poly Vinyl Chloride) bottles	Food grade stretch blow molded PVC bottles are used for packaging edible oil in the country. PVC bottles have good clarity and excellent oil resistant properties.	IS: 12883 –1989 gives specifications for PVC bottles for edible oil packaging.

6	Flexible plastic pouches	Flexible pouches may be made from laminates or multi-layered films of different compositions. The pouches may be in the form of pillow or as stand-up-pouches. They are more economical than any other packaging system.	The selection of a laminate or a multi-layer film is governed primarily by the compatibility of the contact layer, heat sealability, heat seal strength and required shelf-life besides machineability and physical strength parameters.

7	Asceptic packaging	Semi rigid packaging materials are translucent leave the oil open to attack from light, and are permeable to oxygen. A carton package provides an excellent solution to arrest light transmission.	-

Conclusion

The long-term effect in the economic and national development in India can be achieved through the development of small scale food processing techniques, and these culminate to rapid food processing and industrialization. Thus, this chapter focuses on entrepreneurship development in groundnut processing through efficient technologies for various unit operations. The development of entrepreneurship represents one way of diversifying agricultural products. It has the potential to increase farmers' income, open up new job opportunities in rural areas, and support the eventual development of advanced food processing industries. Setting up of rural agro processing complex for groundnut which involve lower capital investment and rely on traditional food processing technologies are crucial to rural development. However, processing enterprises are facing with significant challenges that compromise their ability to function and to contribute optimally to the economy. Financial constraints, market difficulties and lack of management skills and skilled workers are the major constraints, among others, that hamper the efficient performance of such enterprises. The promotion of food processing and preservation demands more technological inputs in terms of mechanization of unit operations, development of new and existing technologies, design and development of machinery for processing different agricultural produce of high target.

References

Desai, B.B., P.M. Kotecha & D. K. Salunkhe, 1999. Groundnut Protein Product. In: *Introduction Science and Technology of Groundnut: Biology, Production, Processing and Utilization.* Naya Prokash Publishers, New Delhi, India, pp 546–582.

FAO 2011. *Report- FAOSTAT* Production Year 2011.

Giyarto., T. F. Djaafar., E. S. Rahaya & Utami. 2012. Fermentation of Groundnut Milk by *Lactobacillus acidophilus* SNP-2 for Production of Non-Dairy Probiotic Drink. *Proceedings of 3rd International Conference of Indonesian Society for Lactic Acid Bacteria*: Better Life with LAB: Exploring Novel Function of LAB. 9p.

Jambunathan, R. 1991. Groundnut Quality Characteristics. Uses of Tropical Grain Legumes In : *Proceedings of a Consultants Meetings* held on 27-30 March, 1989. ICRISAT Centre, Patancheru, Hyderabad, India. P: 267-275.

Mercer, L.C., J. C. Wynne & C. T. Young, 1990. Inheritance of Fatty Acid Content in Groundnut Oil. *Groundnut Sci.* 17:17-21.

Nagaraj, G. 1988. Chemistry and Utilization. In: Groundnut (Reddy, PS., ed.). Indian Council of Agricultural Research, New Delhi. P: 554-565.

O'Keefe, S.F., V. A. Wiley & D. A. Knauft, 1993. Comparison of Oxidative Stability of High- and Normal-Oleic Groundnut Oils. *JAOCS* 70(5): 489-492.

Woodroof, J. G. 1983. *Groundnuts: Production, Processing, Products*, 3rd Edition, Avi Publishing Company, Westport, CT, 1983.

Annexure 1

Some of the groundnut processing machinery manufacturers

1. Shreeji Nut Co, #15, Sardar Patel Udyog Nagar, Jam Kandorna 360 405, Rajkot (Dist) Ph:+919825066854, shrijeenut@gmail.com

2. Rajkumar Agro Engineers Pvt Ltd, Naresh Gambhir (Director), Near Union Bank Of India, Ghat Road, Nagpur - 440018, Maharashtra, India. Ph : 08079452968, +(91)-9370925271, +(91)-9225237191

3. John Fowler & Co Ltd, Plot No. 6 & 6P Bommasandra Industrial Area, Hosur Road Bengaluru 560099

4. Forsberg Agritech (I) Ltd., 123, GIDC Estate, Makarpura, Vadodara-390 010, Gujarat. Ph: 09825072483, 9825082483, 0265-2636752, 6590584

5. Brimco Engg. Works, M24/1, Street No 9, Anand Parbat Indl. Area, New Delhi 110 005, Phone: 25726347, 6178 Fax:22145040

6. Osaw Agro Industries Private Limited, Sanjeev Sagar(Managing Director) Post Box No. 5, Agrosaw Complex Jagadhri Road, Ambala - 133001, Haryana, India. Ph 08071865016, +91-9416027043,Telephone: +91-171-2699547, +91-171-2699167

7. Fowler Westrup India (P) Ltd., Plot No. 60 - 63, KIADB Industrial Area, Malur – 563160, Kolar District (INDIA) Phone : + 91 8152 282500 Fax : + 91 8152 282555

8. Durai Industrial Works, 1143, Mettupalayam Rd., Coimbatore-641043.Tel No. 2442380/2444429

9. Harvest Sortmac Shosha Pvt Ltd, Nutech Vikas, No.6, 1st Avenue, 100 Feet Road, Ashoknagar, Chennai 600 083. Ph: 24717588, Fax: 24717688

10. Sagar Industries, GIDC Estate, Naroda, Ahmedabad

11. Yash Industries, Aji Indl. Estate, Rajkot

12. Sahyog Steel Fabrication, 28, Bhojrajpara, Gondal, 360311. Tel No. 224075

13. Mekins Agro Products (PVT) Ltd., 6-3-866/A Begumpet, Greenlands, Hyderbad 500016

14. Rajan Universal Exports (MFRS) PVT. Ltd., "Raj Buildings", Post Bag No. 250 162,Linghi Chetty Street, Chennai-600 001. Tel: (044) 5341711, 5340731, 5340751Fax: (044) 5342323 E-Mail: rajeximp@vsnl. com

Annexure 2

Case study: Economic analysis of roasted groundnut enterprise

1) Capacity of the plant: 300 kg/ day (Assuming 300 days of working, plant capacity would be 90 T)

S.No.	Description	No	Rate (Rs.)	Amount (Rs.)
	FIXED COST			
	Cost of land 2000 sq.m			7,00,000
	Cost of Buildings 1000 sq.m			30,00,000
	Tools and Plants			
1.	Groundnut decorticator cum grader, 300-350 kg/h capacity	1	75,000.00	75,000.00
2.	Groundnut seed cleaning unit comprising of Elevator 10ft, Round size grader, Destoner (2 No), cyclone blower with cleaner of capacity 500 kg/h (3 HP motor)	1	1,75,000.00	1,75,000.00
3.	Electric roaster, 75 kg/h	1	1,25,000.00	1,25,000.00
4.	Wholenut blancher, 200 kg/h	1	1,25,000.00	1,25,000.00
5.	Chamberless vacuum packaging machine	1	1,50,000.00	1,50,000.00
6.	Electrical connections			10,000.00
7.	Furniture and Packaging equipments			50,000.00
	Total			**38,35,000.00**
	VARIABLE COST			
	Raw materials required for month			
1.	Groundnut (Rs 5000/Quintal) (4 Quintal per day, 25 days)	100	5,000.00	5,00,000.00
2.	Electricity 200 kwh/day @ 25 days	5000	10.00	50,000.00
3.	Water charges			5,000.00
4.	Labour charges - Operator 2 No @ Rs.10,000 /month	2	10,000.00	20,000.00
5.	Helper	4	7,000.00	28,000.00
6.	Other contingencies			20,000.00
	Working capital /month			**6,23,000.00**
	Working capital/Day			24,920.00

Contd.

Payback amount in 10 yrs on fixed assets per day			1,278.00
Depreciation on machinery @ 10% per day			1,278.00
Interest on capital investment @12% per day			1,534.00
Total capital investment on machinery per day			4,090.00
Interest on working capital @ 16%			3987.00
Total Cost of Production per day			28,907.00
Sales realization per day:			
Cost of grade I roasted groundnuts	300	250.00	75,000.00
Profit per day before taxes			46,093.00

Analysis of the project

The project needs a fixed capital of (Rs.)	38,35,000.00
Working capital requirement per day (Rs.)	24,920.00
Working capital cycle is 15 days	
Working capital required for 15 days (Rs.)	3,73,800.00
Total Project cost (Rs.)	42,08,800.00
Let equity share be 1/3 and 2/3 will be procured from Financial institutions.	
Equity share (Rs.)	14,02,933.00
Profit per day before tax (Rs.)	46,093.00
Annual profit before taxes (Rs.)	138,27,900.00
Amount payable towards taxes (Rs.)	46,09,300.00
Annual earning (Rs.)	**92,18,600.00**

Costs and sales realization for Different Production capacities

Qty (kg)	Variable Cost	Total cost (Rs)	Sale @Rs.250/kg	Diff (Rs)
0	0	4,090	0	-4,090
26.6	2554	6,644	6650	6
30	2880	6,970	7500	530
50	4800	8,890	12500	3,610
75	7200	11,290	18750	7,460
100	9600	13,690	25000	11,310
125	12000	16,090	31250	15,160
150	14400	18,490	37500	19,010
200	19200	23,290	50000	26,710
250	24000	28,090	62500	34,410
300	28800	32,890	75000	42,110

Payback period: 0.83 years (Assuming that only 50% of net annual profit is realized due to unforeseen circumstances). The economic analysis of the above project proves to be viable.

11

Entrepreneurship Development in Jaggery Processing

Jagannadha Rao P V K

Introduction

India is the largest consumer and second largest producer of sugar in the world next to Brazil. Among the sugar yielding crops like sugarcane, sugar beet, palms and sorghum, sugarcane is the second most important agro-industrial crop of the country. It is the only raw material for nearly 538 sugar mills (2014-15), producing about 28 million mt of sugar annually from a sugarcane production of 366 million tons in an area of 5.307 mh. Apart from this, it is the predominant source of potable alcohol, industrial alcohol and fuel-ethanol. In India, 35 million farmers are engaged in sugarcane cultivation and another 50 million depend on employment generated by these 538 sugar factories and other related industries using sugar. Sugarcane occupies a very prominent position in the agricultural crops of India covering large areas in UP, Punjab, Haryana and Bihar (sub tropical) and Maharashtra, Karnataka, Gujarat and Tamil Nadu (tropical). On an average, white sugar production accounts for 74% of total cane produced, 14-15% is utilized for jaggery and khandsari production and the rest is utilized for other purposes including seed.

Jaggery and khandsari, the traditional Indian sweeteners are the natural mixture of sugar and molasses. If pure clarified sugarcane juice is boiled, what is left as solid is jaggery, which contains 65-85% sucrose. Khandsari sugar is a finely granulated and crystallized which contains 94 to 98% sucrose. In the early 1930's, nearly 2/3rd of sugarcane production was utilized to manufacture these two alternate sweeteners. With the introduction of sugar mills and their multiple growth and better standard of living and higher per capita income, the demand for sweetener has shifted to white sugar which contains purely sucrose (99.7%).

Commercially available sucrose has very high purity (>99.9%) making it one of the purest organic substances produced on an industrial scale. To obtain such a pure product from sugarcane, complex isolation and purification processes are

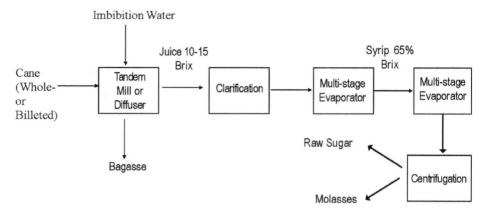

Fig. 1. Basic scheme of raw sugar manufacturing process in sugar factory

followed. Industrial sucrose production consist essentially a series of separations of non-sucrose compounds (impurities) from sucrose (Fig.1). Juice is first extracted from sugarcane (sucrose yields range between 10%–15% weight of sugarcane) by tandem milling or diffusion and converted to raw sugar (97.5% – 99.5% pure sucrose; golden yellow/brown crystals) at factories. Secondly, after raw sugar has been transported to a refinery, it is refined using similar unit processes used in raw sugar manufacture, as well as additional decolourization steps such as ion-exchange resins and activated carbon, to the familiar white, refined sugar (>99.9% sucrose) (Gillian and Isabel, 2015).

Sugarcane by-products based industries in India

Sugarcane is considered as one of the best converter of solar energy into biomass and sugar. The biomass which contains fiber, lignin, pentosans and pith can be converted into value added products by application of suitable chemical, biochemical and microbial technologies. Sugarcane is a versatile crop being a rich source of food (sucrose, jaggery and syrups), fiber (cellulose), fodder (green leaves and tops of cane plant), fuel and chemicals (bagasse, molasses and alcohol) and fertilizer (press mud and spent wash).

Almost all the countries in the world which produce cane sugar have realized that though the production of sugar from sugarcane is undoubtedly the most paying proposition, it is better to produce many value added products by diversification and utilizing the by-products of sugar industry, instead of depending on just one product i.e. sugar (Solomon and Singh 2005; Yadav and Solomon, 2006; Rao, 2008).

The main by-products are (a) Bagasse (b) Molasses (c) Press mud. There are other residues which are produced from sugarcane and have less commercial value such as trash, green tops, wax, fly ash and spent wash.

The sugar industry by-products are vast potential reserves for human and animal consumption as well as capable of providing energy as renewable source. Sugarcane and its by-products are useful raw materials to over 25 industries; some important ones are shown in Table 1. In India, on an average, processing of 100 tons of sugarcane in a factory yields 10 tons of sugar, 30–34 tons of bagasse (of which 22–24 tons is used in processing and 8–10 tons is saved), 4.45 tons of molasses, 3 tons of filter mud (press mud), 120 tons of flue gases (at 180°C) and 1,500 kWh of surplus electricity. The by-products like bagasse, molasses, press-mud, etc., being rich in carbon compounds and minerals provide ample opportunity for physico-chemical transformation or microbial fermentation to value added products like building and structural materials, pharmaceuticals, new food/feed products and low calorie sweeteners, energy options like cogeneration/fuel/bio-diesel and ethanol, medicines, pesticides, etc.

At present, sugarcane is used as food, fiber, fuel and fodder crop. In addition to these residues, large quantities of spent wash and yeast sludge are produced continuously from distilleries, which are rich in minerals and proteins.

Table 1: Agro-industrial value of sugar industry by-products

Sector of economy	Value added products from sugarcane & sugar industry residues
Food	Sweeteners (traditional, modern, synthetic), vitamins, acids, beverages, fats and oils, edible proteins (SCP and mushroom)
Health	Chemical, antibiotics, anti-cholesterol (policosanol), lingo-meds, enzymes, vaccines, juice
Agriculture fertilizers, compost, food, feed, fodder, forages, pesticides	A range of food, feed, fodder, fertilizer and forages
Industry	Solvents, plastics, bio-plastic, alcohol based chemicals, anti-corrosive compounds, tenso-active compounds, biocides
Energy electric power, biogas, bagasse fuel, fuel alcohol	Bagasse as fuel, biogas, co-generation of power, ethanol from bagasse
Transportation	Ethanol-petrol/diesel blends (gasohol), bio-diesel
Education and culture	Text books, note books, newsprints, writing and printing paper
Housing/construction	Particle boards, hard boards, ac ducts, decorative laminate
Light industry	Textile, polish, bitumen, carbon paper and chemicals
Communication	Insulating materials
Heavy industry	Resins for casting molds
Human resource development	Employment generation in rural areas.

Source: Singh and Solomon 1995.

Bagasse based value added products

Bagasse is the fibrous residue left over after sugarcane is crushed. When it comes out from milling plant it is called Mill Wet Bagasse because of high moisture (50%) content. The approximate composition of Mill Wet Bagasse is as follows: (a) Fiber: 46–52% (b) Water: 43–53% (c) Sugar: 2.5–3.0% (d) Mineral constituents: 0.55%. Because of high combustibility of bagasse, it is used as captive fuel in sugar factories. The approximate composition of dry bagasse is as follows (a) Cellulose: 45% (b) Pentosans: 28% (c) Lignin: 20% (d) Sugar: 5% (e) Minerals: 1% (f) Ash: 2%. The cellulose content of bagasse is useful as fibrous raw material in the pulp and paper industry, whereas its pentosan content is used in the manufacture of chemicals. The details of value added products from bagasse are as follows:

Paper

Many bagasse based plants produce different kinds of papers such as kraft paper, white writing and printing papers, newsprint, etc. Bagasse pulp is used in the manufacture of bags and card board of corrugated wrapping and writing paper; and toilet papers.

Panel or insulating board

It is made by treating bagasse with hot water or steam under pressure in rotary digester. The pulp is then washed and fed into board machines from which it emerges as a continuous wet mat. The finished product is an efficient insulator of heat and can be made resistant to insects and fire by impregnation with appropriate chemicals. It is used as ceiling material and wall partitioning.

Particle board

It is much denser and harder than insulating board, made in dry process by binding the bagasse fiber with a resin. It can be made water proof and used for making all kinds of furniture, cabinets, cupboards, racks, almirahs, partition boards, table tops, etc.

Molded products

The molded products are articles manufactured with ligno-cellulosic material with hot-press organic binders in different shaped molds, depending upon the article desired. The molded particle boards have a broad range of potential uses, such as cabinet for radio, television and recorder, kitchen furniture, cabinet doors, boxes, suitcase, trays, and coffins.

Rayon grade pulp

Rayon grade pulp also known as dissolving pulp used for making fabrics and absorbent sanitary products is manufactured from bagasse.

Bio-electricity (Co-generation)

The bagasse co-generation can make substantial contributions to national power generation in several countries including Brazil, India, Thailand, Mexico, Cuba, Pakistan, Colombia and the Philippines. These countries produce 70% of the world's sugarcane and, as an average, bagasse based cogeneration could satisfy over 7% of total national demand. It is possible to generate about 450 kWh electricity from 1 mt of bagasse. At present, as many as 147 sugar mills in India have co-generation projects to generate total power to the extent of 3,067 MW. It is estimated that the sugar mills in India have potential to generate as much as 7,000 MW of electric power and have potential to supply about 3,000 MW of surplus electric power to the National Grid in the future years.

Biogas

Bagasse can be used for the production of sludge or biogas (a mixture of carbon dioxide and methane) by fermentation through inoculation or addition of farm manure. The gas produced usually contains 60–65% methane, 30–35% of carbon dioxide and small quantities of hydrogen. It can be used to drive both petrol and diesel engines.

Furfural

Furfuraldehyde, known as furfural, a colourless oil of pleasant smell is produced from bagasse

Animal feed

Bagasse and bagasse pith are often used as fillers in compound diets and as carriers for molasses or molasses–urea mixture.

Ethanol

Bagasse could also be used for ethanol production by simultaneous saccharification-cum-fermentation (SSF) route using enzymatic or acid hydrolysis technology. This route could yield approximately 200 liters of ethanol/ton bagasse.

Miscellaneous uses of bagasse

Bagasse ash is used as fertilizer and for making cheap glass. Bagasse can be used in the manufacture of biodegradable plastic (PHB). It can also be used as cheap and light filler in building material, soil conditioner, compost and emergency cattle feed and for growing mushrooms. Dehydrated fresh bagasse is used as agriculture mulch. Lignin content of bagasse is an important source of medicine popularly known as ligno-med and sucrolin.

Molasses based industries

Blackstrap molasses is a by-product (end product) in the manufacture of sugar. It is a heavy, viscous liquid separated from the final low grade massecuite from which no further sugar can be crystallized out by the usual methods.

It is generally not used for human consumption on account of its colour and chemical composition. Approximately 23–28 l of molasses/ton of ground cane is produced, but the quantity of molasses increases when stale, burnt or poor quality cane is processed. Around 10–12 million tons of molasses is produced annually in India and used for ethanol and alco-chemicals production.

The main constituents of molasses are chiefly sucrose (30–35%), glucose and fructose (10–25%), non-sugar compounds (2–3%), water and mineral content. The total fermentable sugars in molasses range from 45–55%.

Commercial products made by fermentation of molasses are ethyl alcohol, carbon dioxide, citric acid, baker's yeast, monosodium glutamate, itaconic acid, acetone, butyl alcohol, etc.

Alcohol production

The production of ethyl alcohol from molasses is catalyzed by the action of yeast, and with improved strains and process, yield is about 230 l from one ton of molasses.

Ethanol as automobile fuel

Ethanol or anhydrous ethyl alcohol is used as an automobile fuel up to 5% as oxygenator, up to 5–20% as blend with gasoline (Gasohol) and higher blend F-95% as fuel extender/fuel replacement. It can be used in gasoline up to 5–20% without engine modification. In Brazil, USA, Canada, Sweden and other parts of Europe, ethanol is widely used for gasoline blending. It has helped in reducing carbon monoxide and hydrocarbon emission to a great extent.

Carbon dioxide

Carbon dioxide as well as alcohol is produced when molasses is fermented by *Saccharomyces cereviseae*. The yield of CO_2 is 16% by weight of molasses and between 70 and 75% of it can be recovered. The carbon dioxide produced is used as cooling agent and in the manufacture of carbonated drinks.

Citric Acid

The annual world production of citric acid is roughly 2, 00,000 tons. It is mostly used in soft drinks and food industries. Earlier, citric acid was extracted from citrus fruits, and now, its main source is molasses. A fungus viz., *Aspergillus niger* is used for citric acid production. The yield is 1 ton of citric acid from 3 tons of molasses.

Baker's Yeast

Baker's yeast is now mainly produced by molasses using *Saccahromyes cereviseae*.

Monosodium Glutamate (MSG)

The flavoring compound can be produced from the fermentation of cane molasses using *Micrococcus glutamicus*.

Itaconic acid

It is used as plasticizer and as a chemical intermediate. It is produced by the fermentation of high test molasses by *Aspergillus terreus*.

Acetone and butanol

These are produced by the bacterial fermentation of molasses by *Clostridium acetobutylicum*. The yield of butanol, acetone and ethanol per ton of molasses is about 150 liters in the proportion of 65:30:05.

Dextran

Dextran is polymer of D-glucose produced by *Leuconostoc mesenteroides* using sucrose as substrate. Dextrans are used as in oil drilling, making tooth paste, and exploration of solid minerals, paints and glues

Ephedrine

Ephedrine is a drug used as principal ingredient in all cough syrups. In India, Nellore plant of Kreb's Biochemicals Ltd., produce Ephedrine from sugarcane molasses.

Molasses as feed stuff

Molasses can be used as animal feed. It contains carbohydrates, mainly as sugars. Molasses itself is not a balanced ration because it is deficient in protein content. It is now known that urea can be mixed with molasses to form a safe feed which can provide one third of the protein requirement of cattle and sheep.

There are several advantages of using molasses as animal feed. Molasses stimulates the microflora of the rumen and allows cattle and sheep to deal effectively with low grade roughage such as straw.

Molasses is rich in potash (0.51%) and nitrogen (0.5%) and is widely used as a fertilizer for sugarcane. It is applied in furrows in the fields, two weeks before planting. It can also be mixed with cane filter cake and used as fertilizer.

Press mud cake

In the manufacture of cane sugar, the precipitated impurities present in the cane juice after filtration forms a cake. It is also known as press mud. It is produced in sulphitation and carbonation factories.

Filter cake contains about 50–70% moisture, 5–14% of crude wax and fat, 15–30% fiber, 5–15% sugar and 5–15% crude protein. Disposal of filter cake in the cane fields is most ideal to improve cane yield. Filter mud or filter cake is a useful fertilizer, especially when applied to phosphate deficient soil and to fields in which top soil has been removed or redistributed. A mixture of filter cake and molasses can also be used as fertilizer.

Sugarcane green tops and trash

Green tops of sugar cane are mainly used as cattle feed in India and fetches good price during the sugar cane harvesting period. The trash is used as filler in the plastics and linoleum industry to replace wood flour.

Sugarcane juice bottling

Sugarcane juice is a nutritious product containing natural sugars, minerals and organic acids. Sugarcane juice of 100 ml provides 40 Kcal energy, 10 mg iron and 6 µg carotene. It is rich in enzyme and has many medicinal properties. It contains water (75%-85%), reducing sugars (0.3-3%) and non-reducing sugar (10-21%). Sugarcane juice is a great preventive and healing source for sore throat, cold and flu. It has low glycemic index which keeps the body healthy. It is an excellent substitute to artificial aerated drinks. It refreshes and energizes the body instantly as it is rich in carbohydrates.

Healthy sugarcane is used to extract the sugarcane juice. The skin and node of the cane is peeled using suitable knife. The peeled cane is washed with water to remove dust and dirt. Sugarcane juice is extracted by power operated horizontal crusher and then filtered through filter and muslin cloth. During crushing, lemon juice and ginger extract are added. About 20 ml of lemon juice and 100 g of ginger extract can be added to 20 kg of sugarcane juice. After thorough mixing of sugarcane juice with lemon and ginger extract, juice is preheated at 50°C for 10 minutes. During preheating, the scum floating on the juice should be removed from time to time using steel ladle. The clear juice is then cooled to ambient temperature without any contamination. Class II preservative, potassium/sodium benzoate @ 180 ppm can also be added to the juice. After thorough mixing, the juice is again filtered through muslin cloth. The cooled juice is then filled in sterilized bottles and sealed with caps using sealing machine. The bottles are then pasteurized at 80°C for 10 minutes to prevent contamination (Fig 2).

Fig. 2: Pasteurized sugarcane juice in bottles

Jaggery

Jaggery or *Gur* is a golden-yellow to dark brown colour, coarse, wholesome, traditional, unrefined sugar obtained by concentrating sugarcane juice. India is the largest producer of jaggery under unorganized agro processing sector (Alam, 1999), sharing 55% of the world production, while Colombia is the second largest producer, contributing 11% of the world production. About 14.2% of the total sugarcane produced in India is being utilized for making jaggery (Anonymous, 2016a). Among all the states in India, Uttar Pradesh is the major producer of sugar cane jaggery followed by Tamil Nadu (Fig. 3).

Jaggery, traditionally referred to sugarcane jaggery, is prepared from sugarcane juice. Besides sugarcane juice, jaggery is being made from the sap collected from different palm trees (Pattnayak and Misra, 2004). All these jaggery have its own characteristic taste and aroma. Sixty to seventy percent of the total jaggery comes from sugarcane; remaining 30% comes from palms (Kamble, 2003).

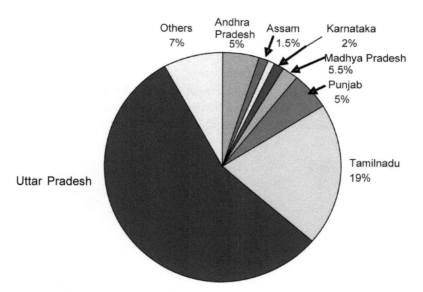

Fig. 3: Jaggery and Khandsari production in India (2014-15)

Jaggery manufacturing process

Preparation of quality jaggery is mainly influenced by agronomic management practices of sugarcane cultivation and jaggery manufacturing process. Various practices like selection of cane variety, soil conditions, time of planting, fertilizer application, time of irrigation, crop protection from pest and diseases, harvest of cane and post harvest care of cane influence jaggery production. Identified sugarcane varieties with high sucrose content are to be selected to prepare good quality jaggery. Well drained and salt free loamy soils are well suited for sugarcane production to obtain good quality jaggery. It is preferable to plant the crop between January and March and harvest the crop at its peak maturity as per the duration of the variety. Inferior quality with less recovery of jaggery will be obtained due to delayed planting and also due to crushing of immature or over mature cane used for preparation of jaggery. Fertilizers (N,P,K) should be applied as per recommendation, excessive nitrogen application reduces both quality and quantity of jaggery. The crop should not be subjected either to moisture stress or water logged condition. Dead canes, water shoots, spoiled and rat infested canes should be discarded before crushing to prepare quality jaggery. Care must be taken to harvest the cane at the base of the clump, as bottom portion of the cane is rich in sucrose. Cane should be crushed immediately after harvest for extracting juice. Otherwise reduction in cane weight of about 5-8% and juice recovery of 2% will occur in two to three days and thus results in deterioration in quality and recovery of jaggery.

The sugarcane juice is an opaque liquid and varies in colour from grey dark green to light yellow depending upon the colour of cane. In addition to various nutritional constituents, it also contains mud, wax and several other soluble and insoluble impurities. To maintain proper quality in jaggery, all these undesirable fractions should be removed. However, the manufacturing process depends on the ultimate form to be produced. Also, minute details of the process vary widely from state to state, in the state from one district to another, and in some cases within a district also.

Solid Jaggery

In general, as presented in Fig.4, jaggery manufacturing process consists of the following unit operations i.e., juice extraction, juice clarification, juice concentration by boiling, cooling of concentrated juice followed by molding and storage.

Extraction of juice by crushing the sugarcane is the first step in jaggery manufacture. Three roller cane crushers (vertical/horizontal) are used to extract the juice. Vertical three roller crusher has the juice recovery efficiency of 50-55%, whereas the same for horizontal crusher is 55-60%. The extracted juice is collected in a masonry settling tank and rested for few minutes for separation of light and heavy particles. The clear juice is drawn from a middle port and transferred to boiling pan to fill only $1/3^{rd}$ of its capacity. Boiling of sugarcane juice is the second important step. Boiling is usually done using bagasse of the cane itself as the fuel. However, furnace design and pan capacity varies from place to place. Overall heat utilization efficiency of traditional type furnaces used by farmers is very low (around 35%) and needs drastic improvement.

In general, jaggery quality, storability and its acceptability depend on the clarity of the juice used in preparation. Juice collected from settling tank is further clarified during the boiling stage. It is mostly done by adding lime (calcium hydroxide). This addition of lime simultaneously increases the normal pH of juice i.e., 5.2 to 5.4 (which depend on harvesting status, variety of cane and soil condition) to around 6.0 to 6.4. Addition of lime also improves the consistency of jaggery by increased crystallization of sucrose, but at the same time it darkens the colour if added in excess. One kg of lime (with purity of 80-90%) is mixed with four liters of water, and about 60-70 ml of the resulting solution i.e., milk of lime is proportionate to every 100 kg of cane juice. Among other chemical clarificant, hydros is the next preferred by farmers. However, hydros being a bleaching agent have a decolourisation effect. Addition of super- phosphate, phosphoric acid, chemiflocks and alum are also reported. Use of these chemical clarificants is specific depending on the juice as they may function as bleaching agent, electrolyte or pH adjusting agent. Problem of using chemical clarificants

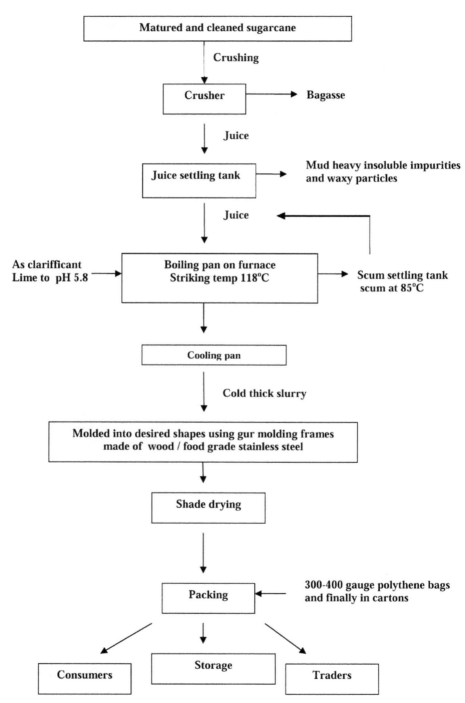

Fig. 4: Process flow chart for manufacturing jaggery cubes

is that in most of the cases, limit of addition exceeds the permitted level. Presently, emphasis is given on natural clarificants from vegetable sources like mucilage's of bhendi, chikani, kateshevari, moranga leaf juice extract, lemon juice etc., which were used during earlier periods.

Cane juice is boiled continuously until it becomes thick syrup with timely removal of scum (Fig.5). Small quantity of edible oil is added to prevent frothing during boiling after removal of entire scum. At this point, temperature plays a major role for molding the syrup into different forms of jaggery. The temperature at which cane juice slurry is modified into jaggery is known as striking point. The striking point varies from product to product. The required striking point for jaggery in cube/lump form is 118°C and for liquid jaggery it is 106 - 107°C

After attaining desired striking point, consolidated syrup of jaggery (jaggery charge) is brought down from furnace, allowed to cool for at least 15 – 20 minutes. After this, the consolidated syrup is molded into desired forms. After solidification, the moisture content of solid jaggery reduces to 10-12% on dry weight basis.

Jaggery is prepared in different shapes and sizes in solid form in different parts of India. Some of the common shapes are balls (250 gms -2 Kg), bucket shaped lumps (10-20 Kg), trapezoidal lumps (4-5 Kg), rectangular blocks etc. Jaggery prepared in any of the above forms gets deteriorated fast because of its high initial moisture content (10 to 12%) and hygroscopic nature. It is difficult to store jaggery during monsoon season especially in coastal areas. The hygroscopic nature of jaggery is due to its non sucrose constituents like glucose, fructose, proteins etc. which absorbs moisture from the atmosphere and run into liquid there by making the product unfit for consumption. About one third to half of the jaggery produced every year in our country is stored for consumption during and after rainy season. It is estimated that about 5 to 10% of jaggery stored is getting spoiled every year causing colossal loss to the nation.

The traditional lump form of jaggery is inconvenient for consumers to purchase for daily requirement. During the marketing process from retail shops, the lump is made into pieces according to consumer need leading to loss of jaggery of about 10% resulting in unhygienic environment. To mitigate the loss, making jaggery in the form of cubes is introduced. Jaggery cubes are made in different sizes viz., 25 g, 125 g, 250 g, 500 g and 1 kg which reduce the moisture content to 5-7% resulting in better shelf-life. The farmer could get benefit of Rs, 40,000/- per ha by preparation of jaggery in the form of small blocks instead of traditional lump form. The gur molding frames (Fig.6) developed by Indian Institute of Sugarcane Research, Lucknow, made of either wooden or food grade stainless steel are used for molding jaggery in different sizes.

Fig. 5: Jaggery preparation by open boiling

Fig. 6: 250 g food grade stainless steel gur molding frame

Storage

Storage of jaggery varies from region to region such as earthen pots, wooden boxes, metal drums etc. Sometimes, without any container, heap of jaggery is just kept covered with cane trash, bagasse, wheat straw, cotton seed, furnace ash, palmyra leaf mat, rice husk etc., to protect the jaggery from ambient humidity.

In respect of the shape and size, jaggery deteriorates fast and become watery within 3 or 4 months because of presence of moisture, invert sugar and its hygroscopic nature. For good keeping quality, moisture content of jaggery should be kept below 6% and at a relative humidity in the range of 43–61% (Singh *et al.*, 1978).

Health benefits of jaggery

Jaggery is consumed directly or used for preparation of sweet confectionery items and traditional (ayurvedic) medicines (Pattnayak and Misra, 2004). It contains moderate amount of important minerals like calcium, phosphorus, potassium, iron, and copper as well as B complex vitamins. The micronutrients present in jaggery possess anti-toxic and anti-carcinogenic properties. Magnesium found in jaggery strengthens the nervous system and potassium maintains electrolyte in the cells. Jaggery is rich in iron and helps in preventing anemia in women (Anonymous, 2016b). Intake of jaggery in the diet can prevent the atmospheric pollution related toxicity and may have a role to reduce the chance of lung cancer (Sahu and Paul, 1998). The nutritive value of jaggery in comparison with sugar is presented in Table 2.

Table 2: Nutritive value of jaggery and sugar (per 100g)

S.No.	Particulars	Jaggery	Sugar
1.	Sucrose (g)	65-85	99.5
2.	Glucose, Fructose (g)	10-15	—
3.	Proteins (g)	0.4	—
4.	Fats (g)	0.1	—
5.	Calcium (mg)	8.0	—
6.	Phosphorous (mg)	4.0	—
7.	Iron (mg)	11.4	—
8.	Copper (mg)	0.8	—
9.	Total minerals (g)	0.6-1.0	0.05
10.	Moisture (g)	3-10	0.2-0.5
11.	Energy (Kcal)	383	398

Forms of jaggery

Jaggery is available in the market mainly in solid, liquid and granular forms (Fig.7). Of the total production of jaggery in India, approximately 80% of the jaggery is prepared in solid form and the remaining 20% is prepared in liquid as well as granular form. Liquid jaggery is a part of diet in most parts of Maharashtra and is gaining commercial importance. The liquid jaggery is being utilized as sweetening agent in foods and drinks in Maharashtra, Gujarat, Kerala, Andhra Pradesh and Tamil Nadu. It is also used in pharmaceutical formulations.

Fig. 7: Forms of jaggery

Granular jaggery

The inherent moisture content in the traditional solid jaggery leads to the development of microbes causing spoilage and reduce its shelf life for a period of 3 to 4 months. To overcome this, process technology and machinery to prepare granular jaggery was developed at Regional Agricultural Research Station, Anakapalli, Visakhapatnam District, Andhra Pradesh. Matured cane is selected for making granular jaggery and these are crushed using three roller horizontal crusher. The juice is allowed to settle in settling tank and clear juice is transferred to boiling pans. The pH of the juice should be maintained in between 6.0 – 6.2 with the addition of lime.

Juice, after adjusting the pH is boiled in a pan on furnace for high heat use efficiency. It takes around 2 hours for one boiling of 350 – 400 Kg juice. Boiling is continued with removal of scum from time to time till juice attains a striking point of 120°C (2°C above normal striking point of solid jaggery). Small quantity of groundnut oil may be sprinkled to control frothing during boiling. The flow chart of making granular jaggery is presented in Fig. 8.

After attaining the striking point, boiling is stopped and concentrated juice is allowed to cool with occasional stirring for few minutes. Allow the mass to cool for few more minutes without stirring for good granular formation. Then, transfer the thick mass from pan to aluminium trays or on to a cement platform and allow to cool, so as to become semisolid mass. At this solidifying stage, the mass must be made into granular using wooden scrappers or stainless steel scrappers by shearing action. The shearing action breaks the mass into particles exposing more surfaces for atmospheric cooling.

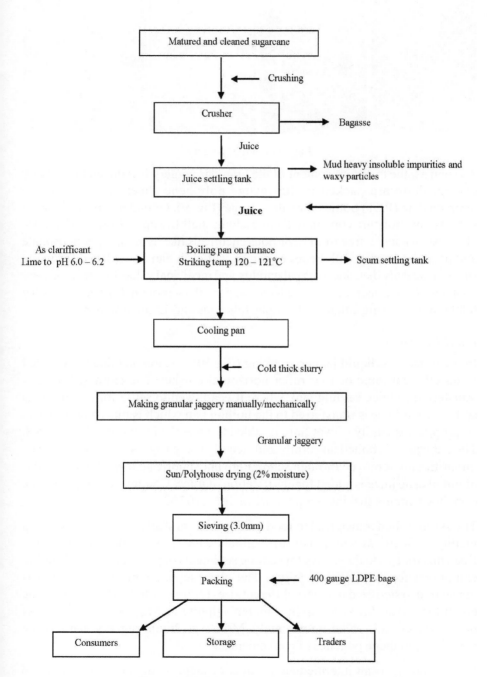

Fig. 8: Process flow chart of granular jaggery

Fig. 9: Granular jaggery

As soon as, the moisture content of the granular jaggery is reduced to 1-2% by drying, these are packed in 400 gauge polythene sheet or polyethylene terephthalate (PET) bottles for safe storage (Fig. 9). Granular form of jaggery with its low moisture content will have a long shelf life up to 2 years. These are also easy to handle, free from hazardous chemicals like hydros, super phosphate and phosphoric acid with better export potential. It can be replaced for white sugar in the daily diet, due to its palatability and medicinal value. Granular jaggery contains 80-90 g sucrose, 5-9 g reducing sugars, 0.4 g protein, 0.1 g fat, 0.6-1.0 g total minerals, 9 mg calcium, 4 mg phosphorous and 12 mg iron per 100g.

Liquid jaggery

In the process of liquid jaggery making (Fig.10), sugarcane juice is extracted using efficient three or four roller horizontal crusher. The extracted juice is transferred to juice settling tank where it is allowed to settle for few minutes and the clear juice is transferred to the boiling pan which is placed on furnace. Appropriate quantity of wet lime is added to raise the initial pH of 5.4 to 5.8. Then, the juice is boiled uniformly and steadily using bagasse/trash as fuel. The impurities are removed through scum. Pure phosphoric acid @ 150 ml/1000 L of initial cane juice is added for clarification of juice. Continue boiling until the cane juice attains striking temperature of 106 - 107°C,

The syrup is then transferred to food grade stainless steel settling tank with tap facility at 5–10 cm above base to facilitate easy flow of supernatant syrup. Keep this for two to three days for settling sediments or passed through filtration equipment for getting clear syrup. The clear supernatant jaggery syrup is transferred to sterilized bottles of different sizes (100 ml–1000 ml) under hygienic conditions. Add 0.1% Potassium metabisulphite (KMS) and citric acid as preservatives and sealed immediately. Moreover, liquid jaggery can also be stored in polythene pouches of convenient size.

Liquid jaggery is an intermediate product of jaggery making process. While concentrating sugarcane juice, when solid content reaches 60-70° Brix with a corresponding temperature of 105°-106°C, the juice is collected. This is popularly known as liquid jaggery (Fig.11). The composition of 100 g liquid jaggery is: water 30-35 g, sucrose 40-60 g, invert sugar 15-25 g, protein 0.5 g, fat 0.1 g, energy 300 Kcal, total minerals 0.75 g, calcium 300 mg, phosphorous 3.0 mg and iron 8.5-11 mg.

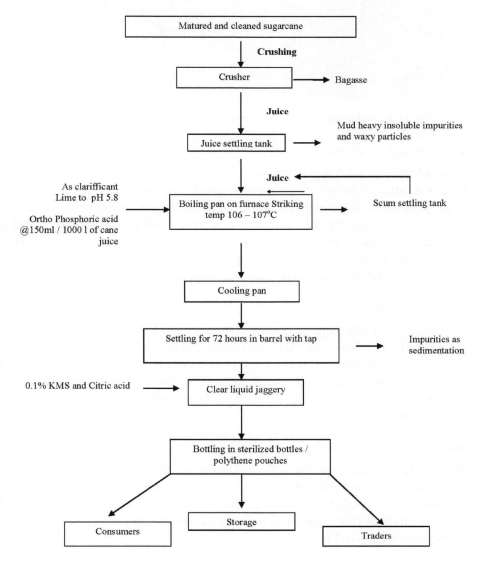

Fig. 10: Process flow chart of liquid jaggery

Fig. 11. Liquid jaggery   **Fig. 12:** Jaggery based extruded snacks

Value added products of jaggery

Value addition to solid jaggery using puffed rice, gram, sesame, nuts, vitamins, iron, and taste enhancers like chocolate powder will increase the demand of jaggery. Nutritive value and palatability can be enhanced by preparing jaggery patti by mixing jaggery with puffed rice, gram or ground nut. Jaggery-wheat flour extruded snacks are prepared by mixing wheat flour and jaggery (Fig.12). Jaggery-besan snacks are prepared with gram flour and jaggery. Procedure to prepare different value added jaggery products are detailed below.

Jaggery cookies

Take multigrain atta, jaggery powder, oat meal and butter in a proportion of 1:0.4:0.4: 0.5 in a bowl. Add $1/4^{th}$ tea spoon of baking powder to this and mix well and prepare dough. Cut the cookies of dough in required shapes on the tray keeping some distance between each. Bake the cookies in a pre heated oven at 180°C for 20 minutes. Remove the tray from oven and cool down to room temperature (Fig.13).

Jaggery cake

Mix multi grain atta, jaggery powder and butter in a proportion of 1: 0.5: 0.5 and $1/4^{th}$ teaspoon baking powder in a bowl. Add water into the mixture to form a smooth batter. Sprinkle some crushed nuts on top of the batter. Bake in a pre heated air fryer at 160°C for 20 minutes. Once baked, remove it from fryer and allow it to cool down (Fig.13).

Sugarcane Juice Jelly

Mix ½ cup sugarcane juice with gelatin (16 mg/ml) and keep aside for 30 minutes to dissolve the gelatin in juice. Separately, take half cup of juice and mix with jaggery powder. Slight warming is necessary if gelatin is not dissolved. Pour sugarcane juice with jaggery powder mixture into the designed molds and cool it until solidification in the form of jelly (Fig.14).

Jaggery chocolates

Grease an aluminum tray with butter and set aside. Melt jaggery in a double boiler and then add butter and mix the contents. As soon as jaggery melts, add crushed peanuts and cocoa powder and mix well. Place the mixture on to silicon jaggery molds and spread evenly. Set aside the molds for 15-20 minutes, and cool down the chocolates to room temperature and then keep in the refrigerator for 5 minutes (Fig.15).

Jaggery powder paper sweet (Pootharekulu)

Paper sweet also known as *Pootharekulu*, is a popular traditional sweet of Andhra Pradesh. It is a wafer like sweet made from rice starch or finely ground soaked black gram and fortified with jaggery/sugar (Fig.16). Preparation of this sweet requires special skills and is a labour intensive process.

Fig. 13: Jaggery cakes and cookies

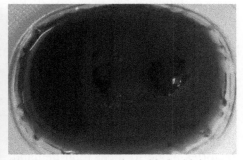

Fig. 14: Sugarcane Juice added with jaggery powder jelly

Fig. 15: Jaggery chocolates

Jaggery coated mouth freshners

A process to prepare mouth fresheners using a thin film of jaggery as an alternative to sugar, over fennel seeds/amla/ginger has been standardized (Fig 17).

Jaggery cubes

Using sugar as a sweetener is harmful to diabetic patients. Instead of artificial sweeteners it is better to promote the usage of nutritive compressed jaggery powder cubes in beverages. The solid jaggery cubes of diameter 9.45 mm and thickness of 3.75 mm with an average weight of 300 mg could be prepared using granular jaggery without binder using semi automatic tablet machine (Fig 18).

Conclusion

Jaggery industry has been one of the most ancient and important rural-based cottage industries in the country. It provides job to the unemployed rural people in their vicinity with minimum capital investment. It also have higher medicinal and nutritional values and easily available to the rural people. Also, it may not be possible for sugar factories alone to meet total demand of sweeteners with increase in the population. It is, therefore, essential to safeguard the interest of jaggery manufacturing unit and improve and modernize the activities for more purposeful ends. Development and modernization of jaggery sector need support in terms of statutory provisions and execution of a clear-cut plan for

Fig. 16: Jaggery powder paper sweet

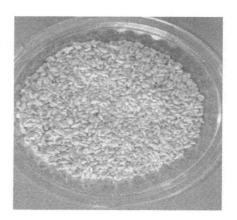

Fig. 17: Jaggery coated mouth fresheners

Fig. 18: Jaggery cubes

setting up of jaggery units in a group of sugarcane producing areas, quality control laboratory at the District level and liaison with mega stores for marketing and supply to consumers

References

Alam, A. 1999. Industrial and Policy Issues Including Export Potential of Jaggery and Khandsari, *Proceedings of the National Seminar on Status, Problems and Prospects of Jaggery and Khandsari Industry in India*, Lucknow, 7-15.

Anonymous, 2016$_a$. State Showing Figures of Area, Yield, Production of Sugarcane, *Indian Sugar*, 17 (8): 40.

Anonymous, 2016$_b$. Jaggery, [online], *Available to: http://www.sugarindia.com*, (Accessed 11, January, 2017).

Gillian, E. & L. Isabel. 2015. Sustainability Issues and Opportunities in the Sugar and Sugar-by-product Industries. *Sustainability*, 7: 12209-12235.

Kamble, K. D. 2003. Palm Gur Industry in India. *Indian Journal of Traditional Knowledge*, 2(2): 137-147.

Pattnayak, P. K. & M. K. Misra. 2004. Energetic and Economics of Traditional Gur Preparation: A Case Study in Ganjam District of Orissa, India. *Biomass and Bioenergy*, 26: 79-88.

Rao, M. P J. 2008. Working of Jiangmen Sugar and Chemical Complex in China. Paper presented at the *Sugar Asia 2008*: *An International Exclusive Exhibition on Sugar & Down Stream Industries*, July 25–26, New Delhi, India.

Sahu, A. P. & B.N. Paul. 1998. The Role of Dietary Whole Sugar- Jaggery in Prevention of Respiratory Toxicity of Air Toxics and in Lung Cancer. *Toxicology Letters.*, 95(1): 154.

Singh, G.B. & S. Solomon. 1995. *Sugarcane: Agro-Industrial Alternatives*. Oxford & IBH Publishing Co.Pvt.Ltd, New Delhi, India.

Singh, M ., K. M. Bharadwaj & M. L. Agarwal. 1978. Storage of Jaggery, *Co-Operative Sugar*, 3.

Solomon, S. & G. B. Singh. 2005. Sugarcane Diversification: Recent Developments and Future Prospects. In *Sugarcane: Agro-Industrial Alternatives*, (ed.) S. Solomon & G.B. Singh, 523–541. New Delhi, India: Oxford IBH Publishing Co. Pvt. Ltd., New Delhi

Yadav, R. L. & S. Solomon. 2006. Potential of Developing Sugarcane by Products Based Industries in India, *Sugar Tech.*, 8 (2-3): 104–111.

12

Entrepreneurial Opportunities in Milk Processing

Aswin S Warrier and Aparna Sudhakaran V

Introduction

Consumption of milk is deeply rooted in our traditions, and it is an inevitable part of the Indian diet and even has ritualistic and medicinal uses. Milk is the first and one among the most nutritious foods that every mammal consumes on earth. It supplies all the nutrients needed to ensure proper growth and development in the postnatal period. Dairy foods in general are commonly considered balanced and nutritive, being frequently included as important components of a healthy diet (Pereira, 2014). Four components of milk are dominant in quantitative terms: water, fat, protein and lactose; while the minor components are minerals, enzymes, vitamins, and dissolved gases (Guetouache *et al.*, 2014). On an average, bovine milk is composed of 87% water, 4 to 5% lactose, 3% protein, 3 to 4% fat, 0.8% minerals, and 0.1% vitamins. Milk and milk products contribute not only nourishment to the consumers, but also livelihood opportunities to farmers, processors, shopkeepers and other stakeholders in the dairy value chain.

India is the largest milk producing country in the world since 1998, with a production of 155.5 million tonnes in 2015-16. Milk production in India has tremendously increased from 22 million tonnes in 1970, which shows a growth of 700 per cent during last 46 years. The per capita availability of milk was at 130 g/day in 1950-51. It sharply increased to 377 g/day in 2015-16, as compared to average world per capita availability of 229 g/day. India had a growth rate of 6.28% in milk production in last two years, three times more than the world average growth of 2.2%. (DAHDF, 2016; Business Standard, 2017).

Despite being the world's largest producer, the dairy sector is by and large in the primitive stage of development and modernization. Being a significant provider to Indian economy, variety of notable programmes has been taken up by the governments to attract more people to this sector. Furthermore, a variety of import restrictions are imposed by the government to protect its domestic dairy market. Hence, there is ample scope for milk processing in Indian market, both for local and multinational players, because of its size as well as potential. Dairy food processing holds immense potential for high returns if proper value addition and branding is done. Therefore, more and more private players are getting into marketing value added dairy products, where all are finding their own space and opportunity. Even global dairy majors are looking at India as a lucrative investment destination (Ajita, 2016).

Health benefits of milk

Milk, in its various forms, has been part of the human diet for millennia, because it is high in a range of nutrients. Milk and dairy foods are considered to be one of the main food groups important in a healthy balanced diet. Indian ancient Vedic texts describe the virtues of milk and dairy products, as is authenticated by modern scientific principles and proofs (Nagpal *et al.*, 2012). The completeness of milk as food will be appreciated when it is seen that it has to serve in the new-born animal as the sole substitute, at least for a period, for the prenatal nourishment that the foetus was receiving through the placental circulation. In addition to the maintenance of the off-spring, the mammary secretion is called upon to supply the energy required for the newly-started functions of respiration, digestion, and even perhaps in some species, locomotion (Rangappa and Achaya, 1948). Milk contains various components with physiological functionality. Milk contains high levels of immunoglobulins and other physiologically active compounds for warding off infection in the newborn. Similarly, colostrum is important for newly born mammals as it provides necessary immunity against infections (Shah, 2000). Research continues to expand the positive role milk and dairy foods play in an individual's health.

Milk is one of the most nutrient-dense foods, filled with a unique blend of carbohydrates, proteins, fats, vitamins and minerals. Milk makes a significant contribution to meet the body's needs for calcium, magnesium, selenium, riboflavin, vitamin B_{12} and pantothenic acid. However, milk does not contain enough iron and folate to meet the needs of growing infants, and the low iron content is one reason animal milk is not recommended for infants younger than 12 months old (Muehlhoff *et al.*, 2013). The gross composition of milk from different animals is given in Table 1. Many factors affect the gross composition of milk, the most significant ones being breed, feed, season, region, and herd health. The average nutritional composition of whole, low fat and skim milk is given in Table 2.

Table 1: Gross composition of milk (Bylund, 1995)

Constituents	Human	Horse	Cow	Buffalo	Goat	Sheep
Protein total (%)	1.2	2.2	3.5	4.0	3.6	5.8
Casein (%)	0.5	1.3	2.8	3.5	2.7	4.9
Whey protein (%)	0.7	0.9	0.7	0.5	0.9	0.9
Fat (%)	3.8	1.7	3.7	7.5	4.1	7.9
Carbohydrate (%)	7.0	6.2	4.8	4.8	4.7	4.5
Ash (%)	0.2	0.5	0.7	0.7	0.8	0.8

Table 2: Average nutritional composition of whole, low-fat, and skim milk (Pereira, 2014)

Composition (100 g)	Whole	Low fat	Skim
Energy (kCal)	62	47	34
Water (g)	88.1	89.1	90.5
Protein(g)	3	3.4	3.3
Fat (g)	3.5	1.6	0.2
Carbohydrates (g)	4.7	4.9	4.9
Cholesterol (mg)	13	8	1
Vitamin A (mg)	59	22	0
Vitamin D (mg)	0.05	0.05	0
Vitamin B_1 (mg)	0.04	0.04	0.05
Vitamin B_2 (mg)	0.14	0.11	0.05
Sodium (mg)	43	41	41
Calcium (mg)	109	112	114
Magnesium (mg)	9	9	10

Milk, yoghurt, and cheese are excellent sources of high-quality proteins. Individual milk proteins have a wide range of potential health benefits and functional properties. Approximately 80% of the protein is casein and 20% is whey, in bovine milk. Both casein and whey proteins are rich sources of peptides that significantly lower blood pressure in those with hypertension and may contribute to satiety and regulate food intake. Whey protein has the highest biological value of any protein, which means it is highly usable by the body. Whey protein is one of the richest sources of leucine, an essential amino acid that triggers initiation of muscle protein synthesis. Other milk proteins also exhibit a wide range of bioactive properties. For example, caseinophosphopeptides, glycomacropeptides, and lactoferrin help to reduce the risk of dental caries. In light of protein's health benefits, manufacturers are isolating various milk proteins, such as milk protein concentrates (MPCs), whey protein isolate (WPI), hydrolyzed whey protein, and whey permeate. These are used as ingredients in dairy and non-dairy foods, and beverages such as milk, yoghurt, energy bars, and cheese to boost their protein content (Smith, 2013).

Milk fat contains approximately 400 different fatty acids, which make it the most complex of all natural fats. About 98% or more of the lipid is triacylglycerol. Phospholipids are about 0.5 to 1% of total lipids and sterols are 0.2 to 0.5%. Consumption of milk and milk products—regardless of fat level—is associated with a lower blood pressure and a reduced risk of CVD and Type 2 diabetes. Moreover, milk fat is never consumed in isolation; dairy foods also contain protein, calcium, and other components that may modulate the effect of fat on health. The milk fat globule membrane (MFGM) has gained a lot of attention due to the growing interest in its nutritional and technological properties. The whole membrane as well as the separate lipid and protein components has great potential for new product applications with unique nutritional and technological properties (Dewettinck et. al., 2008). Research on dairy fat has come to some surprising conclusions: People who eat full-fat dairy are no more likely to develop cardiovascular disease and type 2 diabetes than people who stick to low-fat dairy. When it comes to weight gain, full-fat dairy may actually be better. (Kratz et. al., 2013).

The main carbohydrate in dairy is lactose and insufficient activity of the enzyme β-galactosidase causes intolerance to lactose with all its consequences including diarrhoea. About 75 % of the world's population is unable to break down lactose, a phenomenon called lactose intolerance. The individuals that are intolerant to lactose show better tolerance to fermented milk than to native milk (Ebringer et al., 2008). Lactulose, a disaccharide composed of D-galactopyranose and D-fructofuranose originates during heat processing of milk. Lactulose has beneficial health effect mainly by selective stimulation of the growth and/or activity (prebiotic effect) of probiotic bacteria including bifidobacteria and lactobacilli (Gibson et al., 2004)

Typically, milk has been recognized as a privileged calcium source but in its mineral fraction, several other elements can be demarcated such as phosphorus, magnesium, zinc, and selenium. The vitamin fraction is composed by liposoluble vitamins A, D, and E and also by water-soluble B complex vitamins such as thiamine and riboflavin. The concentrations of fat-soluble vitamins in milk depend on fat content and thus, low-fat and skim milk varieties have lower amounts of A, D, and E vitamins. In some countries, skim milk is fortified with vitamin A and D to improve its nutritional richness (Pereira, 2014).

The most frequent claim favourable to milk consumption has been its richness in calcium and this mineral's role in bone density. Low bone mass is the main risk factor for osteoporosis and it is known that bone mass in later life is quite dependable from the peak bone mass achieved during growth. Milk consumption has been previously associated with a higher bone density, which is protective (Kim et al., 2013). A number of studies suggest that milk consumption is largely anti-cariogenic when combined with a typical routine of oral hygiene. This effect can be mostly attributed to several factors like tooth remineralization, inhibition of bacterial colonization, and biofilm inhibition, which are likely due to numerous proteins found in milk (Merritt *et al.*, 2006).

There is a long history of health claims pertaining to living microorganisms in foods, particularly lactic acid bacteria. The term probiotic, meaning "for life," are "live microorganisms which when administered in adequate amounts confer a health benefit on the host". Probiotics are normally consumed in the form of capsules or through dairy or non dairy based foods. Some of the reported benefits of probiotic consumption include: (i) improving the health of intestinal tract; (ii) enhancing the immune system, synthesizing and enhancing the bioavailability of nutrients; (iii) reducing symptoms of lactose intolerance, decreasing the prevalence of allergy in susceptible individuals; and (iv) reducing risk of certain cancers (Parvez *et al*, 2006).

Dairy products are currently the most common platform for the delivery of probiotics. Dairy probiotic products available in the market can be broadly divided into either (a) fermented, that includes yoghurt, cheese, dahi, butter milk, kefir etc. or (b) non-fermented like ice creams and other dairy desserts, milk powder, normal and flavoured liquid milk.

Value added products of milk

The value added dairy products can be categorized as shown below:

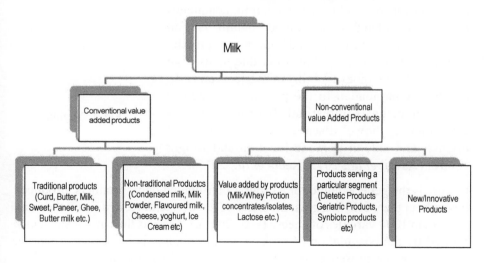

As value addition can include any operation (including selection, processing, packaging, branding etc.) intended to increase the value of the raw material, a lot can be written about the asw-topic. However, considering the space constraints, some of the conventional value added products only are mentioned in this topic.

Fermented dairy products

Milk products prepared by fermentation by lactic acid bacteria or yeasts, moulds or other bacteria are called fermented or cultured milk products. Dahi (curd) is a very popular traditional product in the Indian subcontinent, made by fermenting milk with lactic acid bacteria. It may be considered as the Indian version of yoghurt. The bacteria used for fermentation are known as 'dahi starter cultures'. Dahi starter cultures act on the lactose portion of milk, converting it to lactic acid, which acts on milk protein to give dahi its characteristic texture and flavour. Dahi is having a lot of nutritional and therapeutic benefits. Dahi, in its stirred form is usually sold in polyethylene pouches, while set dahi is sold in plastic tubs.

Yoghurt is one of the most popular fermented milk products worldwide and has gained widespread consumer acceptance as a healthy food (Fig.1). Yoghurt is a product of lactic acid fermentation of milk by addition of starter culture containing *Streptococcus thermophilus* and *Lactobacillus delbrueckii* ssp. *bulgaricus*. When sufficient quantity of lactic acid is produced, milk coagulates and this coagulated milk is called yoghurt. It is an excellent source of calcium and high-quality protein. It is considered as a healthy food due to its high digestibility and bioavailability of nutrients and also can be recommended to the people with lactose intolerance, gastrointestinal disorders such as inflammatory bowel disease and irritable bowel disease, and aids in immune function and weight control (Aswal *et al.*, 2012; Reeta *et al.*, 2015).

Fig. 1: Yoghurt

Cheese is the generic name for a group of fermented dairy products, produced throughout the world in a great diversity of flavour, texture, and form; there are more than 1000 varieties of cheese (Fox, 2011). Cheese is the curd or substance formed by the coagulation of milk of certain mammals by rennet or similar enzymes in the presence of lactic acid produced by added or adventitious microorganisms (Fig.2). Part of the moisture has been removed by cutting, warming and pressing, which has been shaped in mold and then ripened (also unripened) by holding for sometime at suitable temperature and humidity

Fig. 2: Cheese varieties (http://flexibristol.org/the-impact-of-cheese/)

(ecourses.iasri.res.in). Cheese manufacture essentially involves concentrating the fat and casein of milk 6–12 fold by coagulating the casein, enzymatically or isoelectrically, and inducing syneresis of the coagulum which can be controlled by various combinations of time, temperature, pH, agitation and pressure (Fox *et al.*, 1993).

Fat rich dairy products

Cream is a fat rich milk product in the form of 'oil-in-water' emulsion, obtained by physical separation of milk. The physical separation of a fat-rich portion (cream) from a nearly fat-free portion (skim milk) by centrifugation relies on the difference in the density between the fat globules and the aqueous phase in which the fat globules are dispersed. There are different varieties like full cream, half cream, table cream, light cream, coffee cream etc available in market.

Butter is a fat rich dairy product made by churning fresh or fermented cream to separate butter fat from buttermilk. It is an emulsion of 'water-in-oil' type and is made in both sweet and sour varieties. Butter is manufactured in four varieties, namely, sweet cream unsalted, sweet cream salted, cultured unsalted and cultured salted. Traditionally, butter was manufactured in small quantities using manually operated churns. As a result of advancement in technology and industrialization, organized manufacture of butter has moved to continuous butter making machines. According to the Codex Alimentarius Commission under the Joint FAO/WHO Food Standards Programme, butter is a fatty product derived exclusively from milk. A 100 g portion of butter must contain a minimum of 80 g fat and a maximum of 16 g water and 2 g nonfat milk solids.

Ghee is the name used in India for the clarified milk fat, produced by heating cream or butter, until the water content is boiled off. In addition to its use as a food ingredient, it is an important ingredient in religious ceremonies and in traditional medications. During heat treatment, considerable browning of non-fat milk solids occurs that develops a special taste and a strong antioxidative effect. Today, ghee is often factory made by more industrialized methods. The standard specifies ghee to have a minimum of 96% milk fat, maximum moisture content of 0.3%, maximum free fatty acids of 0.3% oleic Acid, and a peroxide value less than 1 meq/kg (Sserunjogi *et al.* 1998).

Indigenous dairy products

Buttermilk is a dairy ingredient widely used in the food industry because of its emulsifying capacity and its positive impact on flavour. Commercial buttermilk is sweet, a by-product from churning sweet cream into butter (Sodini *et al.*, 2006). This liquid phase contains most of the water-soluble components of cream. After disruption of fat globules, milk proteins, lactose, minerals and some lipids are recovered in buttermilk as well as milk fat globule membrane (MFGM) fragments. MFGM is composed mainly of proteins, phospholipids and minerals (Morin *et al.*, 2007). Chhas, a diluted buttermilk based drink, in its various forms, is a very popular traditional beverage in India.

Paneer, a popular Indian dairy product, is similar to an unripened variety of soft cheese which is used in the preparation of a variety of culinary dishes and snacks. It is obtained by heat and acid coagulation of milk, entrapping almost all the fat, casein complexed with denatured whey proteins and a portion of salts and lactose. Paneer is marble white in appearance, having firm, cohesive and spongy body with a close-knit texture and a sweetish-acidic-nutty flavour (Kumar *et al.*, 2014).

Milk based sweets are an integral part of Indian cuisine. Majority of milk sweets in India are made from two base materials, namely, khoa and chhana. Khoa is a concentrated whole milk product obtained by open pan condensing of milk under atmospheric pressure. It is used in the preparation of sweets like pedha, burfi, gulabjamun and kalakand. Chhana is a product of acid coagulation of hot milk and draining out the whey. Chhana based sweets are very popular in Eastern part of India, particularly in Bengal. Sandesh and rasgulla are the most famous Bengali sweets made from chhana.

Concentrated & dried milk products

Condensed milk is a dairy product obtained by partially removing the moisture content of milk, by evaporating it under vacuum, so as to minimize heat induced changes. It may be sweetened or unsweetened, even though the term condensed milk is often associated with sweetened condensed milk. Unsweetened condensed milk is usually called as 'evaporated milk'.

Dried milk or milk powder is the product obtained by removal of water by heat or suitable means, to produce a solid containing 5 per cent or less moisture. The purpose of drying is to stabilize milk constituents for their storage and later use.

Drying ensures an increased shelf life at atmospheric temperature, due to its low water activity. Moreover, it significantly reduces the bulk, making the packaging, storage and transportation easier. Nowadays, milk powders are mostly produced commercially by spray drying.

Frozen dairy products

Ice cream is one of the widely accepted dairy products among children and adults all over the world. It is a frozen dairy product made by suitable blending and processing of cream and other milk products, together with sugar and flavour, with or without stabilizer or colour, and with the incorporation of air during freezing process. Ice cream generally contains seven categories of ingredients: fat, milk solids-not-fat, sweeteners, stabilizers, emulsifiers, water, and flavour. Manufacturing is performed by the preparation of a suitable liquid mix by blending, pasteurizing, homogenizing, cooling, and aging at 4°C; subsequently, freezing that mix to -5°C through a scraped-surface freezer while under shear (which incorporates air and produces small, discrete ice crystals); optionally incorporating any flavouring materials that will remain discrete in the product (fruits, nuts, candy, or bakery pieces); packaging or shaping (as in the case of 'novelty' or 'impulse' products); and then finally blast freezing these products to a temperature of -25° to -30°C (Goff, 2011).

Safety and quality aspects of milk products

The excellent nutritional profile and health value of dairy products make it a suitable medium for growth of microorganisms. Presence of carbohydrates, proteins, fats and abundant water combined with neutral pH supports and encourages a microbial ecology that can be both diverse and highly variable. Number and type of microorganisms in raw milk determines the final quality of milk products. Safe milk concept covers both the quality and safety aspects. The milk when secreted in udder is sterile but gets contaminated in subsequent stages like milking, processing etc. Milk contains both good and bad microbes and major challenge in processing is to reduce the undesirable microbes including pathogens and spoilage microbes to a safer limit. Spoilage microbes can degrade milk and deteriorate the shelf life of processed products and also its yield. The environmental contaminants represent a significant percentage of spoilage microflora. They are ubiquitous in the environment from which they contaminate the animal, equipment, water, and milker's hands. Spoilage microorganisms include psychrotrophic gram-negative bacteria, yeasts, moulds, heterofermentative lactobacilli, and spore-forming bacteria. Psychrotrophic bacteria can produce large amounts of extracellular hydrolytic heat stable enzymes, and the extent of recontamination of pasteurized fluid milk products with these bacteria is a major determinant of their shelf life. Fungal spoilage of dairy foods is manifested by the presence of a wide variety of metabolic by-products, causing off-odours and flavours, in addition to visible changes in colour or texture. Coliforms, yeasts, heterofermentative lactic acid bacteria, and spore-forming bacteria can all cause gassing defects. Raw milk also contains somatic cells which also play a significant role in deteriorating quality of milk products and reflects the health status of the mammary glands. Presence of pathogens and their metabolites has implications on human health. Some of the reported pathogens in milk and milk products include *E. coli, Salmonella, Staphylococcus aureus, Bacillus cereus, Clostridium, Listeria mono-cytogenes* etc.

Depending upon the processing and nature of the products, the spoilage microorganisms will be different. Table 3 highlights some of the predominant spoilage microorganisms in different dairy products.

Table 3: Predominant spoilage microorganisms in dairy products

Type	Spoilage microorganisms/enzymes
Raw milk	A wide variety of gram positive and negative bacteria, yeast and moulds
Pasteurized milk	Psychrotrophs, sporeformers, thermophiles, heat stable enzymes
Concentrated milk	Spore formers, osmophilic yeast and moulds
Dried milk	Spore formers, heat stable enzymes
Butter	Yeast and moulds, psychrotrophic bacteria
Cultured butter milk	Coliforms, yeast and mould, psychrotrophs
Curd/ yoghurt	Coliforms, yeast and mould, psychrotrophs
Soft cheese	Psychrotrophs, coliforms, yeast and mould
Ripened cheese	Sporeformers, lactic acid bacteria, coliforms, yeast and mould

The major food safety issues related to milk and milk products are

- Presence of microbes and their toxins
- Presence of mycotoxins
- Presence of antibiotic residues, drug residues, pesticides
- Presence of heavy metals.

Milk and milk products are highly perishable and are consumed by all age groups. Hence, utmost care has to be taken for the maintenance of safety and quality of milk and milk products. This can be achieved mainly by practicing clean milk production; good hygienic practices throughout production, processing, storage and distribution; cold chain maintenance; and following standard operating procedures. The hygienic production and processing determines the shelf life and safety of the dairy foods. The production of clean and wholesome milk and processing using proper technology are important for improved quality of milk and milk products. The pasteurization (72°C/15 sec or 63°C/ 30 min) of milk take care of almost all spoilage and all pathogenic organisms making it safe for consumption. The effective prevention of post pasteurization contamination helps to sustain the quality of the product and shelf life. The quality evaluation of milk and milk products, processing, environment, handler etc. for microbiological as well as chemical quality are essential for ensuring safety.

The quality of milk and milk products can be assessed at all stages of production and processing by different tests like

- Physical tests - organoleptic properties, quantity
- Chemical tests - composition, detection of adulterants, drug residues
- Microbiological tests - hygienic conditions, cleanliness and quality, packaging material.

Raw milk received in dairy plant is first subjected to a series of rapid platform tests which include: Organoleptic (taste, flavour, colour) test, Titratable Acidity, Clot-on-Boiling, Alcohol test, Alcohol Alizarin test, 10 minutes Resazurin test (RRT), Direct Microscopic Count etc. These tests serve the purpose of acceptance or rejection of milk. Several chemical tests can be done to detect adulterants, pesticide residues, heavy metals etc. The other tests like Methylene Blue Reduction Test (MBRT), Standard Plate Count (SPC), Coliform Count etc. may be conducted to get a better idea of the microbiological quality. Besides, a number of microbiological standards/limits are prescribed to serve as a guideline for controlling the quality of raw materials and processing factors as well as the finished products. The regulations for milk and milk products are given by Food Safety Standards Authority of India (FSSAI) and organized dairy sector has been paying increased attention to improve quality of products as per standards.

Packaging of milk products

Packaging is a fundamental part of the food supply chain. Packaging can delay and protect food from chemical, biological, and physical deterioration, extend shelf-life, maintain, and assure the quality and safety of products.

Distribution in bottles was the most universally accepted milk distribution system before the advent of single-serve containers. However, it had lot of limitations like expensive infrastructure requirement, larger operational area, large store space, large labour requirement, high transportation and handling cost along with the issue of bottle breakages. Nowadays, most of the dairies in India are using polyethylene pouches for milk distribution. The plastic pouches are single-trip packages, very light in weight, and hence distribution costs are less compared to glass bottles. Losses during pouch filling are less than bottle filling and less floor space is required for packing section and cold storage. Another packaging which has become popular recently is the carton system like Tetrapak Tetrabrik (Fig 3). Such cartons made of paper board/aluminium foil/PE, in the absence of ambient air provide best protection against loss of nutrients and maintains best organoleptic properties and can be stored for months at ambient temperature conditions. The equipment necessary to produce tetrabrik is expensive to install, but its advantages include a PE-lining which is strong and does not leak, and a lower price per unit of milk in running costs (Kumar, 2005). Sterilized flavoured milk is presently marketed in glass bottles, cartons and PET bottles. Products like stirred dahi, butter milk, spiced butter milk etc are also packed in polyethylene pouches or cartons.

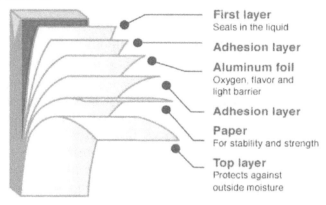

Fig. 3: Tetrabrik Milk Carton (www.nestle.tt)

Often, in dairy product packaging, there is little difference in filling and closing between solid and fluid products. The difference comes later in distribution after the product has set. Thus, from an initial packaging standpoint, packaging is the same, but from a package selection standpoint, it is important to choose structures that will contain the final product and be useful to the consumer. Products such as yoghurt and pudding are handled from a packaging operation standpoint as if they were fluids, but from a consumer standpoint, their packages must take account of spoonability. Butter is formed, pressure pumped, and cut into appropriate shapes for pats which may be dispensed onto paperboard trays or overwrapped in aluminum foil laminations in mini-molds or greaseproof paper in full size molds. Soft cheeses are generally pumped into either thermoformed polystyrene or injection molded polypropylene cups or tubs. These are then closed by a combination of aluminum foil, heat sealed to the flange, friction fit thermoform, with or without tamper resistant ring around the rim. Ice cream is packaged in bulk in paperboard coated with polyethylene (Brody, 2008). The retail packaging of ice cream is done in cups made of paperboard or plastic, cones and cartons along with various novelties. Dehydrated dairy products are generally hygroscopic in nature and even slightest increase of the moisture content will decrease the shelf life considerably. Hence, the packaging material should be impervious towards water vapour and such property is to be considered important for packing dehydrated dairy products (ecourses.iasri.res.in). Milk powders are usually bulk packed in sacks lined with polyethylene, while retail packaging is done in lacquered tin cans, flexible laminates such as metallized PET/BOPP/aluminium foil/polylaminates, HDPE bottles, laminated lined cartons or in bag-in-box form. Traditional Indian sweets like khoa, burfi and peda can be packed in vegetable parchment paper, cellophane along with polyethylene, aluminium foils or in laminates. Gulabjamun, rasgulla etc. are packed in lacquered tin cans, injection molded/thermoformed containers or stand up laminated pouches. Ghee is usually packed in lacquered tins, glass bottles or HDPE pouches.

A number of other packaging materials, forms and technologies are being utilized and are being developed for packaging dairy products. However, elaborating all of these is beyond the scope of this chapter.

Scope of by-product utilisation/effluent treatment/utilisation

A by-product is a secondary product obtained while a raw material is transformed to a finished commercial product. It may be a useful product, may be made useful by some modifications or it can be a waste. Effluent, on the other hand, refers to the liquid waste or sewage that flows out of a commercial establishment into a water body such as a river, lake, ocean or a sewer system.

By- products of milk processing

As the by-products of dairy processing are rich in nutrients and having high commercial value, their proper utilization is essential. It is often said that 'the by-product of today can be a main product of tomorrow.' Utilization of dairy by-products improves plant economy, makes valuable nutrients available for humans and reduces environmental pollution originating from dairy waste. In India, most of the problems associated with the production and utilization of dairy by-products are: a low per capita availability of milk, higher proportion of buffalo milk, poor quality of raw milk, lack of organized manufacture of products, lack of adequate technology, high cost of new technologies, lack of in-house R & D, lack of proper infrastructure, lack of indigenous equipment and plants etc. Before setting up a by-product factory, it is important to consider the economical aspects of the plant, because sometimes the cost of manufacture of the by-product may be more as compared to the by-product, and therefore, the setting up of the by-product plant is not economically feasible. But, with advancement in science and technology as also the automation of plants, the economic feasibility of these can be improved (ecourses.iasri.res.in). The major dairy by-products, along with some edible ways of utilization are given in Table 4.

Table 4: Dairy by-products and utilization

S.No	Main product	By-product	Processing method	Products made
1.	Cream	Skim milk	Pasteurization	Flavoured milk
			Sterilization	Sterilized flavoured milk
			Fermentation	Cultured buttermilk
			Fermentation and Concentration	Concentrated sour skim milk
			Concentration	Plain and sweetened condensed skim milk
			Drying	Dried skim milk or skim milk powder or Non Fat Dry Milk NFDM)
2.	Butter	Buttermilk	Coagulation	Cottage cheese, Quarg, edible casein
			Fermentation and Concentration	Condensed buttermilk
			Concentration and drying	Dried buttermilk
			Coagulation	Soft cheese
3.	Cheese, Casein, Channa, Paneer	Whey	Fermentation	Whey beverage, Yeast whey
			Concentration	Plain and sweetened condensed whey, whey protein concentrate, whey paste, lactose
			Drying	Dried whey
			Coagulation	Ricotta cheese
4.	Ghee	Ghee residue	Processing	Sweetmeat, toffee, sweet paste

Fig. 4: Whey (http://www.brighthealing.com)

Effluent Treatment

The dairy industry, like most other agro-industries, generates strong waste water characterized by high biological oxygen demand (BOD) and chemical oxygen demand (COD) concentrations representing their high organic content (Demirel *et al.*, 2005). Dairy effluents originate due to spillage of milk and dairy products and cleaning of product contact as well as non contact surfaces, combined with the sanitary waste water generated. The dairy industry is one of the most polluting industries, not only in terms of the volume of effluent generated, but also in terms of its characteristics as well. It generates about 0.2–10 L of effluent per liter of processed milk with an average generation of about 2.5 L of waste water per liter of the milk processed (Shete and Shinkar, 2013). Generally, dairy waste contains large quantities of milk constituents such as casein, lactose, fat, inorganic salts besides detergents and sanitizers which contribute largely towards high BOD and COD (Marwaha *et.al.*, 2001). The high values of suspended solids and dissolved solids show its high pollution potential. Discharge of such waste into inland surface water will lead to depletion of oxygen in water bodies, affecting aquatic life and creating unaesthetic anaerobic conditions (Noorjahan *et al*, 2004). The characteristics of dairy waste water as reported by Deshannavar *et. al.*, (2012) is given in Table 5.

Table 5: Characteristics of dairy waste water

Parameter	Range
Ph	7.2 -8.8
Colour	White
Total Suspended Solids (mg/L)	500 – 740
Volatile Suspended Solids (mg/L)	400 – 610
Chemical Oxygen Demand (COD) (mg/L)	1900 – 2700
Bio-chemical Oxygen Demand (BOD) (mg/L)	1200 – 1800

The volume, concentration, and composition of the effluent arising in a dairy plant are dependent on the type of product being processed, the production programme, operating methods, design of the processing plant, the degree of water management being applied, and, subsequently, the amount of water being conserved. Dairy cleaning water may also contain a variety of sterilizing agents and various acid and alkaline detergents. Thus, the pH of the waste water can vary significantly depending on the cleaning strategy employed. The highly variable nature of dairy waste water in terms of volume and flow and in terms of pH and suspended solid (SS) content makes the choice of an effective waste water treatment regime difficult (Britz *et al.*, 2006). Currently, there are a lot of technologies available for the on-site treatment of dairy effluent so as to lower its reduction potential and to make it compliant with the legal requirements. Some dairy plants dispose the treated effluents to municipal sewers or water bodies, while some prefer reusing it for irrigation. The on-site effluent treatment includes some or all of the steps given below:

Pretreatment

- Physical screening using wire screens and grit chamber
- pH control for making the pH optimum for biological treatment.
- Flow and composition balancing using an equalization tank
- Fat, oil and grease removal using gravity traps, air floatation or enzymatic hydrolysis.

Biological Treatment

- Aerobic treatment methods like trickling filters, rotating biological contactors, activated sludge process or treatment lagoons
- Anaerobic systems like anaerobic lagoons, stirred tank reactors, anaerobic filter digestors or UASB digestors

The residual sludge after biological treatment is dewatered using a decanter to separate the liquid portion for re-use (irrigation) and the solid sludge for disposal. The conventional wastewater treatment process in dairy industries is as shown in Fig. 5.

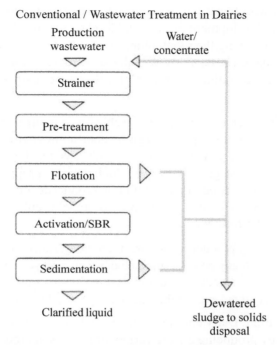

Conventional / Wastewater Treatment in Dairies

Fig. 5: Conventional wastewater treatment process (www.oilgae.com)

Scope of entrepreneurship development in milk processing sector

In India, the dairy sector plays an important role in the country's socio-economic development, and constitutes an important segment of the rural economy. Dairy business has been practised as a rural cottage industry in India for centuries. Dairy industry provides livelihood to millions of homes in villages, ensuring supply of quality milk and milk products to people in both urban and rural areas. From a largely unorganized condition earlier, India's dairy sector witnessed a spectacular growth between 1971 and 1996; the period was known as the Operation Flood era. The Operation Flood was one of the world's largest rural development programmes which ran for 26 years and eventually helped India to emerge as the world's largest milk producer. As part of the programme, around ten million farmers were enrolled as members of about 73,000 milk cooperative societies (Jagadish, 2013). However, it has remained a boring market as the per capita consumption is very much skewed and also as most of the milk was consumed in its basic, liquid form, or at best as ghee and some butter. The growth potential of this sector is enormous and it is expected that the consumption of the value added dairy and food products would grow at a very fast pace. Economic liberalization and rising consumer prosperity are opening up new opportunities for diversification in the dairy and food processing sectors.

Liberalization of world trade will open up new vistas for growth. In this scenario, entrepreneurship development in dairy and food sectors will be a key driver for promoting and sustaining the momentum of growth. Government, institutions and individuals have increasingly visualized entrepreneurship as a strategic intervention for accelerating the pace of development in any economy (Pal *et al.*, 2006).

Entrepreneurship development is indispensable for capitalizing the full potential of dairy industry as well as in providing job opportunities to the ever-increasing population of the country. As mentioned already, two third of the dairy market in India is still in the unorganized sector, which requires proper channelization. There is a phenomenal scope for innovations in product development, packaging and presentation. Some of the major opportunities for entrepreneurs in dairy processing are in the following areas:

- Ensuring quality products which are microbiologically and chemically safe. The quality of many products marketed by some players in the unorganized sector is below acceptable standards. Customers are always looking for good quality items even if they have to spend a little more.

- Value addition: Dairy entrepreneurs have the opportunity to add value to raw milk by producing a wide variety of processed products, to increase the return above the base price of raw milk. Moreover, there is substantial scope for innovations in product development, packaging and presentation.

- Export potential: Some organized players have already started exploiting the export potential of dairy products. There is a lot of scope for export of traditional Indian dairy products due to the significant presence of Indian Diaspora across the globe.

- Branding and Certification: Selecting and branding the products properly can make a huge difference. There is a niche market for items like 'farm fresh milk', 'A2 milk', ready-to-cook and ready-to- eat foods, fortified foods, probiotic foods, diabetic products, geriatric foods, low calorie foods, nutraceuticals etc. and also for certifications like 'Organic', 'Halal' etc.

Success stories in milk processing sector

The most popular success story in the field of milk processing from India is that of Amul. A brand managed by the Gujarat Co-operative Milk Marketing Federation (GCMMF) Ltd., Amul is a household name throughout India. In the dusk of colonial era, the condition of dairy farmers in Anand, like other farmers across the country, was miserable, as the milk marketing system was controlled by contractors and middlemen. At that time, Polson Dairy was the only dairy processor in Anand, who had a monopoly over the income levels of poor farmers.

Angered by the unfair trade practices, the farmers of Kaira (the district to which Anand belonged, at that time) approached Sardar Vallabhbhai Patel under the leadership of local farmer leader Tribhuvandas K. Patel. Sardar Patel reiterated his advice that they should market their milk through a co-operative society of their own. In 1946, a trusted deputy of Sardar Patel, Mr.Morarji Desai held a meeting of farmers which resolved that milk producers' co-operative societies should be organized in each village of Kaira District to collect milk from their member-farmers. All the milk societies would federate into a Union which would own milk processing facilities. The Government should undertake to buy milk from the Union. If this wasn't done, the farmers would refuse to sell milk to any milk contractor in Kaira District. The Government turned down the demand. The farmers called a 'milk strike'. It lasted for 15 days. Not a drop of milk was sold to the milk merchants. After 15 days, the milk commissioner of Bombay, an Englishman, and his deputy visited Anand, assessed the situation and accepted the farmers' demand. This marked the beginning of the Kaira District Co-operative Milk Producers' Union Limited, Anand, popularly known as Amul Dairy, which was formally registered on December 14, 1946. The cooperative was further developed and managed by Dr.Verghese Kurien with H.M. Dalaya. Dalaya's innovation of making skim milk powder from buffalo milk (for the first time in the world) and a little later, with Kurien's help, making it on a commercial scale, led to the first modern dairy of the cooperative at Anand. The trio's (T. K. Patel, Kurien and Dalaya) success at the cooperative's dairy soon spread to Anand's neighbourhood in Gujarat. Within a short span, five unions in other districts were set up. To combine forces and expand the market while saving on advertising and avoid competing against each other, the GCMMF, an apex marketing body of these district cooperatives, was set up in 1973. The Kaira Union, which had the brand name Amul with it since 1955, transferred it to GCMMF. Late Shri Lal Bahadur Shastri, the then Prime Minister of India who visited Anand on 31st October for inauguration of Amul's Cattle Feed Plant, having spent a night with farmers of Kaira and experiencing the success wished and expressed to Mr.Kurien, the General Manager of Amul at that time to replicate Amul model throughout our country. In order to bring this dream into reality, during 1965, The National Dairy Development Board (NDDB) was established at Anand and by 1969-70 NDDB came out with the dairy development programme for India popularly known as "Operation Flood" or "White Revolution". The Operation Flood programme, even today, stands to be the largest dairy development programme ever drawn in the world. This saw Amul as model and this model is often referred in the history of White Revolution as "Anand Pattern". Replication of "Anand Pattern" has helped India to emerge as the largest milk producing nation in the world. Now, GCMMF, is India's largest food product marketing organisation with annual turnover (2015-16) of US $ 3.5 billion. Its daily milk procurement is approximately 16.97 million lit from 18,545 village milk cooperative societies, 18 member unions covering 33 districts, and 3.6 million milk producer members. It operates through 56 Sale Offices and has a dealer network of 10000 dealers and 10 lakh retailers, one of the largest such networks in India. Its product range comprises milk, milk powder, health beverages, ghee, butter, cheese, pizza cheese, ice-cream, paneer, chocolates, and traditional Indian sweets, etc. GCMMF is India's largest exporter of dairy products, exporting its products to USA, Gulf Countries, Singapore, Philippines, Japan, China and Australia (www.amul.com, www.amuldairy.com)

Other notable co-operative dairies in India include Sudha from Bihar and Nandini from Karnataka.

A noteworthy story of success in the field of dairy processing from the private sector is of Mr. R. G. Chandramogan, who started his ice cream business in 1970 with a limited capital of Rs. 13,000 and a manual production of about 20 litres of ice cream per day. Chandramogan can well be touted as the ice cream man of South India as his company –Hatsun Agro, commands 60% share of ice cream market in Tamil Nadu and 35% in the entire southern region. The packaged milk sold under 'Arokya' brand contributes 65% to its revenue and value-added milk products contribute the rest. The company's revenue expanded remarkably to about 3445 crores in 2015-16, its net profit being around 134 crores. From being a small player in ice-cream, Hatsun Agro today is a listed company that constitutes the largest dairy in the private sector (without considering the MNCs) and is the largest exporter of milk powder and derivatives to 26 countries across the globe (The Hindu, 2012; Business Line, 2007; Economic Times, 2014, Hatsun Agro, 2016).

Maharashtra based Parag Milk Foods Ltd., led by Mr. Devendra Shah, is another remarkable private venture in the dairy industry. The company which started its operations in a small way in 1992, is one of India's elite private sector dairy company, with a diverse portfolio in over 15 consumer centric product categories. They manufacture products under brand names such as Gowardhan, Go, Topp Up & Pride of Cows. Product portfolio includes ghee, fresh milk, skim milk powder, whole milk powder, paneer, an array of processed and natural cheese, cheese spreads, butter, dahi, dairy whitener and gulabjamun mix under the brand names of 'Gowardhan' and 'Go'. Pride of Cows is a brand of fresh farm-to-home milk and Topp Up is flavoured milk in many variants. In 2008, the company established one of India's largest cheese plants and started manufacturing value-added milk products. Parag also owns one of the largest modern dairy farms in India with rotary milk parlour. Cheese consumption has zoomed over the years and Parag has a lock on institutional clients such as Domino's, Pizza Hut and

Papa John's. Unlike other dairy companies, which see 25-30 per cent of their top line coming from value-added products, Parag generates 75 per cent of revenue from cheese, flavoured yoghurt and ghee, all of which are value added products (Nayak, 2016; Srivastava, 2013).

The Managing Director of Milky Mist, Mr. T Sathish Kumar, who dropped out of school after Class Eight to take over his father's struggling milk business, selling around 3000 liters of milk daily, and turned it into a Rs.200-plus crore company, is another daredevil entrepreneur in the dairy sector. Based in Erode, the company makes a range of value added milk products including Milky Mist Paneer, its flagship product. The product portfolio includes fresh paneer, fresh cream, lassi, Mozzarella cheese, butter, yoghurt, cheddar cheese, ghee, shrikhand, cheese slices, khoa, payasam, cheese spread, curd and fruit yoghurt_(The Weekend Leader, 2015).

New technologies/innovations in value addition of milk

Innovations in product diversification

Product diversification is critical for the expansion of dairy industry. The impulse behind value addition is new product development thereby expanding the product portfolio, changing consumer demands, developing convenience products, extending shelf life etc.

Functional Foods: The captivating market for health/functional foods leads to a wide variety of products as detailed in Table 6.

Table 6. Functional foods (Siro *et al.*, 2008)

Sl.No	Type	Description	Example
1.	Fortified product	A food strengthened with additional nutrients.	Milk powder fortified with Vitamin D or A, Iron.
2.	Enriched products	A food added with new nutrients or components not normally found in a particular food.	Prebiotics, Probiotics.
3.	Altered products	A food from which a deleterious component has been removed, reduced or replaced with another substance with beneficial effects.	Lactose hydrolyzed milk.
4.	Enhanced commodities	A food in which one of the component has been naturally enhanced through special growing condition, new feed composition, or genetic manipulation.	CLA enhancement in milk through feeding of green fodder.

By combining nutritional and genetic interventions, researchers are developing 'designer milk' tailored to consumer preferences which has dual advantage with health benefits as well as in processing (Sabikhi, 2007). Applications of designer milk in diet and human health is that it generates a greater proportion of unsaturated fatty acids in milk fat, reduce lactose content that benefits lactose intolerant individuals and removal of â-lactoglobulin from milk. However, its applicability in processing and technological developments include alteration of primary structure of casein to improve technological properties of milk, production of high-protein milk, accelerated curd clotting time for cheese manufacture, increased yield and/or more protein recovery, milk containing nutraceuticals and replacement for infant formula etc. (Umaraw *et al.*, 2015)

Organic milk is collected from organic farms in which livestock are raised without giving antibiotics or hormonal treatment and fed with organic feeds which are free from pesticides, synthetic fertilizers or antibiotics. The cattle are raised on pastures that comply with organic standards and milk is collected at excellent hygienic conditions so as to avoid any chances of contamination.

Among different caseins, beta casein has two genetic variants A1 and A2 which differ at amino acid position 67 with histidine (CAT) in A1 and proline (CCT) in A2 milk as a result of a single nucleotide difference. This polymorphism leads to a key conformational change in the secondary structure of expressed â-casein protein. A1 protein is predominant in crossbred cows where as indigenous cow milk contains A2 protein. It is reported that A1 proteins can adversely affect health and nowadays A2 milk is gaining market momentum. Genetic test helps to determine whether a cow produce A2 or A1 type protein in its milk. (Sodhi *et al.*, 2012).

Innovations in dairy processing

Emerging technologies like Membrane processing, High hydrostatic pressure technology, Pulsed electric field, High pressure homogenization, Ohmic and microwave heating, Ultrasound and Pulsed light are used for processing different value added milk products.

Automation in quality analysis

Newer technologies are adapted by dairy industry to make up with the fast growing pace of dairy production and processing. Automation with latest technologies like NIR spectroscopy for the in/online monitoring, image sensing technology, fibre optic sensors, Lateral Flow Assay etc. have been introduced for rapid process monitoring.

Innovations in packaging

New concepts in packaging which gains momentum are eco-friendly packaging, edible packaging, active packaging, controlled atmospheric packaging, intelligent packaging etc.

Challenges in dairy processing

In the modern era of health consciousness, consumers are demanding foods which are natural, nutritionally better, free from chemical preservatives and microbiologically safe with extended shelf-life. Hence, a vast potential area is open to the dairy technologists to improve the current scenario. Some of the challenges are:

- Un-organized milk production system
- Insufficient awareness about clean milk production
- Inadequate power supply/ or difficulty in the maintenance of cold chain
- Lack of rapid field level kits for monitoring of different microbial contaminants, adulterants in milk
- Lack of enforcement of rules in the unorganized dairy sector
- Limited mechanization for traditional products
- Inadequate laboratory infrastructure facilities and resource persons
- Inadequate inspections and monitoring of product quality and safety
- Lack of problem based traceability system.

Future scope of research

The milk and milk products are the major source of animal proteins to the large vegetarian segment in Indian population. Researches in the dairy field are mainly focused on developing cost efficient technologies which will be accessible to farmer level. Some of the key areas that can be exploited are:

Farm level

- Increasing productivity of animals, better health care and breeding facilities and management of dairy animals
- Rapid on-line analysis of milk for contaminants and adulterants during milking

Processing level

- On-line measurement of product quality in dairy processing – physical, chemical and microbiological parameters
- Newer packaging technologies
- Initiatives to establish an industry–academic collaborative research centre
- A zero emission environment friendly dairy industry

Product diversification

- Designer milk
- Value addition of by-products
- New flavour generation in dairy products
- Value addition by diversified functional dairy products
- Buffalo milk based dairy products
- Development of personalized dairy products with specific therapeutic benefits

Conclusion

India has moved a long way ahead from a country deficient in milk to one that leads the world, in a span of few decades. The prime mover of this accomplishment has been the entrepreneurial and managerial spirits of a few determined people. Small scale entrepreneurs looking forward to enter the milk processing sector have a lot of lessons to learn and implement, which not only make him/her aware about the operations, opportunities and hurdles in the business, but also the connections and interdependencies between various factors affecting these. In this era of 'Make in India', plunging into entrepreneurship is valuable not only for working in one's own terms or creating profits, but also by being a part of the region's development by creating jobs and the conditions for a flourishing society.

References

Ajita, S. 2016. The Rs 80,000 Crore Milk Business. *Business Today*. http://www.businesstoday.in/ magazine/cover-story/indian-dairy-market-is-on-a-tear-due-to-new-players/story/ 232545.html. Accessed on 27.03.2017.

Aswal, P., S. Priyadarsi & S. Anubha, 2012. Yoghurt Preparation, Characteristic and Recent Advancements. *Cibtech Journal of Bioprotocols*, 1(2): 2319-3840.

Britz, T. J., C. van Schalkwyk & Y. T. Hung, 2006. Treatment of Dairy Processing Wastewaters. *Waste Treatment in the Food Processing Industry*. CRC Press.

Brody, A L. 2008. Packaging Milk and Milk Products. In: R. C. Chandan, (Ed.) *Dairy Processing & Quality Assurance*. John Wiley & Sons, USA.

Business Line. 2007. Be Different, Determined, Success will follow. http://www.thehindubusinessline.com/ todays-paper/tp-economy/article1668621.ece . Accessed on 29.03.2017.

Business Standard. 2017. Per Capita Availability of Milk in India. http://www.business-standard.com/article/news-cm/ 117011300843_1. html . Accessed on 27.03.2017.

Bylund, G. 1995. *Dairy Processing Handbook.* Tetrapak Processing Systems AB, Sweden.

DAHDF. 2016. Basic Animal Husbandry Statistics. Department of Animal Husbandry Dairying & Fisheries. http://dahd.nic.in/sites/default/files/BAHS-2016%20Updated%20on%2016.08.16.pdf. Accessed on 27.03.2017.

Demirel, B., O. Yenigun & T.T. Onay, 2005. Anaerobic Treatment of Dairy Wastewaters: A Review. *Process Biochemistry,* 40(8): 2583-2595.

Deshannavar, U.B., R. K. Basavaraj & N. M. Naik, 2012. High Rate Digestion of Dairy Industry Effluent by Upflow Anaerobic Fixed-bed Reactor. *J. Chemical & Pharmaceutical Research,* 4(6): 2895-2899.

Dewettinck, K., R. Rombaut., N. Thienpont., T.T. Le., K. Messens & J. Van Camp, 2008. Nutritional and Technological Aspects of Milk Fat Globule Membrane Material. *International dairy journal,* 18(5): 436-457.

Ebringer, L., M. Ferenèík & J. Krajèoviè, 2008. Beneficial Health Effects of Milk and Fermented Dairy Products—Review. *Folia Microbiologica,* 53(5): 378-394.

Economic Times. 2014. ET 500: Hatsun Agro Product, country's largest private dairy company, set to become a formidable player. http://economictimes.indiatimes.com/markets/stocks/news/-articleshow/ 42535607.cms . Accessed on 29.03.2017.

Fox, P. F. 2011. Cheese: Overview. In: J.W. Fuquay., P.F. Fox & P. L. McSweeney (eds.) *Encyclopedia of Dairy Sciences.* Academic Press.

Fox, P.F., J. Law., P. L. H. McSweeney & J. Wallace, 1993. Biochemistry of Cheese Ripening. In: *Cheese: Chemistry, Physics and Microbiology.* Springer US.

Gibson, G.R., H. M. Probert., J. Van Los., R. A. Rastall & M. B. Roberfroid, 2004. Dietary Modulation of the Human Colonic Microbiota: Updating the Concept of Prebiotics. *Nutrition Research Reviews,* 17(2): 259–275.

Goff, H D. 2011. Ice Cream and Frozen Desserts: Product Types. In: J. W. Fuquay., P.F. Fox & P.L. McSweeney, (eds.) *Encyclopedia of Dairy Sciences.* Academic Press.

Guetouache, M., B. Guessas & S. Medjekal, 2014. Composition and Nutritional Value of Raw Milk. *Issues in Biological Sciences and Pharmaceutical Research,* 2(10): Pages 115-122.

Hatsun Agro Product Ltd. 2016. 31st *Annual Report.* http://hap.in/pdf/annualreport/2015. pdf . Accessed on 29.03.2017.

http://www.amul.com/m/organisation . Accessed on 29.03.2017.

http://www.amuldairy.com/index.php/about-us/history. Accessed on 29.03.2017.

http://ecourses.iasri.res.in/ . Accessed on 29.03.2017.

http://www.oilgae.com/algae/cult/sew/new/dai/dai.html . Accessed on 29.03.2017.

Jagadish, T. 2013. *An Economic & Financial Analysis of Dairy: A Case Study of Guntur District Milk Producers' Mutually Aided Co-operative Union Ltd., Vadlamudi.* PhD Thesis Submitted to Acharya Nagarjuna University, Andhra Pradesh.

Kim, S.H., W. K. Kim & M. H. Kang, 2013. Effect of Milk and Milk Product Consumption on Physical Growth and Bone Mineral Density in Korean Adolescents. *Nutrition Research and Practice,* 7(4): 309-314.

Kratz, M., T. Baars & S. Guyenet, 2013.The Relationship between High-Fat Dairy Consumption and Obesity, Cardiovascular, and Metabolic Disease. *European Journal of Nutrition,* 52(1): 1-24.

Kumar, K. R. 2005. *Packaging Aspects of Milk & Milk Based Products*. In: G.C.P. Rangarao, (Ed.) Plastics in Food Packaging. A monograph prepared by Central Food Technological Research Institute, Mysore for Indian Centre for Plastics in Environment, Mumbai. ICPE, Mumbai.

Kumar, S., D. C. Rai., K. Niranjan & Z. F. Bhat, 2014. Paneer-An Indian Soft Cheese Variant: A Review. *Journal of Food Science and Technol.* 51(5): 821-831.

Marwaha, S.S., P.S. Panesar., V. Gulati & J. F. Kennedy. 2001. Development of Bench Scale Technology for the Treatment of Dairy Waste Waters by *Candida parapsilosis* MTCC 1965. *Indian Journal of Microbiology*, 41(4): 285-287.

Merritt, J., F. Qi & W. Shi, 2006. Milk Helps Build Strong Teeth and Promotes Oral Health. *Journal of the California Dental Association*, 34(5): 361-366.

Morin, P., R. Jiménez-Flores & Y. Pouliot, 2007. Effect of Processing on the Composition and Microstructure of Buttermilk and its Milk Fat Globule Membranes. *International Dairy Journal*, 17(10): 1179-1187.

Muehlhoff, E., A. Bennett & D. Macmahon, 2013. *Milk and Dairy Products in Human Nutrition*. Food and Agriculture Organization of the United Nations, Rome.

Nagpal, R., P. V. Behare., M. Kumar., D. Mohania., M. Yadav., S. Jain., S. Menon., O. Parkash., F. Marotta., E. Minelli., C. J.K. Henry & H. Yadav, 2012. Milk, Milk Products, and Disease Free Health: An Updated Overview. *Critical Reviews in Food Science and Nutrition*, 52(4): 321-333.

Nayak, P. 2016. How Akshali Shah is giving her Father's 1,500cr Milk Business a Modern Makeover. https://yourstory.com/2016/09/akshali-shah-parag-milk-foods/. Accessed on 29.03.2017

Noorjahan, C.M., S. D. Sharief & N. Dawood, 2004. Characterization of Dairy Effluent. *Journal of Industrial Pollution Control*, 20(1): 131-136.

Pal, D., A. A. Patel., A. Jha., S. Singh & S. K. Kanawjia, 2006. *Proceedings and Recommendations: National Seminar on Value Added Dairy Products & National Workshop on Entrepreneurship Development in Dairy and Food Industry*. Dairy Technology Society of India, National Dairy Research Institute, Karnal.

Parvez, S., K. A. Malik., S. Ah Kang & H. Y. Kim, 2006. Probiotics and their Fermented Food Products are Beneficial for Health. *Journal of Applied Microbiology*, 100(6): 1171-1185.

Pereira, P. C. 2014. Milk Nutritional Composition and its role in Human Health. *Nutrition*. 30: 619-627.

Rangappa, K. S. & K. T. Achaya, 1948. *The Chemistry & Manufacture of Indian Dairy Products*. The Bangalore Printing and Publishing Co., Bangalore.

Reeta, K. S., J. Ankita & N. Ramadevi, 2015. Fortification of Yoghurt with Health-Promoting Additives: A Review. Research & Reviews: *Journal of Food and Dairy Technology*. 3(3): 9-17.

Sabikhi, L. 2007. Designer Milk. *Advances in Food and Nutrition Research*, 53: Pages 161-198.

Shah, N.P. 2000. Effects of Milk-Derived Bioactives: An Overview. *British Journal of Nutrition*, 84(S1): 3-10.

Shete, B. S. & N. P. Shinkar, 2013. Dairy Industry Wastewater Sources, Characteristics & its Effects on Environment. *International Journal of Current Engineering and Technology*, 3(5): 1611-1615.

Siro, I., E. Kapolna., B. Kápolna & A. Lugasi, 2008. Functional Food. Product Development, Marketing and Consumer Acceptance—A Review. *Appetite*, 51(3): 456-467.

Smith, K. G. 2013. Milk Proteins: Packing a Powerful Nutritional Punch. *Today's Dietitian*. 15(3): 26.

Sodhi, M., M. Mukesh., R.S. Kataria., B. P. Mishra & B. K. Joshii, 2012. Milk Proteins and Human Health: A1/A2 milk hypothesis. *Indian J Endocrinol. Metab*, 16(5): 856.

Sodini, I., P. Morin., A. Olabi & R. Jiménez-Flores, 2006. Compositional and Functional Properties of Buttermilk: A Comparison between Sweet, Sour, and Whey Buttermilk. *Journal of Dairy Science*,89(2): .525-536.

Srivastava, S. 2013. How Parag Milk Foods Got it Right with Cheese. http://www.forbesindia.com/article/big-bet/how-parag-milk-foods-got-it-right-with-cheese/35431/1. Accessed on 29.03.2017.

Sserunjogi, M.L., R. K. Abrahamsen & J. Narvhus, 1998. A Review Paper: Current Knowledge of Ghee and Related Products. *International Dairy Journal*, 8(8): 677-688.

The Hindu. 2012. The Ice Cream Man. http://www.thehindu.com/features/metroplus/the-ice-cream-man/article3428751.ece . Accessed on 29.03.2017.

The Weekend Leader. 2015. Turning dad's old business around by infusing a new flavour, taste and brand to the products. http://www.theweekendleader.com/Success/2083/milky-boom.html. Accessed on 29.03.2017.

Umaraw, P., A. K. Verma & D. Kumar, 2015. Designer Milk- A Milk of Intrinsic Health Benefit: A Review. *Journal of Food Processing and Technology*, 6(3): 426.

13

Entrepreneurship Ventures in Fish Processing

George Ninan and Ravishankar C N

Introduction

Fish is a source of valuable animal protein and is considered a health food. This has resulted in increased consumer demand. Fish is now more expensive than meat and other animal foods. Being a highly perishable commodity, fish require immediate processing and various options are available for the value addition of fish. Fish processing, particularly seafood processing and marketing have become highly complex and competitive and exporters are trying to process more value added products to increase their profitability. Value can be added to fish and fishery products according to the requirements of different markets. These products range from live fish and shellfish to ready to serve convenience products. In general, value added food products are raw or pre-processed commodities whose value has been increased through the addition of ingredients or processes that make them more attractive to the buyer and/or more readily usable by the consumer.

According to the recent statistics, the annual capture and culture based fish production in India is around 90, 00,000 MT. Seafood export sector is one of the major foreign exchange earners in India. In 2015-16, India exported 9, 45,892 MT of seafood worth Rs.30,420 crores (Anon, 2016). USA and South East Asia are continued to be the major importers of Indian seafood. Frozen shrimp continued to be the major export item followed by frozen fish. Marketing of value added products is completely different from the traditional seafood trade. It is dynamic, sensitive, complex and very expensive.

Technology developments in fish processing sector offer scope for innovation, increase in productivity, increase in shelf life, improve food safety and reduce waste during processing operations. Large number of value added and diversified products both for export and internal markets, based on fish, shrimp, lobster, squid, cuttlefish, bivalves etc. have been identified. However, the commercialisation of fish products still pose lot of challenges to the entrepreneur and researcher in terms of optimization of technologies and ultimately developing the technologies into a commercially viable business plan. In this regard, the Indian Council of Agricultural Research (ICAR) has started a Business Incubation Unit at the Central Institute of Fisheries Technology (CIFT) exclusively for the fisheries sector through the World Bank funded National Agricultural Innovation Project (NAIP). It is designed to accelerate the growth and success of entrepreneurial start-up efforts through the mobilization of an array of business resources and services. Later, in 2016 an Agri-Business Incubation Centre (ABI) was established in CIFT under the XII plan scheme of National Agriculture Innovation Fund (NAIF) of ICAR. The role of ABI Centre is to facilitate the innovator and the researcher to turn their ideas into commercial ventures with focus on incubation and business development programme, including entrepreneurship, skill development and grass root innovators activities.

Health benefits of fish

As a rich source of nutrients, fish provides a good balance of proteins, vitamins and minerals, and relatively low calorie content. In addition, fish is excellent source of Omega-3 poly unsaturated fatty acids (PUFA) which appear to have beneficial effects in reducing the risk of cardio- vascular diseases and are linked with positive benefits in many other pathological conditions particularly, certain type of cancers and arthritis.

Fish represents an excellent option as a major source of nutrients. On a unit calorie basis, fish can provide a broad range of nutrients. A high intake of fish is compatible with a reduction in the intake of both calorie and saturated fatty acids. Coronary heart disease, hypertension, cancer, obesity, iron deficiency, protein deficiency, osteoporosis and arthritis are contemporary health problems for which fish provides a number of nutritional advantages and some therapeutic benefits.

Conventional finfish and other fishes potentially provide 100 to 200 kcal/100g, which is mainly attributed to the protein and fat content of fish. The amount of carbohydrates in fish is very small. Finfish usually contains less than 1% carbohydrate whereas shellfish have very low fat content. Compared to other muscle food, they contribute very low fat calories to the average diet. For example, each gram of fish muscle provides only 0.05 – 0.2 g of fat compared to 0.25 – 0.5 g fat per gram of red meat. The most important constituent of fish muscle is protein. The protein content in fish varies from 17 to 25%, though values as low as 9% are sometimes encountered as in the case of bombay duck. Fish protein is highly digestible because of very low stroma protein and has an excellent spectrum of essential amino acids. Like milk, egg and mammalian meat proteins, fish protein has a high biological value. Cereal grains are usually low in lysine and/or the sulfur containing amino acids, whereas fish protein is an excellent source of these amino acids. In diets based mainly on cereals, fish as a supplement can therefore, raise the biological value significantly.

Fish oil contains primarily the Omega-3 series of fatty acids. The PUFA components of fish lipids can be effective in reducing plasma lipids. Epidemiological data from Japan and Netherlands indicate that frequent consumption of fish even in quantities as low as 30 g/ day may have beneficial effects in reducing heart disease. Consumption of medium (100 g) to large amounts prevents thrombosis and ameliorates ischemic heart disease. These effects are mediated by the Omega -3 PUFA of fish lipids which alter the production of certain biologically important components called eicosanoid. The efficiency of the Omega -3 PUFA components is influenced by the amount ingested and the concentration of other unsaturated fatty acids in the diet, especially Omega -6 PUFA. Squalene, an isoprenoid molecule present in shark liver oil in higher quantities has been reported to possess antilipidemic, antioxidant and membrane stabilizing properties. Fish and shellfish, particularly anchovies, clams, oysters and sardines are rich sources of vitamin B_{12}.

Fish consumption is compatible with optimum dietary practices/ recommendations and that substitution of fish for other foods can help to maintain a balanced nutrient intake compatible with a low fat consumption. In addition, the consumption of fish- or more precisely, fish lipids–provides significant health benefits.

Value added products from fish and shellfish

Live fish

A recent development in international fish trade is the growing demand for live fish and shell fish. Live fish and shellfish command premium price in export as well as domestic markets. Transportation of fish, crustaceans and molluscs in live condition is the best method to ensure that the consumer is supplied with fresh product. In India, traditional mode of live transport in open earthen containers and metal containers was practised. In terms of the range of species and the distance shipped, tropical fishes stand first in live fish transport. Waterless transportation of live fish is also practised for many species where the animals are kept in moist conditions under optimal cold temperatures.

Chilled fish

The most common means of chilling fish is by the use of ice. Although, ice can preserve fish for some time, it is still a relatively short-term means of preservation. Chilled fish is an important value added item of international trade. The most prominent among this group is sashimi grade tuna. Indian major carps like, catla, mrigal and rohu are packed in boxes in iced condition and exported. Up to 35% yield of high value products can be expected from fish processed within 5 days of storage in ice, after which a progressive decrease in the utility will be observed. With increase in storage days and beyond 9 days of ice storage no high value products could be processed (Venugopal and Shahidi, 1998).

Frozen fishery products

Freezing is the most accepted method for long term preservation of fish and fishery products. Freezing reduces the spoilage activity and extends the shelf life of the product. Freezing represents the main method of processing fish for human consumption, and it accounted for 55.2% of total processed fish for human consumption and 25.3% of total fish production in 2010 (Anon, 2012). During freezing process, the temperature of the fish should be lowered to -30°C before it is transferred to the cold store. Most of the commercial freezers operate at temperatures of -40°C to -35°C. The thermal centre of the fish should attain -20°C prior to its removal from the cold store. The time taken to lower the temperature of the thermal centre to -20°C is termed as the freezing time.

Individually quick frozen (IQF) products made from fish fetch better price than block frozen products. However, for the production of IQF products raw materials of very high quality need to be used, as also the processing has to be carried out under strict hygienic conditions. The products have to be packed in attractive moisture proof containers and stored at -30°C or below without fluctuation in storage temperature. Thermoform molded trays have become accepted containers for IQF products in western countries. Utmost care is needed during transportation of IQF products, as rise in temperature may lead to surface melting of individual pieces causing them to stick together forming lumps. Desiccation leading to weight loss and surface dehydration are other serious problems met during storage of IQFproducts.

Skinless and skin on fillets from lean/medium fat white meat fish have enormous market potential. Fillets from fish can be packed as block frozen in 2 kg or 5 kg packs. Many varieties of fresh water fishes viz., catla, rohu, common carp, grass carp are suitable for making fillets both for domestic market and for export to developed countries in block frozen and IQF forms. In the importing countries, these fillets are mainly used for conversion into coated products.

Fish fillets can also be used for the production of ready to serve value added products such as fish in sauce and fish salads.

Similarly, different products from prawn are packed in blocks as whole, headless (HL), peeled and cooked (PC), peeled and deveined (PD) and peeled and undeveined (PUD). The commercially important prawn species from fresh water aquaculture is the giant fresh water prawn (Scampi) which is packed normally in whole or head less condition.

Thermally processed fish products

Canning and retort pouch processing are the common methods employed for thermal processing of fishery products. However, fish canning industry in India is declining due to high cost of cans. Recent innovations like polymer coated Tin Free Steel (TFS) cans provide a cheaper alternative. Studies conducted at Central Institute of Fisheries Technology (ICAR-CIFT) showed that polyester-coated TFS cans are suitable for processing ready to serve fish products, which can be stored at room temperature for long periods. The industry can utilize these cans for processing ready to eat fish and shell fish products for both domestic and export markets. This will help in reviving the canning industry in India.

Retortable flexible containers are laminate structures that are thermally processed like a can. These are shelf stable and have the convenience of keeping at room temperature for a period of more than one year without refrigeration. The process relies on heat sterilization and in many respects is analogous to canning with the imported tin can being replaced by a cheaper indigenous heat resistant flexible pouch. The most common form of pouch consists of a 3 ply laminated material. The flexible pouches manufactured indigenously, employing the configuration recommended by ICAR-CIFT has opened the way for commercialization of fish curry in retortable pouches (Fig.1).

Fig .1: Ready to Serve fish curry in flexible pouches *Cured fish products*

Exhaustive work has been carried for the process optimization of retort pouch packed ready-to-serve fish preparations from commercial species. Test marketing of mackerel curry conducted by MPEDA have shown that the product had good acceptability and there is good demand for fish curry in flexible pouches.

The traditional methods of processing fish by salting, drying, smoking and pickling are collectively known as curing. In India, roughly 20% of the fish caught is preserved by curing. Considerable quantities of cured fish are also exported, mainly to Singapore, Sri Lanka and to the Middle East.

Some of the commercially important dry fish products are

i. *Masmin:* It is the main smoked dried fish product of Lakshadweep islands. It is prepared from skipjack tuna. The meat is boiled in sea water and alternately dried and smoked till the characteristic flavour and colour is got. The finished product is a hard-smoked and hard dried one with a shelf life of more than a year in ambient storage conditions.

ii. *Dried squid:* The whole squid is cleaned, slit open, dipped in salt solution and washed in clean water. It is dried on ropes, hung by the anterior side to a moisture level of 18%. The mantle is stretched and kept flat by passing through rollers.

iii. *Dried jellyfish:* Both unsalted and salted dried jellyfish are produced for export. The salted jellyfish has final moisture content of 60% and unsalted about 20%. They are graded based on the size of the umbrellas.

iv. *Dried bombay duck:* Fresh fish is gutted and washed thoroughly. The fish is then dried on a scaffold by interlocking the jaws of two fishes. The head and fins are removed and it is split open, longitudinally. A dip treatment in 1% brine for 20 minutes is given and the fish is dried again on mesh trays to a moisture content of about 16-17%. It is then flattened out in rollers and trimmed to required shape. The product is again dried until a moisture content of 10% is reached.

Battered and breaded products

The most prominent among the group of value added products is the battered and breaded products processed out of a variety of fish and shellfish. Battered and breaded products offer a 'convenience' food widely valued by the consumer. These are products, which receive a coat or two each of a batter followed by coating with bread crumbs, thus increasing the bulk and reducing the cost element. The pick-up of coating can be increased by adjusting the consistency of the batter or by repeating the coating process. By convention, such products should have a minimum fish component of 50%. Coated products viz., fish fingers, squid rings, cuttlefish balls, fish balls and prawn burgers form one of the major fish and shellfish based items of trade by the ASEAN countries (Chang *et al.*, 1996).

Fish mince and mince based products

Mechanically deboned fish meat is termed as fish mince. Fish mince is more susceptible to quality deterioration than the intact muscle tissue, since mincing operation cause disruption of tissue and exposure of flesh to air which accelerates lipid oxidation and autolysis. The quality of the mince is dependent on the species, season, handling and processing methods. Also, low bone content in the mince (01-0.4%) is desirable for better functional and sensory properties. Depending on the type of raw material, fish mince can have a frozen storage life up to 6 months without any appreciable quality deterioration (Ciarlo *et. al.*, 1985). Generally, minced fish is frozen as 1-2 kg blocks at -40°C in plate freezers and stored in cold store at -18°C. Lipid oxidation and protein denaturation during frozen storage of mince can be prevented by the incorporation of spices, cryoprotectants and hydrocolloids.

Fish mince is a major source of raw material for the preparation of traditional products such as patties, balls, wafers, loaves, burgers, fish fingers, dehydrated fish minces, cutlets and pickled products. The mince from different species could be combined to prepare composite fillets.

Surimi

Surimi is stabilized myofibrillar protein obtained from mechanically deboned flesh that is washed with water and blended with cryoprotectants. Washing not only removes fat and undesirable matters such as blood, pigments and odoriferous substances but also increases the concentration of myofibrillar protein, the content of which improves the gel strength and elasticity of the product. This property can be made use of in developing a variety of fabricated products like shellfish analogues. Presently, India has eleven surimi plants and the main raw materials are threadfin bream, croaker, ribbon fish and big eye snapper. Presently, 5,00,000 tons of surimi are marketed globally per annum of which India's share is about 9% or 45,000 MT (Shamasundar & Chandra, 2012).

Pickled products

Fish pickle makes use of the non-fatty variety of low cost fish having good meat content. Major ingredients are: fish, garlic, green chilly, ginger, chilli powder, turmeric powder, gingelly oil/ground nut oil, salt, vinegar and sugar. The method of preparation of pickle is simple, the preservative being oil, salt and vinegar.

The traditional packing is in glass bottles. Modern packing materials like pouches and stand packs made of 12 micron polyester laminated with 118 micron LD/HD co-extruded film can also be used for packing pickles.

Utilisation of secondary raw material from fish processing

Fish processing operations generate more than 60% of the raw material as waste comprising of skin, head, viscera, trimmings, liver, frames, bones and roes. In India, these are disposed or converted into animal feed, fish meal and fertilizer. The disposal of fish processing waste is under strict regulations due to environmental issues and it adds to the operational cost of seafood industry. Hence, effective utilization of fish processing waste is gaining importance. Presently, the fish processing waste is considered as a secondary raw material due to richness of proteins, lipids and minerals. Developing appropriate technology to recover or isolate the valuable components could be of paramount importance.

Fishmeal

Fishmeal is a highly concentrated nutritious feed supplement consisting of high quality protein, minerals, B complex vitamins and other unknown growth factors. It is produced by cooking, pressing, drying and grinding the skeletal remains along with the adhering proteinaceous tissues of fish from filleting or canning operations or by processing whole miscellaneous fish mainly caught along with prawns, which include jew fish, sole, silver bellies, ribbon fish and the like. The composition of fishmeal differs considerably due to the variations in the raw materials used and in the processing methods and conditions employed. Fishmeal is important as a feed supplement for poultry and cattle.

Fish body oil

The main source of fish body oil in India is oil sardine. Apart from sardine oil, fish body oil is also obtained from the fishmeal plants. Production of fish oil has achieved considerable improvement both in quality and quantity in recent times due to the establishment of large organized modern fishmeal plants.

Fish liver oil

The therapeutic value of fish liver oil was discovered in the 18th century and fish liver oil became a common medicinal product. Both vitamin A and D are found in certain fish liver oils. The most important fish liver oils are obtained from cod, haddock and shark. Halibut and tuna livers are also rich sources of vitamin A and D. The weight of liver, fat content and presence of vitamins are dependent on a number of factors like species, age, sex, nutritional status, stage of spawning, and area from where it was caught.

Fish silage

Fish silage can be defined as a product made from whole fish or parts of the fish to which no other material has been added other than an acid and in which liquefaction of the fish is brought about by enzymes already present in the fish. Almost any species of fish can be used to make fish silage though cartilaginous species like shark and rays liquefy slowly. There are no problems in storage of fish silage if the correct acidity is maintained. During storage, protein becomes more soluble and there is an increase in free fatty acids if any fish oil is present. If silage is made from oily fish it is desirable to separate the oil after liquefaction.

Squalene

Squalene ($C_{30}H_{50}$) is an unsaturated hydrocarbon found in the unsaponifiable fraction of fish oils, especially of certain species of shark. Deep sea shark, *Centrophorus moluccensis* is the richest source of squalene and the compound got its name squalene as it was identified from the oil of deep sea shark belonging to the genus Squalus. Squalene is widely used in pharmaceuticals and cosmetics.

Fish calcium

Fish bone is an important source of calcium in the form of dicalcium phosphate with high bio availability. Calcium powder processed from fish bone can be used to combat calcium deficiency. The fish calcium capsules "CALCIFIT' developed by ICAR-CIFT (Fig.2) is being test marketed for commercial adoption.

Fig. 2: Calcium capsules from bone

Chitin and chitosan

The body peelings from shrimp processing plants are a major and economical source of chitin. Lobster and crab shell waste also contain sizeable quantities of chitin. The shells are deproteinised with alkali and demineralized with dilute hydrochloric acid. The fibrous portion obtained after washing is chitin. Chitin can be deacetylated with caustic soda to give chitosan. The deacetylation is achieved by treatment of chitin with (40% w/w) aqueous potassium or sodium hydroxide at about 100°C. The product obtained is dried in hot air dryer to a temperature not exceeding 60°C. Chitosan finds extensive applications in many industries such as pharmaceutical, textiles, paper, water purification etc.

Glucosamine hydrochloride

Chitin can be hydrolysed to glucosamine hydrochloride by adding concentrated hydrochloric acid and warming until the solution no longer gives opalescence on dilution with water. The excess acid can be distilled off under vacuum. The crude glucosamine hydrochloride is diluted with water and clarified with activated charcoal. The solution is filtered and evaporated under vacuum. The crude glucosamine hydrochloride coming as the residue can be separated from mother liquor by adding alcohol. Glucosamine hydrochloride is used as a nutraceutical for osteoarthritis, rheumatoid arthritis, glaucoma, joint pain, back pain, and weight loss. Glucosamine hydrochloride is applied to the skin in combination with chondroitin sulfate, shark cartilage, and camphor for osteoarthritis.

Fish maws/Isinglass

The word isinglass is derived from the Dutch and German words which have the meaning sturgeon's air bladder or swimming bladders. The air bladder of deep water hake is the most suitable for production of isinglass. In India, air bladders of eel and cat fishes are used for the production of isinglass. It is used as a clarifying agent for beverages like wine, beer, vinegar etc. by enmeshing the suspended impurities in the fibrous structure of swollen isinglass. India exports dried fish maws, which form the raw material for production of isinglass and such other products. Export is mainly confined, at present, to Hong Kong, Singapore and Germany.

Fish Protein Concentrate

Fish Protein Concentrate (FPC) is any stable fish preparation, intended for human consumption, in which the protein is more concentrated than in the original fish. Fish Protein Concentrate (FPC) is a mixture of cross linked and or aggregated molecules of various muscle proteins.

Collagen peptide

The collagenous materials discarded in fish processing include skin, swim bladder, bones, fins and scales. Controlled hydrolysis of collagen gives biologically active peptides, which have great potential in pharmaceutical, health food, nutraceutical and cosmetic industries. Collagen peptide is also known as "Collagen Hydrolysate", "Gelatin Hydrolysate", "Hydrolysed Collagen". Collagen peptide contains amino acids, such as proline and cysteine, which serve as the building blocks of keratin (the main structural protein of hair, nails, and stratum corneum). Supplementation with collagen peptide not only increases levels of collagen in the skin — which makes skin more firm and supple — but also reduces the activity of the enzyme collagenase, a matrix metalloproteinase that breaks down collagen in the extracellular matrix .

Gelatin

Gelatin from marine sources (warm- and cold-water fish skin, bones and fins) is a possible alternative to bovine gelatin (Rustad, 2003; Kim and Mendis, 2006; Wasswa *et al.*, 2007). One major advantage of marine gelatin sources is that they are not associated with the risk of Bovine Spongiform Encephalopathy (BSE) and are acceptable to many religious groups. Although, fish gelatin will be unable to completely replace mammalian gelatin, in the future, it might become a niche product offering unique and competitive properties to other biopolymers, as well as meeting the demand of global halal/kosher market (Karim and Bhat 2009).

Packaging of fish products

Fish is one of the most perishable of all foods. The best package cannot improve the quality of the contents, so the fish must be of high quality prior to packaging. Different products have different packaging requirements and it is important to choose suitable packaging material accordingly. It is important to know the intended storage conditions of the product, i.e., temperature, relative humidity and expected shelf life. This is why multilayered plastics are very popular since properties of different films can be effectively used. The basic function of food packaging is to protect the product from physical damage and contaminants, to delay microbial spoilage, to allow greater handling and to improve presentation.

Fresh fish

A suitable package for fresh fish should keep the fish moist and prevent dehydration, retard chemical and bacterial spoilage, provide a barrier against moisture and oxygen to reduce fat oxidation and prevent permeation of external odours. In India, baskets made of split bamboo, palmyrah leaf and similar plant materials were traditionally used for packing fresh iced fish. However, they do not possess adequate mechanical strength and get deformed under stacking. The porous surface of these containers will absorb water and accumulate slime, creating an ideal breeding ground for spoilage bacteria, which can contaminate the fish. Used tea chests provided with 2.5 cm thick foamed polystyrene (in polyethylene sleeving) slabs inside have been found extremely beneficial for transport of fish over long distances up to 60 hours duration. Modern insulated containers are made of HDPE or polypropylene with polyurethane insulation sandwiched between the inner and outer walls of the double walled containers. They are durable and in normal use have a life span of over 5 years. Materials such as aluminium, steel and fibreglass are also used in the construction of insulated containers. Insulation properties of these containers depend on the integrity of the layer of insulation. Contamination of insulation layer with water drastically reduces insulation properties of the medium. A recent development is an insulated corrugated plastic container, which is the lightest of all packages available in the country for iced fish transport. It lasts for 5 trips with collapsible design and lightweight and return of empty container is very easy. For cycle hawkers U shaped box (100 kg capacity) made of high molecular weight high density polyethylene is ideal.

Frozen fish

World trade in frozen fishery products has been increasing every year. In India, frozen fishery products form a major constituent of the export earnings. About 80 per cent of the total aquatic products exported from India to Europe, USA, Japan and Middle East countries are frozen in different forms and packs. Fish being highly perishable, transportation and storage of frozen fishery products requires a cold chain. Hence, these fishery products are stored at temperature below −18°C. They are transported by sea in refrigerated freight containers of the reefer type. The kind of packaging and the material used depends on the value of the product and the end user or customer. Both consumer and bulk packaging are used to facilitate trade. The retail pack of smaller quantity is used for household purposes whereas bulk packs with larger quantities are used in restaurants, catering services or for repacking into consumer packs.

The packaging used for frozen fish products should be of good quality. It should protect the product from transit and storage hazards as well as provide a barrier for retention of moisture. Packaging material used for frozen fresh fish is mainly polyethylene, either as pre made bags or wraps which are then packed into waxed duplex cartons. Frozen fish are also over wrapped in polystyrene trays for display. Individual fillets are packed in cellophane or PVC. Vacuum packaging of fish fillets and steaks in laminate films, and PVDC copolymer system is also practised. The corrugated fiberboards are usually waxed or provided with a liner. Sometimes an inner slab of polystyrene foam is effective in providing increased insulation. Laminated packs are usually used for IQF products. The sequence of packing starts with the primary inner wrap and finishes at the master carton.

Block frozen products

A block frozen fish product is primarily wrapped with Low-density polyethylene (LDPE). This can be in the shape of a bag or a film. Usually 2 Kg material is packed along with 10-20% glaze. Glazing should be optimum at the recommended level, since this will add to packaging and transportation. Alternately, films of HM-HDPE, which is not as transparent as LDPE film can be used, being more cost effective. 100 gauge LDPE is used for wrap while 200 gauges is used for bag. The corresponding values for HDPE are 60 and 120 gauges. Polyethylene films should be of food grade conforming to IS: 9845 specifications.

The frozen blocks are wrapped in film and then packed in duplex cartons. A number of such blocks are packed in a master poly bag and then packed into master cartons. The carton should have details like net weight, type and size, name and address of the producer and the country of origin. In the case of frozen shrimps about 6 units of 2 kg each or 10 units of 2 kg each are packed into master cartons. Corrugated fiberboards are used for the packaging of frozen fish.These may be of virgin material and having three or five ply with liners. The cartons may be wax coated or supported with liner paper with higher wet strength to make it moisture resistant.

Individually Quick Frozen (IQF) shrimp products

Packaging requirements of IQF shrimps are different from block frozen products. IQF shrimps are mainly packed for retail marketing in consumer packs ranging from 100 g to 5 kg (Fig 3). IQF shrimps are filled in primary containers along with code slip and weighed. The product is filled into primary pack and further is packed in master cartons for storage and transportation. The primary pack may be plastic film pouches (monofilm coextruded film or laminated pouches). The unit pouches may be provided with unit/intermediate cartons or directly packed into master cartons. The unit/intermediate cartons are made of duplex or three ply corrugated fibreboard laminated with plastic film on the inside and outside to improve the functional properties as well as aesthetic value of the pack. The most functional cost effective film has been identified as 10 μ biaxially oriented polypropylene (BOPP). Some duplex cartons are also wax-coated. Compression strength of 500 kg is the minimum recommended specification, which might give reasonable safety to the product.

Fig. 3: QF Fish fillets in re-sealable packages

Battered and breaded fish products

Thermoform trays produced from food grade materials like PVC, HDPE are suitable for packaging battered and breaded fishery products. These trays are unaffected by low temperature of frozen storage and provide protection to the contents against desiccation, discolouration and oxidation during prolonged storage.

Dried and smoked fish

It is estimated that about one third of the dry fish manufactured in India is lost due to poor handling and processing techniques, packaging and storage. Packaging of dry fish has many functions to reduce the loss. The properties of a suitable packaging material are inertness, leak proof, impermeability to water and oxygen and resistance to mechanical abrasion and puncture. High density polyethylene (HDPE) woven gusseted bags laminated with 100 gauges LDPE is suitable for packaging dried fish. HDPE is impervious to microbial and insect attack. HDPE is a material which will not spoil even if it gets wet. It is hard and translucent and has high tensile strength.

In the consumer market, the dried fish is packed in low-density polyethylene or polypropylene. Due to high moisture content of about 35% in certain salted fishes they are often attacked by microbes. Hence, fish should be dried to a moisture level of 25% or below. Packets of different sizes and weights ranging from 50 g up to 2 kg and bulk packs are available. Nowadays monolayer and multilayer films, combination and co-extruded films are used for bulk and consumer packaging of dry fish. Cleaned and processed, ready to-fry condiment incorporated and ready to serve products are also found in super markets.

Polyester polyethylene laminates and thermoform containers are used to pack dried prawns and value added dried products.

Accelerated freeze dried products

The final moisture content of AFD products generally is about 2%. Low moisture content and large surface area make these foods extremely hygroscopic. Most dried products deteriorate when exposed to oxygen. Changes in colour also may take place as a result of bleaching. Light accelerates oxidative reactions and hence contact with light should be prevented. As fish contains fat there will be chances of taking up the taints from the packaging material. The particular structural properties of freeze-dried products lead to damage by mechanical means. Freeze dried products are also liable to damage caused by free movement within the package. Measures must be taken to fit the product compactly in the container, while leaving the minimum headspace for filling inert gas. Rigid containers both glass and cans were used earlier to pack freeze dried products. However, it can now be seen that metalized polyester laminated with polyethylene or aluminum foil /paper/ are used since they have low oxygen transmission rate and water vapour transmission rate. Most of the packages are filled with an inert gas. The product can also be packed under vacuum to give better protection against damage.

Success stories of business incubation programme at CIFT

PRAWNOES –the extruded snack product

Mrs. Omana Muralidharan is a homemaker at Ernakulam district of Kerala. One of the extension programmes conducted by the ICAR-Central Institute of Fisheries Technology (CIFT), Kochi, changed her destiny. Many new fish processing and packaging technologies were discussed during the programme. Her attention developed into interest and furthered her desire to start a small business enterprise with CIFT technologies. However, the challenges were (a) no finances, (b) competition from big firms, (c) no infrastructure, (d) no machineries, (e) no skilled manpower, and (f) no trainings. At that juncture, the Agri-Business Incubator (ABI) attached to the CIFT has come to her rescue. Presently, she is one of the most successful women entrepreneurs of Kerala with the brand of "PRAWNOES" — the extruded snack products in different flavours. Before registering as an Incubatee at ZTM-BPD Unit of ICAR-Central Institute of Fisheries Technology (CIFT), Mrs Omana Muraleedharan was running a small-scale metal industry named Amruta Metal Works. She approached ICAR-CIFT with the idea to develop the extruded snack food flavoured with prawn. A brand named 'Prawnoes' was created and registered for trademark protection by ZTM-BPD Unit. CIFT developed and standardized three varieties of Fish Kure for the Incubatee, 'Spicy Shrimp', 'Shrimp n Onion' and 'Prawn Seasoning'. The BPD Unit also helped the entrepreneur to carry out feasibility studies, prepare Business Plan and DPR (Detailed Project Report) and helped her in mobilizing seed funding from Canara Bank to start her own production facility in the Industrial area at Aroor, Kerala. The production facility was designed and machines were sourced through the BPD Unit. Some of the machines were indigenously designed and manufactured as per the suggestions from ICAR-CIFT. CIFT gave her technical guidance in developing the product, standardization of process parameters, testing, packaging solutions, ideas for branding, assistance in trademark filing and setting up their own production unit at Aroor. The unit was inaugurated on 28 June 2014. Presently Prawnoes (www.prawnoes.com) is marketed in seven flavours and the produce is sold in four districts in Kerala. Mrs. Omana Muraleedharan received the best woman entrepreneur award from the Government of Kerala State. Prawnoes received excellent product reviews during its test marketing period and Mrs. Omana is planning to add more snack foods to her product range. With the support of all government institutions like the District Industries Centre (DIC), Ministry of Microm Small and Medium Enterprises (MSME), Banks and CIFT, she is now promoting a healthy snack food brand with a campaign "Save Children, Eat healthy snack".

Spicy Shrimp   Shrimp 'n' Onion   Prawn Seasoning

Ready to serve fish curry in retort pouches

With the changing socio-economic pattern of life and the increasing number of working couples, the concept of convenience food products is fast becoming popular in Indian market. To meet the ever changing and diverse customer demands, CIFT has developed ready-to-eat fish products in flexible retortable pouches. The superior product quality with well-preserved natural colour, flavour and texture characterizes the CIFT technology. The technology was successfully adopted by two Incubatees.

Ideal caterers, Cochin

Mr. Hisham Kabir of Ideal Caterers belongs to the first generation of entrepreneurs to complete business incubation from the Business Incubation Unit at CIFT, Cochin. He joined CIFT's business incubation initiative in 2011 and his products hit the market by early 2013. Three ready-to-cook gravies, Kumarakom fish curry, Nadan chicken curry and Kerala chicken curry were the first to roll out from the Incubation Centre under the brand name "Freedom Kitchen". CIFT provided him solutions and scientific expertise related to food processing and machinery for setting up his own production unit.

Monsoon bounty foods manufacturing Pvt. Ltd., Chennai

Mr. Sunil Ravi, an IT professional, joined as Incubatee at CIFT during 2012, after being introduced to the Convenience Food sector by CIFT Business Incubation Unit. He is now successfully running his Chennai based company, manufacturing ready-to-eat fish, meat and vegetable products. The wide product range includes curries, soups, stews, snacks etc. which were launched to the market in early 2013.

Vitagreen-Organic nutrient extract from fish

Organic agriculture is based on a holistic viewpoint that perceives nature as more than just the separate, individual elements into which it may be split. The principles of organic farming are found in ecology, a science concerned with the interrelationship of living organisms and their environment. With the aim of supporting and strengthening biological processes without recoursing to synthetic products, CIFT introduced fish based organic fertilizers which contains natural enzymes, protein, minerals and other plant foods beneficial in the growth process of plants. Green Allies Organics Private Limited, an Incubatee Company working towards delivering quality organic products and services has adopted the CIFT technology for producing Organic Nutrient Extract from fish. The product has a perfect balance of trace minerals required for plant growth derived from highly nutritious sea food through acid ensilation process. This is a natural fermented blend of whole fish and fish waste products and entrails with supplemental products to produce one of the best liquid fertilizers in the market today. The products, marketed under the brand name "Vitagreen" with a registered trademark, are 100% organic and have received certification from INDOCERT, a nationally and internationally operating certification body accredited by National Accreditation Body (NAB), Government of

India, as per National Programme for Organic Production (NPOP). The products were found to be very effective in promoting and enhancing plant growth, increasing germination and sprouting, and boosting plant vigour.

Swadish-fish gravy paste

The technology for ready-to-cook instant fish gravy paste was taken by Promise Food Products through the Business Incubation Unit of CIFT. The product was labelled as Swadish fish gravy paste and was launched in January 2014 and is currently marketed through retail kiosks in Kerala.

Safety and quality aspects of fish products

The variety and species of aquatic organisms used as food are too many. Some of these organisms by virtue of their genetic makeup or food habits are found to contain some toxic substances. Thus, ciguatoxin, paralytic shellfish poison (saxitoxin), diarrhetic shellfish poison (okadaic acid), amnesic shellfish poison etc. are species related toxins, which encounter health hazards. Similarly, scombroid or more generally fishes with red meat on temperature abuse produce histamine and cause the commonly reported scombroid toxin poisoning. Some species, particularly bivalves are known to accumulate certain heavy metal residues from the eco system to toxic levels. The cephalopods, squid and cuttle fish are found to accumulate lead and cadmium depending on pollution of the environment as well as age of the organism. On a similar fashion, large fishes like tuna, marlin etc. is known to accumulate mercury to toxic levels with increase in age and size. While most of these problems are common in case of wild catches, the presence of antibiotic residues, pesticide residues, hormone residues, muddy moldy flavour (geosmin) and certain heavy metal residues are common health hazards encountered in products of aquaculture. All these residues pose various kinds of health risk to consumers. This problem of contamination is highly significant in products of aquaculture than that of wild catches especially from marine water.

In spite of all these health hazards, fish and shell fish continue to be in great demand as a food material in the developed world on account of the better taste, nutritional quality and medicinal properties of aquatic food. To avoid public health problems in using fish and shell fish as a food of mass consumption several quality assurance programmes were evolved and enforced from time to time. Today, various kinds of quality standards like Codex standards, US FDA standards, EU Norms, BIS standards etc. are in operation at national and international levels. To achieve these standards several quality assurance programmes were also developed and practised in different parts of the world. The HACCP system of USA, the European Council directives, the QMP of Canada and TQM of Japan are such quality assurance programmes aimed to ensure safety and quality of fish and fishery products consumed in these countries.

With change of time, all these quality standards and quality assurance programmes became more and more stringent and mandatory and pose severe challenges to the developing countries, which used to export a major share of their fish and fishery products to the developed countries. As a result, many developing countries including India had to face trade ban on fish exports. The imposition of HACCP in the early nineties by US FDA, the EU ban of fishery products from India on account of sanitation and hygiene and rejection of several consignments on account of antibacterial substances, antibiotic residues, heavy metal residues, muddy moldy flavour etc. by USA, EU and Japan played havoc in Indian seafood industry. Even though, TQM of Japan aimed total quality management, it failed to ensure both safety and quality, probably due to certain lacuna like calibration, good laboratory practice, good personnel policy etc. There were several attempts to improve the TQM concept of Japan to make it suitable to tackle all problems of safety and quality of a given product.

Today, Total Quality Management (TQM) is a widely used technique for quality assurance in a wide range of production industries. The use of TQM is gaining more and more important in Indian seafood industry perhaps due to the higher incidence of health risks/hazards in fish and fishery products as well as to tackle the challenges in the name of quality and safety posed by major importing countries. In India, the Central Institute of Fisheries Technology, Kochi played the key role to equip the Indian Sea Food Industry to squarely face these formidable challenges by introducing a total quality management concept. The action plan adopted by CIFT consisted of plant inspection, identification of deficiencies, rectification of deficiencies, guidance for preparation of HACCP manual, defects and defect action plan, validation, auditing, installation of GMP, SSOP, GLP and good personnel policy. These combination procedures are termed TQM. Even though, TQM is widely used, depending on various factors, the approach and concepts for implementation of TQM vary from industry to industry and from person to person. Obviously there is a need for consolidation of all relevant aspects to attain uniformly assured quality for products of mass consumption.

Emerging technologies for value addition of fish

Irradiation
Radiation processes that can be applied to fishery products include radurization (pasteurization of chilled fish), radicidation (sanitization of fresh and frozen products including fish mince by elimination of non-spore forming pathogenic bacteria) and disinfestations. Irradiation at doses in the range of 0.1 to 1.0 kGy can prevent development of beetle larvae and adults in packaged, salted, dried fishery products

Accelerated freeze drying
Accelerated freeze drying is now being increasingly used for the preservation of high value food products. In India, this technique is now applied for processing shrimp, squid rings etc. The possibilities for various ready-to-eat products based on fish and shellfish employing this technique are immense.

High pressure processing

High pressure processing (HPP) is a non-thermal food processing method with application of high pressures (87000 psi or 600 MPa) with or without heating. This method provides an environment friendly fresh-like product. HPP with 250 MPa did not inactivate *L. monocytogenes*, but significant lag phases of 17 and 10 d were observed at 5°C and 10°C respectively in cold smoked salmons (Lakshmanan and Dalgaard, 2004). Studies carried out at ICAR-CIFT have shown that HPP of tuna extended the shelf life for ~10 days when compared to conventional processing methods.

Enzyme treatment

In commercial food processing, transglutaminase (TGase) is used to bind proteins together. TGase catalyzes the cross-link of side chains of two amino acids (glutamine and lysine) and thus yielding peptide bond. Examples of foods made using transglutaminase include imitation crab meat and fish balls. TGase can be used as a binding agent to improve the texture of protein-rich foods such as surimi or ham. Transglutaminase is one of the several forms of "meat glue" which help in improving the texture of low-grade meat, whose characteristics are attributed to stress and a rapid postmortem pH decline.

Extrusion technology

It is a technique used to form shapes by forcing a material through a region of high-temperature and/or pressure, and then through a die to form the desired shape. ICAR-CIFT has worked on the production of extruded products by incorporating fish mince with cereal flour. The fish mince is mixed with cereal flour, spices and vegetable oil and extruded using a twin-screw extruder. The product obtained is finally coated with spice mix to provide a delicious snack that has been christened as "Fish Kure".

Pulsed light technology

Pulsed light acts as a sterilization method for products with uniform surface, packaging materials, pharmaceutical and medical products. As fish has non-uniform surfaces, it cannot sterilize the fish products but reduces microbial load. Pulsed light is effective in inhibiting the growth of *Listeria monocytogens* in vacuum packed smoked salmons. Pulsed light technology has also been shown to be effective in inactivating *L. innocua* from the surface of fish products such as cold smoked salmon.

Hurdle technology

The concept of hurdle technology is based on the application of combined preservative factors to achieve microbiological safety and stability of foods. Most important hurdles used in food preservation are temperature, water activity, acidity, redox potential, antimicrobials, and competitive microorganisms. A synergistic effect could be achieved if the hurdles hit at the same time at different targets that disturb the homeostasis of the microorganisms present in foods. For fish products manufactured in industrialised countries, hurdle technology has been employed for two groups of products viz. convenience products and lightly preserved fish products.

Conclusion

Fish processing and value addition has evolved over the years as an important sector in Indian Agriculture. Fish and fishery products earn maximum foreign exchange in the category of agricultural produce exported from India. This sector has immense scope for development through diversification and generation of employment for the skilled and unskilled workforce of the country. Many new species are being introduced in the Indian Aquaculture sector; *L. vannamei*, *Monosex tilapia*, is to name a few. A comprehensive study on the suitability of these species for value addition has to be carried out to propose optimized utilization protocols. Functional fish products will be in much demand in future; the challenge will be to retain the functional benefits of fish and shellfish meat by way of adopting product specific processing protocols or alternate delivery systems for sensitive components.

References

Anonymous, 2012. The State of World Fisheries and Aquaculture 2012 Part I *World Review of Fisheries and Aquaculture* pp 63.

Anonymous, 2016. The Marine Products Export Development Authority. *Annual Report* 262p.

Chang, N.M., C.G. Hoon & L.H. Kwang, 1996. *Southeast Asian Fish Products* (3rd Ed.), Southeast Asian Fisheries Development Centre, Singapore.

Ciarlo, A.S., R. L. Boeri & D. H. Giannini, 1985. Storage Life of Frozen Blocks of Patagonian Hake (*M.hubbsi*) Filleted and Minced, *J.Food Sci.*, 50: 723.

Karim, A.A. & R. Bhat, 2009. Rev. Fish Gelatin: Properties, Challenges, and Prospects as an Alternative to Mammalian Gelatins. *Food Hydrocolloids* 23: 563–576.

Kim, S. & E. Mendis, 2006. Bioactive Compounds from Marine Processing by-products – A Review. *Food Res. Int.* 39: 383–393.

Lakshmanan, R. & P. Dalgaard, 2004. Effects of High-Pressure Processing on *Listeria monocytogenes*, Spoilage Microflora and Multiple Compound Quality Indices in Chilled Cold-Smoked Salmon. *.J Appl. Microbiol.* 96:398–408.

Rustad, T. 2003. Utilisation of Marine by-products. Electron. *J. Environ. Agric. Food Chem.* 2: 458–463.

Shamasundar, B. A. & M.V. Chandra, 2012. Surimi and Surimi Based Products. In: *Advances in Harvest and Post-Harvest Technology of Fish* (Nambudiri D.D & K.V.Peter Eds.), New India Publishing Agency, New Delhi, 133-170.

Venugopal, V. & F. Shahidi, 1998. Traditional Methods to Process Under-utilized Fish Species for Human Consumption. *Food Rev. Int.* 14(1):35–97.

Wasswa, J., J. Tang., X. Gu & X. Yuan, 2007. Influence of the Extent of Enzymatic Hydrolysis on the Functional Properties of Protein Hydrolysate from Grass Carp (*Ctenopharyngodon idella*) Skin. *Food Chem.* 104 : 1698-1704.

14

Opportunities and Entrepreneurship Developments in Meat Processing

Renuka Nayar and Vasudevan V N

Introduction

Meat is a highly nutritious animal product relished by majority of the population in India. India produced around 6.2 million MT of meat in 2013 (FAOSTAT, 2013), standing at fifth position in world meat production and accounts for only 3 per cent of total meat production in the world (APEDA, 2017). Contribution of livestock sector to national economy in terms of gross domestic product (GDP) was 4.1 per cent in 2012-13 (Islam *et al.*, 2016). The percentage of meat subjected to processing in India is only 1-2 per cent compared to 70 per cent in developed countries (Sen *et al.*, 2013). Relatively higher proportion of poultry (7-8%) is converted into further processed poultry products. Major proportion of the meat produced in the country is derived from old and unproductive animals. Availability of spent birds is also on the rise due to the rapid growth in poultry industry. Meat from such animals and poultry has less desirable palatability attributes, especially tenderness, which makes it more suitable for processing into value added products. This will provide consumers with more acceptable, convenient, healthy and versatile products while simultaneously increasing the economic returns to the processor. The profitability of meat industry is critically dependent in deriving extra value from lower-value meat cuts. To achieve this, value must be added in the eyes of the consumer across the wide chain of value added products.

Nutritive value and health benefits of meat

As in any other food, meat also contains five classes of nutrients - proteins, lipids, carbohydrates, vitamins and minerals. Meat contains about 72-80 per cent water and the rest solids. Meat of young animals contains more amount of water and that of fatty animals contains less amount of water.

The proteins in meat include sarcoplasmic proteins, myofibrillar proteins and connective tissue proteins. The proteins in meat have high biological value and meet the body's requirement of all essential amino acids. Muscle foods also provide proteins in a concentrated form.

Lipid portion of meat includes triglycerides, phospholipids and cholesterol. The main misconception is that meat fat is totally constituted by saturated fatty acids and vegetable fat is totally constituted by unsaturated fatty acids. But, meat contains both saturated and unsaturated fatty acids and the single most abundant fatty acid in meat is the unsaturated fatty acid, oleic acid. Less than half of the fatty acids in pork and beef and 51 per cent of fatty acids in mutton are saturated. Predominant saturated fatty acids in meat are palmitic and stearic acids. Meat has low levels of atherogenic fatty acids like myristic acid. Linoleic and linolenic fatty acids are also present in meat. Omega-3 fatty acids are present in minute quantities. Recent studies have shown that conjugated linolenic acid (CLA) is present in ruminant meat especially beef and mutton and that it has tumour and atherosclerosis reducing properties. Meat contains very low level of carbohydrates and the only carbohydrate in meat is glycogen. Meat is a rich source of vitamin D but rather poor in vitamin A, E and K. Meat is a rich source of all water soluble B vitamins especially thiamine, riboflavin, niacin, pantothenic acid, vitamin B_{12} and B_6, but a poor source of vitamin C. Pork has got a higher concentration of thiamine when compared to other meat.

Meat is generally a good source of all minerals except calcium. Meat is a good source of haem iron, which is rapidly absorbed from body whereas non-haem iron of vegetables is less absorbed from the body. Meat also enhances the availability of non-haem iron present in plant foods. This effect is known as *Meat factor*. Meat is a good source of zinc and also provides phosphorous, copper, magnesium, iodine and chloride in useful amounts. It has also been observed that various health beneficial biopeptides are released during enzymatic changes in meat post mortem. Many of these peptides have antihypertensive, anticarcinogenic, antioxidant, antithrombotic and immunomodulatory effects (Pighin *et al.*, 2016). However, there are certain limiting factors to nutritive value of meat like absence of dietary fibre. There have also been several reports linking meat consumption with colon cancer, and cardiovascular diseases (CVDs), but no studies could give a fool proof evidence for this. According to McAfee *et al.* (2010), including red meat in diet results in a more balanced diet with improved contributions to consumption of essential nutrients like conjugated linoleic acid and omega 3 fatty acids and they stressed that a wider research is needed before meat can be considered as a risk factor for cancers and CVDs.

Value addition of meat

Meat is processed to produce value added products. Value addition of meat offers various advantages to producers, processors as well as consumers.

- Value added meat products provide variety and versatility to consumers
- Value addition makes the products available to the consumers at all times and at different places
- Value addition increases market demand of products
- Products suitable for changing health concerns and lifestyle paradigms will be made available
- Lower-valued portions of the carcass, offal meat and fat can be utilised which increases profitability of the enterprise
- Non-meat ingredients can be incorporated to value added products which enhances yield and economic returns
- Value addition reduces perishability of fresh meat and make available to a wider population
- Value addition of meat can insulate entrepreneurs from market fluctuations in the live animal/bird prices
- Value addition promotes export of meat products
- Benefits of value addition can attract more potential entrepreneurs to meat processing sector.

Value added meat products include coarse/fine ground products, emulsified products, enrobed products, cured and smoked products, dried products, canned/retort pouched products, restructured products and traditional products. Value addition of meat starts from cutting and deboning of carcasses to chunking or dicing, marination, packing and to further processing. Value added meat products can be roughly classified into whole muscle products and comminuted meat products.

Whole muscle products: These products are not subjected to a higher level of size reduction like cutting or grinding which include

Meat cuts/portions (Bone-in/boneless): Carcass of any species has certain premium cuts/portions which have greater demand and are highly priced and certain low value cuts which are priced low. Sale of meat as specific cuts/portions has become popular especially in the poultry sector eg. chicken breasts, chicken drumsticks, giblets, janata cuts etc. These portions can be marketed as bone-in or boneless, depending on demand eg. Boneless breast fillets. They can be attractively packed in trays with overwrap which offers good consumer appeal. The advantage of this marketing is that consumer can select the carcass/ meat portion depending on the method of cooking and the type of meat preparation and on his economic status.

Marinated meat cuts/portions (Bone-in/boneless): These are cuts/portions which are soaked, tumbled, massaged or injected with a marinade. A marinade may include weak organic acids like lactic acid as in curd, citric acid as in lemon juice or acetic acid as in vinegar, salt and spices. Sometimes the effect of marination is enhanced by addition of phosphates, natural enzymes, natural antioxidants like herbs etc. Marinade uptake is usually limited to 10 per cent of weight of meat portion. Marination improves tenderness, juiciness and the flavour of meat cuts. Phosphates in marinades improve cooking yield and antioxidants prolong the shelf life of meat cuts. Vacuum tumbling or massaging improves the effect of marination by increasing tenderness and juiciness. Marinaded meat cuts are convenient ready-to-cook products and can be retail packed and marketed.

Precooked products: These are whole muscle products which are fully cooked and ready for consumption. Indian meat dishes like meat curries, tandoori chicken etc. are examples for this. These precooked products offer greatest convenience but keeping quality under refrigerated conditions may be reduced by the development of warmed-over flavour (WOF). This off flavour is due to the oxidation of unsaturated fatty acids in meat and is evident when precooked refrigerated meat is reheated before consumption. It is described as a "cardboard-like" flavour. This off flavour can be reduced to some extent by addition of antioxidants especially natural antioxidants like organic acids or vitamin E.

Precooked products can also be prepared as canned/retort pouched products which are shelf stable and can be kept at ambient temperature for months together. The process of retorting subjects the products to high temperature and pressure and makes the products "commercially sterile". The procedure is elaborate and expensive and hence the end products are priced higher. Examples of this are canned/retort pouched chicken curry, mutton curry, *biriyani* etc.

Breaded/coated products: The breadings and coatings are usually starch or flour based and sometimes protein based and the products are usually cooked in oil. This process seals the moisture in and improves juiciness of the product. This also improves consumer appeal and enhances palatability. Poor coating adhesion is a major problem in this process and predusting of the meat cuts with flour will enhance the adhesion of coatings on meat cuts. Breaded and fried chicken drumsticks, breasts etc. are common and quite popular.

Cured whole muscle products: Curing is the process of applying salt, sugar, nitrate/nitrite etc. on to meat. Often phosphates, ascorbates, mono sodium glutamate, spices, flavourings etc. are added to cure. Ham and bacon are the two cured whole muscle products. Curing ingredients can be applied either in dry form or wet form. Based on this, curing process is divided into dry and wet curing.

Dry curing is the oldest curing method. It is done by rubbing the dry cure mix over meat portions/cuts. These salts draw moisture from meat, dissolve in the moisture and then penetrate inside. This is a lengthy process and takes few months to one year for completion. This result in greater amount of shrinkage, but the products have intense flavour and are highly priced. Examples of dry cured meat are country-cured hams and bacons.

In wet curing, curing ingredients are dissolved in water and the pickle formed is either used for immersing meat cuts or the pickle is injected into the meat. Immersion/cover curing takes a long time (up to six weeks). Injection curing involves injecting the curing solutions either as single stitch or multiple stitch injection into meat and is a faster technique. Another technique of wet curing is artery pumping done in meat portions like ham with a well defined artery like femoral artery. The curing solution is directly injected into the artery and is uniformly distributed in the meat through the already existing blood vessels. This is also a fast technique.

Curing is often accompanied by smoking. Smoking is still practised to have variety products in terms of colour and flavour, to aid colour development, to protect from oxidation and to develop a protective skin on emulsion type sausage. During smoking, a Maillard (browning) reaction occurs between amino groups of proteins and the carbonyl groups from the smoke. Smoke retards surface bacteria by both dehydration and the phenol and acid contents.

The common chemical components found in smoke include phenols, organic acids, alcohols, aldehydes, carbonyls, hydrocarbons and gases such as carbon dioxide, carbon monoxide, nitrogen and nitric oxide. Phenols are colour and flavour enhancers. Their bacteriostatic and antioxidant properties also are made use of in smoking. It has been found that the particle phase and not the vapour phase of wood smoke contributes to the antioxidant property. Phenols and carbonyls in vapour phase of smoke contribute to colour development by Maillard type reaction with the amino groups on meat surfaces. Smoked meat has a characteristic flavour due to the phenolic components in vapour phase. Vanillic acid of smoke gives a sweet aroma to the smoked meat. Phenols along with other smoke components like acetic acid, formaldehyde and creosote prevent microbial activity. The role of alcohols especially wood alcohol or methanol appears to be the carrier for other volatile components. Among organic acids in smoke, 1-4 carbon organic acids such as formic, acetic, butyric, propionic etc are found in the vapour phase whereas 4-10 carbon acids such as valeric, caproic, heptylic, nonylic and capric acids are found in the particle phase of smoke. Their main role is the surface coagulation and skin formation along with a red colour development. Carbonyl components also contribute to the colour, aroma and flavour. Hydrocarbons such as benzapyrene and dibenzanthracene in smoke are carcinogenic to humans. Removing the particulate phase of smoke can eliminate these as in liquid smoke. Gases in smoke like carbon monoxide and carbon dioxide are readily absorbed on meat surface and produce bright red pigments namely carbon monoxide-myoglobin and carboxymyoglobin. Oxygen can convert myoglobin to oxymyoglobin or metmyoglobin. An important gas in the smoke namely nitrous oxide can lead to nitrosamine formation.

Bacon is a cured and smoked product prepared from pork bellies. Dry or pickle curing is followed. Pickle curing is practised in commercial units and injection curing is done. Bellies are held in the curing cooler for 10 to 14 days for proper curing and then smoked. The length for which the belly remains in the smoke house depends on size of the belly; smoke house air velocity and temperature. Bacon is usually cooked to an internal temperature of 126-132°F. In air-conditioned smoke houses the relative humidity is controlled between 25 and 40%. Afterwards bacon is chilled and the rind (skin) is removed and sliced.

Ham is a processed product using the ham portion of pork carcass. It is a cured product and usually smoked. Hams can be bone in, semi boneless and boneless. Hams are pickle cured commercially and the pickling solution contains salt, sugar, nitrite, ascorbate, phosphate etc. Pickling solution is either injected into the meat or pumped through the femoral artery. Usually, to get a good cure, the pumped hams are held in a cooler for 24 hours. For maximum cure, the hams are held for 3 to 7 days. In the country cured hams, usually dry curing is practised and hams are generally smoked. In case of 'cooked hams', hams are commercially cooked to an internal temperature of 170-180°F at 30-40% relative humidity. These hams are not usually smoked.

Dried meat products: This includes western products like jerky, biltong, charque etc. Jerky is a popular meat snack in USA and is prepared by making thin strips of beef, mixing with marinades and smoked and dried. In India also in some hilly areas dried meat is popular and is a method of preservation of meat.

Comminuted meat products

These are meat that have been reduced in size by mincing, cutting, chopping etc. Comminution helps to reduce the toughening effect of collagen, helps in better mixing of fat and lean portions, more uniform incorporation of salt and spices. Most comminuted products have extenders/binders added to improve the bind strength, palatability and health benefits of the products and they also lower the cost of formulation. Some common comminuted products include the following.

Meat mince: This is comminuted/ground raw meat and is called as *keema* in India. Minced meat is used in the preparation of patties, cutlets, *koftas*, nuggets, sausages etc.

Cutlets: They are popular meat snacks in Kerala and are prepared from coarsely ground cooked boneless meat, mainly beef and chicken. Potatoes are added as extenders in the product and also sauted condiments, spices and salt are added. The batter is flattened into cutlets, dipped in egg white, breaded and deep fried in oil and is an example of enrobed meat product. Enhanced quality of cutlets in terms of increased phenolic content, reduced lipid oxidation and protein degradation can be brought about by replacement of potato with cooked yam (*Amorphophallus paeonifolius*) without affecting the sensory attributes (Kurian *et al.*, 2016).

Emulsion based meat products: Meat emulsions are prepared by grinding or chopping meat along with added water and salt to a fine matrix in which fat globules are dispersed. Preparation of emulsion allows the processor to utilize lower value meat portions because comminution obviates the effects of meat toughness. Emulsion preparation also allows inclusion of non-meat additives like cereals, milk products, soy products etc., which increases the profitability of the enterprise. The products can also incorporate edible non-muscle portions like edible offal, poultry skin etc. Emulsion based meat products provide consumers with convenience, variety and choice with respect to flavour, healthy options and ease of preparation. Some emulsified meat products are detailed below.

i) *Sausages*: Sausages are one of the oldest forms of processed foods and are chopped or minced meat, seasoned and salted and filled into casings. Casings are sausage containers and can be natural or artificial. Natural casings are prepared by processing gastrointestinal tract and involve the submucosa portion. Oesophagus, intestines, stomach, urinary bladder etc are commonly processed to prepare casings. Artificial casings include those made up of cellulose, regenerated collagen etc. Sausages are prepared using ingredients like meat, fat, ice, spices, salt, curing agents, extenders/binders etc. Sausages are only popular in big cities in India. There are hundreds of sausages available in different countries and due to its diversity no single classification is possible for sausages. The United States Department of Agriculture has classified sausages into fresh, uncooked smoked, cooked smoked, cooked, dry and semidry and luncheon meat.

ii) *Meat patties*: The emulsion is filled in stainless steel container lids or glass petri plates smeared with oil. The molded patties are removed from the molds and dry cooked under grilling for 15-20 minutes in a hot air oven or microwave oven. Cooked patties are placed between bread slices or burger buns and consumed along with cheese, onion, tomatoes etc.

iii) *Nuggets*: The emulsion is filled into tiffin boxes, smeared with oil taking care to avoid air pockets. After closing, the boxes are placed in a pressure cooker and cooked for 20-30 minutes without pressure. The boxes are taken out, cooled and the cooked blocks are taken out of the box and sliced. The nuggets are lightly fried on a pan with a bit of oil before consumption. The nuggets can also be prepared from whole muscle pieces like chicken breast fillets which are battered and breaded category products.

iv) *Croquets*: A portion of the emulsion are taken and mixed with pre-cut onion, garlic, ginger, curry leaves etc., and small portions of this mixture is formed in the form of balls and fried in oil for about 3-4 minutes. The fried golden brown croquets are consumed along with any sauce.

v) *Meat balls or kofta*: These are prepared using meat emulsion which is formed into balls followed by cooking in boiled water for 20 minutes along with spices, condiments, coconut etc. Koftas can be consumed directly after shallow pan frying or used to make curry.

Restructured meat product: This includes any meat product that is completely or partially disassembled and then reformed into the same or different form. This involves chunking or flaking the meat and tumbling or massaging the meat with salt and polyphosphates to extract the myofibrillar proteins and stuffing the blended mix in casings or trays of desired shape and freezing. Sectioned and formed meat product forms a category of restructured meat but does not include the processes of grinding, chopping, slicing or flaking except for the preparation of binders. In this, intact muscles or sections of muscles are the major meat components and the meat particles are used as binding materials. Sectioned and formed meat products have advantages like easy to slice and serve, reduce cooking loss, utilize cheaper cuts to produce more valuable products and there is better control over composition. Some disadvantages are that low quality meat is not improved by this method and additional equipment is needed. Restructured meat products and sectioned and formed meat products are prepared so that the products resemble whole muscle cuts and are often priced between whole muscle cuts and comminuted products. This enables the elimination of excess seam fat (fat in between muscles) and connective tissue.

Traditional Indian meat products: Traditional meat products have high sensory quality with good nutritive value. Some of the traditional meat products are

i) *Kababs*: These are prepared from meat chunks, minced meat or meat emulsions (incorporating extenders and meat by-products) by charbroiling or oven-roasting. The flavour of charbroiled kababs is unique due to combustion of fat that drips on the red hot charcoal. For commercial scale cooking, oven roasting may be more preferable.

ii) *Tandoori chicken*: It is a popular traditional Indian meat dish prepared using marinated tender broiler chicken meat which is baked in a traditional or gas tandoor. Small superficial incisions are made on the deskinned primal cuts or carcasses to facilitate the penetration of marinade and smeared salt and lemon juice and baked in a gas tandoor.

iii) *Meat pickles*: Meat pickles are the shelf-stable, ready-to-eat and convenience meat product of Indian origin, preserved by a simple pickling process by careful selection of ingredients.

iv) *Dried meat*: Dried beef is an ethnic delicacy in many parts of Kerala and there exists great scope for marketing of dried meat products which have assured quality, safety and shelf-life.

v) *Meat balls*: A meat ball is made of ground or minced meat rolled into a small ball, sometimes along with other ingredients, such as bread crumbs, minced onion, egg, butter, and seasoning. Meatballs are cooked by frying, baking, steaming, or braising

vi) *Meat curry*: It is made from different meats, cooked with various spices and usually served with rice. For curried products, the meat to gravy ratio should be 3:2. Pre-salted cooked meat chunks are mixed with the blend of fried condiments and spices and heated till the development of desirable brown colour and flavour. Subsequently, adequate amount of water is incorporated and boiled.

vii) *Meat keema*: Keema can be made from almost any meat, can be cooked by stewing or frying, and can be formed into kababs. Keema can also be marketed as semi-processed meat product, as consumers and housewives can eliminate the mincing or chopping step and can directly proceed with making minced meat products like kabab or cutlets.

viii) *Tikka and samosa*: Tikka is prepared from either meat chunks or ground meat and is characterized by the crispy exterior and a less springy texture than patties, burgers and other comminuted meat products. Samosa is a popular, convenience ready to eat enrobed meat product, prepared with stuffs consisting of precooked/fried meat mince or chicken mince along with spices and seasoning. Deep fat frying is done until the exterior becomes golden colour and crispy.

Apart from the above products, a variety of traditional products of Kerala like *erachi pathiri, kozhi ada, beef ada, alissa, kozhi pidi, kappa biriyani* etc. also require special mention.

Advances in value addition of meat

Advanced processing techniques are adopted to improve the health benefits of meat by incorporation of certain functional ingredients without affecting palatability and other quality characteristics. Functional foods by virtue of physiologically active ingredients prevent certain diseases and promote health rather than just meeting the nutrient requirements. Functional meat products can be processed by either adding compounds that are not naturally present in meat or increasing the concentration of compounds which are present in meat or replacing certain compounds in meat during processing. Low sodium and low fat meat products, meat products enriched with dietary fibre, calcium, unsaturated fatty acids, CLA, bioactive peptides, pre and probiotics, ACE (Angiotensin I converting enzyme) inhibitory peptides, antioxidants etc. are examples of functional meat products. The level of the functional ingredients can be increased to a desired level only during meat processing.

Low sodium meat products are processed by using low-sodium blends (potassium chloride and magnesium chloride), use of flavour enhancers like mono sodium glutamate, yeast extract, hydrolysed vegetable proteins etc. Low fat meat products are processed by trimming of natural fat, using fat replacers, fat mimetics etc. Incorporation of dietary fibre mainly from plant sources helps to make fibre enriched meat products. Other advancement in meat processing is the replacement or reduction of nitrite in cured meat by using nitrite replacers without reducing the antimicrobial effect and not altering the pink colour of cured meat. Use of locally available and cheaper non meat ingredients with health benefits is to be stressed to render the products more palatable and to reduce the processing cost.

Apart from the above, use of irradiation, high pressure and ultra sound waves to improve the microbiological quality of meat, use of hydrodynamic waves for meat tenderization, robotics in meat processing etc are recent advances in the meat sector. Advances have come in packaging techniques like active and intelligent packaging and rapid and accurate analyses of meat products.

Scope of entrepreneurship in meat processing

Venturing into meat processing at small and medium scale is both a lucrative and challenging step for new entrepreneurs. The typical demographic patterns in the post globalisation era in the country including the changing life styles, more working women, faster pace of life, quality consciousness, better brand recognition and wider access to a variety of foods with both ethnic and exotic flavours have created a great opportunity for marketing value-added, convenient and quality assured meat products which are attractive, healthy and safe to consume. The prospects of successful entrepreneurship in meat processing depend on many factors, some of which are listed below.

i) Availability of raw material is of paramount importance. Supply of raw material of uniform and assured quality must be ensured. For meat of large animals like buffaloes, this could be a problem, especially with larger volumes of processing. Processing of poultry meat will be a more viable option, considering the availability of raw material and consumer acceptance. However, fluctuation in live bird price is a stumbling block for the poultry meat sector.

ii) The volume of operations and marketing has to be carefully planned. A better option will be to start with a small scale production unit which handles about 20-30 kg boneless meat for producing popular value added products that are readily acceptable and attractive to the customers. Initial steps should include making tie-ups with local bakeries, catering units and institutional food establishments for marketing the products. One problem at this stage will be the lack of a brand for this new product. When the scale of operation is gradually increased, efforts for brand building also have to be initiated. For brand building affordable methods of advertisements like leaflets, radios, local cable channels and cookery shows can be initially tried.

iii) The point of marketing has to be carefully planned. The lifestyle and meat preferences of the customers around the sales point are to be considered while finalising the list of products and formulations. Fluctuations in the raw material price can be effectively tackled if frozen storage facilities are made available. Birds can be procured when cost is low and frozen carcasses can be stored for product preparation at a later time. Frozen bird carcasses can also be directly marketed to catering units, hotels and restaurants as they do not need meat from the wet market.

iv) Selection of products, their formulations and appropriate facilities for their manufacture are key determinants for the success of the enterprise. As mentioned earlier, it will be better to start with two or three familiar, popular products like cutlet, pickle, kababs etc. Gradually, new and ethnic products can be introduced which caters to different sections of population including emulsion based products and dried meat products.

v) For small scale entrepreneurs, a better option will be to start with poultry carcasses or deboned poultry meat rather than establishing primary chicken slaughter facilities. Those who run poultry farms can initially plan marketing poultry carcasses in the wet or frozen form by establishing small scale primary processing facilities. They can further expand later into poultry products.

vi) Quality and safety of meat products are of utmost concern for the consumers. Traceability of the meat in terms of farm of origin, treatment or vaccination schedule of the animals or birds and other details are also of interest to the consumers. Sustainability of meat processing in future will be largely depend if entrepreneur consistently meet these consumer concerns. Sourcing the live birds or carcasses from existing poultry farms can solve the traceability to a larger extent. Tie up with a local veterinarian for regular ante-mortem and post-mortem inspection with remuneration on a monthly basis can substantially corroborate the consumer's confidence in the poultry meat or meat products.

vii) Development of processed meat sector in India largely depends on strengthening the small scale processing sector which caters to the requirements and preferences of different segments of consumers. Such units use less sophisticated equipment and processes thus cutting down the capital and operational costs. Small scale meat processing ventures with simple technologies and machineries would thus be a viable option than very large automated processing units. However, there are very few manufacturers in the country who are reliably supplying indigenous meat processing equipment. Most of the larger processing units rely on imported equipment.

A meat processing facility can be run in rented or own premises which could be a small unit of 250 sq.ft to much larger facilities. The output from small units can also vary from 20-30 kg per day to several hundreds. Small scale processors can distribute the products on two wheelers fitted with insulated boxes. Simultaneously, effort can be taken to start fast food eateries inside the city for marketing and popularisation of the products.

Some machinery required for small scale meat processing

Meat mincer

Meat mincer is primarily used for particle size reduction in meat processing. It consists of a rotating augur which pushes meat towards a grinder plate available in different sizes suitable for coarse or fine comminution. Minced meat can be directly marketed, or can be used for preparation of products ranging from cutlets to emulsion-based products. Low-cost indigenous meat mincers are being manufactured in the country.

Fig.1. Meat mincer   Fig.2.Planetary mixer   Fig.3.Bowl chopper

Planetary mixer: Planetary mixer is mainly used for preparing coarsely comminuted or ground meat products. The minced meat is mixed with non-meat ingredients and mixed in stainless steel bowl using mixing paddles.

Bowl chopper: Bowl choppers consist of a horizontally revolving bowl and multiple curved knives rotating vertically on a horizontal axle at high speed of up to 5,000 rpm. It is primarily used for preparation of meat emulsions. The minced meat is converted into a fine batter. It also mixes all other ingredients and finally forms the meat emulsion. Both imported and indigenous versions of the equipment of varying capacities are available.

Sausage stuffer: The meat batter is filled into natural or synthetic sausage casings using the sausage filler or stuffer. Hydraulic as well as manually operated fillers are available

Fig. 4: Sausage stuffer　　　　**Fig. 5:** Smoking cum drying chamber

Smoking cum drying chamber: Automated and sophisticated equipment are available which can regulate the smoke and heat generation within the unit. In small scale units, the temperature and humidity can be adjusted for controlled drying of meat, and smoke can be generated from wood at the bottom that enters into the chamber.

Meat slicer: Some of the meat products are marketed in the form of thin slices and slicing is done using meat slicer. The machine consists of a circular blade and the thickness of the slice can be adjusted.

Band saw: Band saws are used in butchery industry, meat shops and restaurant kitchen for cutting or slicing of frozen meat.

Barbeque unit: Barbequing is a dry method of cooking in which heat from coal, hardwood, LPG or electricity is transferred by both air convection and conduction onto marinated tender meat cuts.

Other equipment required for small-scale meat processors including cooking equipment, sealing machines, stainless steel tables etc can be sourced from local manufacturers.

Packaging of meat and meat products

Packaging is done to contain and unitize a product, to prevent contamination, to efficiently and economically distribute a product and for consumer appeal. Materials used in packaging meat include paper based, glass, metal and plastics. Paper materials include kraft paper, grease proof paper etc. They are cheap, have good printability, and disposal is easy. But, they are liable to tear and the strength is reduced when wet. Metal and glass containers are commonly employed in meat products like canned meat, potted meat, corned meat etc. Plastics include polymers which are made by linking together monomers, either same or different types. A combination of properties can be achieved by combining materials of different layers by co-extrusion or lamination. All meat packaging materials must be free from toxic substances and must not impart off colour or off-flavour to meat.

Packaging materials for fresh meat

In developed countries where wholesale and retail distributions of meat are present, there are two different types of packaging in fresh meat itself.

For wholesale cuts usually vacuum packaging is preferred. Flexible high barrier polymer bags made from co-extrusion films are used and these bags are shrunk to fit the shape of meat cut using a tunnel (90°C) or hot dip. These bags are fitted with highly puncture resistant polyolefin patches laminated to the bag at key locations to prevent sharp bones from penetrating the bags. These materials are called as "bone guards". Vacuum packaging reduces bacterial growth especially psychrophilic growth, prevents contamination and desiccation, and reduces lipid oxidation, purge and weight loss. Myoglobin in vacuum packaged meat remains deoxygenated and is purplish in colour, but on opening it will immediately oxygenate and attains bloom. Ethylene vinyl acetate (EVA), ethylene vinyl alcohol (EVOH) etc are the commonly used materials. Overwraps, pouches etc. are also used for whole sale packing.

For retail cuts, a moisture resistant oxygen permeable packaging material is needed. Overwraps, tray with overwraps, pouches etc. are commonly used. The packaging should be puncture resistant and antifog additives are commonly employed so that the product remains visible. Trays made up of pulp, clear polystyrene (PS), expanded polystyrene (EPS), oriented polypropylene (OPP) etc. are commonly used. In the trays, drip pads are used to absorb the drip and are made up of absorbent cellulose with a thin polyolefin layer on one side to prevent sticking on meat. Plasticised poly vinyl chloride (PVC) with oxygen permeability but with moisture resistance is used as overwraps in trays. Vacuum packaging is not preferred in retail sale of meat due to the dark purplish colour of meat, which makes it unappealing. Modified atmosphere packaging can be done in meat cuts where high barrier containers are used. Air is removed and a mixture of gases is introduced, usually 50-80% oxygen, 20-50% carbon dioxide and sometimes nitrogen. This helps to increase the shelf life of meat.

Packaging materials for frozen meat

For frozen meat, the packaging material should prevent oxidation, desiccation and must be puncture resistant. Common materials used are heat shrinkable polyolefin bags with metal clips or heat seals, low density polyethylene (LDPE), EVA co-extrusions with polyvinylidine chloride (PVdC), EVOH and nylon.

Packaging materials for cured meat products

Materials with heat sealing and oxygen barrier properties like EVA, PVdC, EVOH and nylon are desired for cured meat products

Advances in meat packaging include retort packing, active and intelligent packing. Retort packaging is done in foods to be subjected to high temperature and pressure treatment of retorting/canning and the products are shelf stable and ready-to-eat straight out of the packet. They are made up of laminates containing 3 or more layers. Retort pouch generally consists of an outer layer of polyester or nylon for printability and toughness, a middle aluminium foil layer that functions as the principal oxygen and water vapour barrier and an inner heat sealed polypropylene (Varalakshmi *et al.*, 2014). These pouches are filled with food preparations, vacuum is created in the packets and they are hermetically sealed and retorted. The pouches can be cut open and the food can be consumed directly from the packet.

Both active and intelligent packagings have functions over and above the ordinary functions of packaging like unitization, preventing contamination etc. Active packaging is defined as "packaging in which subsidiary constituents have been deliberately included in or on either the packaging material or the package head space to enhance the performance of the package system" (Robertson, 2006). This includes antimicrobials or antioxidants, either as such in sachets or pads or as nano particles so as to enhance the shelf life of the product. Intelligent packaging is defined as ""packaging that contains an external or internal indicator to provide information about aspects of the history of the package and/or the quality of the food" (Robertson, 2006). Intelligent packaging provides information to the consumer regarding the environmental conditions to which the products have been subjected and whether the products have deteriorative changes, eg. Time -Temperature Indicators (TTIs).

Utilization of by-products from meat industry

By-products and offals are those parts of the animal, which are not included in the carcass. The by-products must be effectively utilized mainly for two reasons 1.) They are valuable sources of potential revenue 2.) Utilisation of by-products prevents environmental pollution and reduces public health problems.

In the animal which is slaughtered, the major portion is formed by carcass, which includes meat, bone, adipose and other connective tissues. In India, in the case of culled cattle, average percentage of saleable meat is 28-33% and quantity of by-products is 67-72%. In the case of pigs, the average percentage of saleable meat is 52-55% and the quantity of by-products is 45-48%. In the case of the culled goats the average percentage of saleable meat is 28-30% and quantity of by-products is 70-72%. The by-products obtained from animals can be grouped into edible and inedible.

The edible by-products from meat animals range from 20-30% of the live weight of cattle, buffalo, sheep, goat and pig. In the case of chicken, it is only 5-6% (Ockerman and Hansen, 1999). Most non-carcass materials are edible biologically if they are properly cleaned and processed. But due to custom, religion, palatability etc. only a few are consumed like liver, heart, casings, tripe, brain etc. The edible products are mainly used as a sausage ingredient, or consumed as such or used as pet food or animal feed or fertilizer (Table 1).

Table 1: Edible by-products from meat industry

By-product	Yield based on live weight	Main use
Blood	2-6%	As a sausage ingredient, for black pudding, blood protein isolate and for pharmaceutical purposes.
Brain	0.08-0.1%	As food after frying or broiling
Edible kill fat	1-7% (cattle)	As shortening or dripping
Feet	1.9-2.1% (cattle)	For jellied food items
Head and cheek meat	0.32-0.4%	As sausage ingredients
Heart	0.3-0.5%	As sausage ingredient or cooked and used
Intestines	-	For casings and as food
Kidney	0.07-0.2% (cattle)	As food and for pharmaceutical purposes
Liver	1.0-1.5%	As food, for liver tonics, liver meal
Lungs	0.4-0.8%	As pet food, for heparin

Contd.

Pancreas	0.06%	For hormones and enzymes
Spleen	0.1-0.2%	As sausage ingredient
Sweet bread	0.03-0.05%	As variety meat
Tail	0.1-0.25% (cattle)	For soup preparation
Tongue	0.25-0.5%	Cured and boiled or used as a sausage ingredient
Pig tail and feet	-	Tid-bits
Poultry giblets	-	As variety meat
Rendered edible fat	2-11% (cattle)	Shortening

Edible by-products due to their higher glycogen content and lower fat covering are highly perishable and must be removed from the carcass and chilled quickly and hygienically. They must also be cooked and served quickly.

Inedible by-products include some blood, hide or skin, bones, hairs, hooves and horns, feathers, rumen contents etc. Some of the important inedible by- products and their uses are listed in Table 2.

Table 2: Inedible by-products from meat industry

By-product	Yield based on live weight	Main use
Blood	Less than 1%	Leather finishing agent, as blood meal as animal feed or as fertilizer.
Hide	6-8%	Leather and gelatine
Bone	18-30%	Glue and gelatine
Intestine	-	Suturing materials and for musical instruments, tennis racquets etc
Horn and hoof	0.5%	For hoof and horn meal, protein hydrolysate
Bile	0.06%	Detergent
Fat	-	For soaps, candles and for grease
Feathers	-	Feather meal, protein hydrolysate, for pillows and mattresses and for ornamental purposes
Rumen contents	-	Biogas

Apart from edible and industrial uses, a few by-products have medicinal and pharmaceutical uses (Table 3).

Table 3: Medicinal and pharmaceutical uses of by-products

Organ	Product
Adrenal cortex	Cortisone
Adrenal medulla	Epinephrine and nor epinephrine
Gall bladder	Bile, bile salts, bile pigments
Brain	Cholesterol, cephalin, thromboplastin
Duodenum	Enterogastrone hormone, Secretin and Intrinsic Factor that aids absorption of vitamin B_{12}
Heart	Valves
Intestines	Heparin, catguts etc.
Liver	Vitamin B_{12}, heparin, catalase and liver extracts
Lungs	Heparin
Ovaries	Progesterone, Oestrogen and Relaxin
Pancreas	Hormones insulin and glucagon. Enzymes like Trypsin, Chymotrypsin and Lipase.
Parathyroid	Parathormone
Pineal	Melatonin
Pituitary (Anterior)	Growth promoting hormone (GH), Thyroid stimulating Hormone (TSH), Prolactin, Gonadotrophic hormone and Adrenocorticotrophic hormone (ACTH).
Pituitary (Posterior)	Vasopressin and Oxytocin
Seminal vesicles	Prostaglandins
Stomach	Pepsin and Rennin
Testes	Hyaluronidase
Thyroid	Thyroxin

In our country, mostly bones, skin and hides and intestines are only processed and the rest are not effectively processed. So, steps must be taken for the effective utilization of by-products as this increases the revenue from meat sector, increases export revenue and tax and provides employment opportunities. Value addition of by-products for edible, medicinal or any other industrial uses can be taken up only in an organized manner on a large scale.

Utilization of abattoir waste

Waste from an abattoir includes excreta of animals, intestinal contents, bedding material from lairage, blood, waste water after cleaning etc. These will create problems as they may attract rats and will become a breeding ground for flies. They can be a source of pathogens and bad odour and become a public nuisance. Therefore, they should be efficiently collected, removed and disposed. The waste from an abattoir can be either solid or liquid. Solid waste includes the faecal matter, partially digested stomach and intestinal contents. The choice of the method for disposal of solid waste depends on the cost, availability of land and labour. Some of the methods of solid waste disposal are:

Dumping: The waste is collected and dumped in low-lying areas. Due to bacterial action, the waste is degraded and converted into humus. But, the waste will be exposed to rodents and flies and there is nuisance of bad smell and unsightly appearance.

Sanitary landfill: This is done by tipping the waste in a trench and covering with earth.

Incineration: This is a hygienic method of disposal and can be adopted when suitable land is not available. This is highly expensive and a source of air pollution.

Composting: This is an effective method of disposal of waste from small slaughter houses, poultry processing units and meat processing units. It is a method of combined disposal of solid waste and sludge or dung. In this, there is breakdown of organic matter by bacterial action resulting in the formation of humus like material, compost. There is alternate layering of solid waste and cow dung in a trench till it rises as a heap over ground level. This can also be done at ground level. The topmost layer should be of solid refuse and is covered by excavated soil. The heat generated in the compost mass destroys pathogenic organisms and parasites. Decomposition is completed by 4-6 months and the resulting substance is odourless and is good manure.

Rendering: This is the most effective method of solid waste treatment in abattoirs. This involves high temperature treatment of solid waste, mainly condemned organs/carcasses, blood, bones etc. There are two types of rendering –wet and dry. In this process, fat is separated and the remaining proteinaceous materials are sterilized, dried and ground to prepare carcass meal or meat cum bone meal. This meal has a protein content of 50-55 per cent and is used as a feed for pets and non ruminants. However, the cost involved in establishing and operating a rendering unit is high and hence is feasible in large processing units.

Liquid waste from an abattoir includes urine, blood, waste water etc. The biochemical oxygen demand (BOD) of the liquid waste depends on the amount of organic matter present in it and it increases as blood is incorporated. The average BOD of abattoir waste is 1500-2000 mg/litre. Liquid waste treatment is divided into primary and secondary. In the primary treatment, solid matter is separated from the sewage partly by screening and partly by sedimentation. Dissolved air floatation is a method for removing suspended solids, fats and grease. In this, small air bubbles are used and they attach themselves to the suspended materials and lift them to the surface to form a scum. In secondary treatment, the sewage is subjected to biologic treatment systems in which a mixed culture of microorganisms is used. It can be anaerobic or aerobic. In anaerobic system, entry of air is prevented. In this, bacteria first break down the complex waste matter into simpler compounds like volatile fatty acids. These volatile fatty acids are further broken down by methanogenic organisms into methane and carbon dioxide. In aerobic oxidation, aeration of the system is achieved using compressed air. The dissolved organic matter is oxidized into carbon dioxide and water. Proteins are broken down into sulphates and nitrates. The product of this process is called as biomass and it is separated out from the treated effluent by a separator/clarifier. A portion of the sludge is taken into aeration tank and the remaining is treated and can be used as a fertilizer. The effluent is chlorinated and disposed by dilution into watercourses. It should not have suspended matter more than 30 mg/L and its BOD should be within 20 mg/L.

Scope for entrepreneurship development in meat sector

Indian meat especially buffalo meat has high demand in other countries especially in the Middle East and other Asian countries. Demand for meat and meat products is also increasing in India due to the increase in economic and social status of the population, double income in a family, increased number of working women and change in life style demanding convenient variety products. Organized development of meat sector is essential for getting full benefits from meat animals. Value addition of meat has to be stressed with development of formulations and processing technologies suiting Indian palette. Many Western meat products like sausages, ham, bacon etc. are relished only by a small section of the population especially in big cities. Certain exotic meat products like Arabian meat dishes are very much appreciated in Kerala especially by the younger generation. Also, our own traditional meat snacks and other preparations which are only locally known need to be popularized all over the state and slowly to other parts of the country. Value addition helps to make full utilization of the carcass, including edible offals. The treatment of waste from processing units to meat cum bone meal, compost etc. needs to be encouraged.

People are still unaware of the potential of meat processing sector and the economic returns from it. Small scale processing units especially by women will go a long way in improving the economics of meat sector. Different units can specialise in different classes of meat products and marketing can be done in retail outlets, where there may be provision for sale of frozen products as well as an eatery section. Supply of high quality products to restaurants, catering units, resorts etc. can also be carried out. Scientific techniques of processing with good hygienic practices are necessary and utmost care must be taken to ensure the quality of the products. Awareness regarding the laws governing the establishment of the units should also be there.

Research can be carried out in making formulations using locally available non meat ingredients which are not only cheaper, but having numerous health benefits. This again helps the farmers when such formulations are popularized. However, there are a few limitations to the meat sector like oppositions from various groups against establishment of meat processing units, problems of waste disposal, misconceptions regarding meat consumption etc.

Given below is an example of economic returns in a small scale processing of chicken cutlet

Expenditure

1. Chicken (deboned) – 10 kg - Rs. 2000/ @ Rs. 160/kg carcass (from 1 kg carcass, 0.8 kg meat. Hence for 10 kg meat, 12.5 kg carcass must be purchased.12.5X160= Rs.2000)

2. Potatoes - 5 kg - Rs. 100/- @ Rs.20/kg

3. Condiments, spices, salt

   oil, rusk, egg, fuel - Rs. 1500/

4. Labour charge   - Rs. 800/-

   Total   - Rs. 4400/-

Income

Minimum sixty packets (200 g each) or 360 numbers of ready-to-fry cutlets will be obtained from the above batter, which can be sold at the rate of Rs. 100/packet (including sales tax). The total income will be minimum Rs. 6000/- .

Success story in meat processing sector

Unlike fruits, vegetables and spice processing units, only very few entrepreneurship developments are taking place in the meat processing sector. This may be due to the unawareness of the people regarding meat processing. One success story in the meat processing sector is that of Mrs. Sherly Sunny, Kulakkattil (H), Mullankolly post, Pulpally, Wayanad, Kerala who started a small meat drying unit at Pulpally, licensed by both FSSAI and Panchayath. She along with her husband is running the unit named Sun taste dried meat in a renovated building in her own agricultural land. Buffalo meat/beef is dried using a fabricated drier. The drier can handle 40 kg of meat at a time and use firewood as fuel. The firewood is obtained mostly from their agricultural land and there are no labourers. Meat is brought from local market, washed and cut into thin strips. They are marinated with salt, turmeric and pepper and kept on the slotted steel shelves of drier. Drying is done for 10-12 hours. Forty kilogram of meat gives around 8 kg dried meat.

The product has huge demand locally especially among people who are working in Middle East countries. They purchase around 5-8 kg of meat per person when they return to the work place after vacation. The product is also supplied to premium restaurants in Ernakulam and Kozhikode districts. Since, the expenditure in processing is only for the purchase of meat, the product is marketed at a comparatively lower price of Rs.1700/ + sales tax per kilogram.

Dr. Dia. S. a Veterinary graduate with M.V.Sc in Livestock Products Technology from College of Veterinary and Animal Sciences, Mannuthy, Kerala, started a ready-to-cook/eat meat products processing unit under the brand name *FABULOUS FOODS* at Perumbavoor, near Kochi, Kerala. The major products are packaged frozen ready-to-fry chicken cutlet, beef cutlet, and ready to eat beef fry and chicken roast. The major marketing points are cold storages and bakeries. The firm had FSSAI registration, health certificate, D&O licence from local bodies, packer registration and Small Scale Industry Registration. The total area of the unit is 600 sq. ft. The initial investment was around Rs. 4 lakhs. There is huge demand for these products in cities.

Safety and quality aspects of meat products

One disadvantage of meat is that it is highly perishable and undergoes rapid spoilage after harvesting. Contamination of meat can occur from animals, during slaughter and handling operations. This not only makes the meat less perishable, but also results in food poisoning out breaks. Various organisms like *Salmonella* sp., *Staphylococcus aureus, Campylobacter jejuni, Yersinia enterocolitica, Listeria monocytogenes, Escherichia coli, Clostridium perfringens* etc. can cause food poisoning in man through contamination of meat.

Meat, once harvested from the carcass needs to be washed in potable water to remove dirt, dung, hairs, feathers and blood clots attached to it. Washing of large portions or cuts of meat is required, since washing of diced/chunked meat results in loss of water soluble proteins and hence loss of nutritive value and flavour. Then, it has to be properly packed and kept in chiller to promote ageing, the process by which the characteristic flavour, tenderness and juiciness of meat develop. After ageing, meat can be cut into retail portions, packed and frozen or used for further processing.

Care must be taken to see that all meat contact surfaces like hands of meat handlers, chopping boards, knives, utensils, refrigerator/freezer shelves etc. are sanitized prior to use. Periodic sterilization of knives and frequent washing of hands are to be practiced. Workers should be provided with proper and clean work clothes and boots and there should be sufficient wash rooms, lockers and changing rooms in the unit for both sexes of workers. Workers should wear plastic/disposable aprons, disposable gloves, masks and head gear during processing. They should be free from any diseases and should undergo frequent health checkups. Any cuts on hands should be well covered/bandaged. Eating, drinking, smoking of cigarettes, betel/pan chewing in processing units should be strictly forbidden.

Processing units should be spacious with proper lighting and ventilation. Walls should be tiled and they must be easily washable. Floors should have non slippery, durable tiles and must be sloped. Management of pests like flies, cockroaches, rats etc. must be given due attention.

In case of value added meat products, non meat ingredients like flour and different extenders, spices, packaging materials etc. also contribute to contamination of products. Hence, care must be taken throughout the processing operations i.e. from harvesting of meat to packing and storing of the products to ensure high quality products.

Conclusion

Indian meat industry is growing at a rapid rate and more stress is put on the production and marketing of fresh meat to suit the preference of local population. Due to the economic gains from export of meat, production of frozen meat, especially frozen buffalo meat is also gaining popularity. More importance should be given in processing of meat into value added products which on marketing results in more economic returns. Large scale processing of meat can be taken up for export purposes, whereas small scale processing can be done to meet local demands. Small scale processing may be tried for processing, marketing and thus to popularise traditional meat products in different parts of India.

References

APEDA, 2017. http://www.apeda.gov.in. Accessed on 20-05-17.

FAOSTAT, 2013. http://www.FAOSTAT. Accessed on 20-05-17.

Islam, M M., A. Shabana., R. J. Modi & K. N. Wadhwani, 2016. Scenario of Livestock and Poultry in India and their Contribution to National Economy. *Inter. J. Sci. Environ. Technol.*, 5(3): 956 – 965.

Kurian, Y., R. Nayar., M. Pavan., K. Rajagopal., C. Sunanda & P. Kuttinarayanan, 2016. Development and Quality Evaluation of Novel Chicken Cutlet Incorporated with *Amorphophallus paeonifolius* (Elephant foot yam). *Ind. J. Natural Sci.* 7(37):11298-11306.

McAfee, A. J., E. M. McSorley., G J. Cuskelly., B. W. Moss., J .M. W. Wallace., M. P. Bonham & A. M. Fearon, 2010. Red Meat Consumption: An Overview of the Risks and Benefits. *Meat Sci.* 84: 1–13.

Ockerman, H. W. & C. L .Hansen, 1999. *Animal By-product Processing and Utilization.* CRC Press, USA.

Pighin, D., A. Pazos., V. Chamorro., F. Paschetta., S. Cunzolo., F. Godoy., V. Messina., A. Pordomingo & G. Grigioni, 2016. A Contribution of Beef to Human Health: A Review of the Role of the Animal Production Systems. *The Scient. World J.* Pages 1-10.

Robertson, G. L. 2006. *Food Packaging: Principles and Practice.* Taylor and Francis, USA.

Sen, A. R., M. Muthukumar & B. M. Naveena, 2013. *Meat Science - A Student Guide,* Satish Serial Publishing Company, N. Delhi.

Varalakshmi, K. P., Devadason., Y. Babji & R. S. Rajkumar, 2014. Retort Pouch Technology for Ready to Eat Products -An Economic Analysis of Retort Processing Plant. *J. Agri. Vet. Sci.* 7(1): 78-84.

15

Using the P's of Marketing to Sell your Product

Sangeetha K Prathap

Introduction

Successful marketing is important to every business. Marketing is a critical tool for establishing awareness, attracting new customers and building lasting relationships. It is the process of planning and executing conception, pricing, promotion and distribution of your ideas, goods or services to satisfy the needs of individual consumers or organisations. It is as important to put adequate efforts on marketing your product or service, no matter whatever time you have dedicated to development of part of it.

We live in an electronic era, where the globe has shriveled into our fingertips. Customers are having lot of options to choose from. Information of what you need is reaching you within a click, that your decisions are often being guided by such information. Marketing uses communication and advertising tactics to persuade customers that your brand, including your products and services are exactly what they need. Marketing is thus crucial to your business which can have a bearing on the success rate and requires research, planning and appropriate budget allocations.

One of the important concepts in marketing is that of marketing mix. The elements of this mix are: product, pricing, physical distribution and promotion. Marketing management is that which optimizes the elements in the marketing mix, so that organisation is able to achieve its objectives while gaining customer utility. This chapter throws light on various aspects of marketing mix that an entrepreneur needs to focus upon to ensure success.

Product

Products refers to the physical product or service what you intent to sell, which includes all features, advantages and benefits that your customers can enjoy from buying your goods and services. When marketing your product, you need to inform the customers about the features and benefits of your product and how it matches customer's needs.

Product development

In the product development phase, the concept is translated into a prototype or trial formulation in case of a food product. It may be noted that technical development of the product does not fall typically under the production domain alone. Marketing personnel need to be involved in every stage of the product development, especially to conduct product tests which will throw light on consumer needs, preference, behaviour and response. Accordingly, based on customer feedback, necessary changes in product formulations or choice between alternative formulations *etc.* can be gauged.

Typically, a new product development process starts from idea generation, idea screening, concept testing, business analysis, technical development and test marketing. Each of these stages in the new product development process acts as milestones before the entrepreneur is to continue or to abandon the business opportunity. A detailed discussion on new product development is beyond the scope of this book. The reader may refer to specific books on product development gaining better insights.

CASE ON PRODUCT DEVELOPMENT

DRIED (UNRIPE) JACKFRUIT

Jackfruit (*Artocarpus heterophyllus*) is the largest edible fruit in the world. It is a seasonal fruit found in almost all the humid tropical regions of the world. The flesh of jackfruit is starchy and fibrous and is a source of dietary fiber. Jackfruit has more protein, calcium, thiamine, riboflavin and carotene than banana. However, the fruit is perishable and cannot be stored for long time because of its inherent compositional and textural characteristics. In every year, a considerable amount of jackfruit, especially obtained in the glut season (June-July) goes waste due to lack of proper post harvest knowledge during harvesting, transporting and storing both in quality and quantity. In order to process the fruit, the skin has to be removed first. Sticky latex from the fruit will ooze out when the tough scaly skin is cut causing difficulties. Proper post harvest technology for prolonging shelf life is, therefore, necessary.

Drying is a process used widely to preserve fruits and vegetables during off season by removing water content by evaporation. This process reduces the weight of the fruit or vegetable and can be preserved in room temperature in an air tight packing. Mature unripe jack fruit is the main raw material for this type of processing. Whole jackfruits are peeled, processed and packed. Drying process reduces the usage of preservatives and also occupies less space to store. It adds diversified and attractive food items in dietary menu as well as contributes to generation of income and employment. Drying of jack fruit involve various unit operations *viz.,* cleaning, grading, cutting, removal of bulb, deseeding, slicing, blanching, drying and packaging.

From the market survey conducted, consumers are found to endorse the following benefits of dried jackfruit.

- The raw material is cheap and easily available
- It is a natural product
- Field problems like pest and diseases are absent
- Product is devoid of harmful chemicals
- Product survives the heat of summer and have better longevity
- Shelf life is high compared to fresh fruit, jackfruit can be made available round the year
- Uniformity in quality can be achieved compared to fresh fruit
- Minimum loss of quality (risk in production is less)
- Product remains eco-friendly and biodegradable
- Low investment is required for establishing a processing unit
- Can be converted into wide range of products like sweets, chips etc.
- Advantage in terms of easier handling of the product, when compared to unprocessed raw product

Branding

Branding add value to a product, forms an important aspect of product management. Most of the industrial products are branded; thus reap the benefits arising from branding such as increased consumer attention towards the products and adds to shopper's efficiency. Brand names tell the consumers of the characteristics of the product and assure them that they will be ensured of the same characteristics while making repeated purchases.

According to American Marketing Association,

'A brand is a name, sign, symbol or design or a combination of them intended to encourage prospective customers to differentiate a producer's product from those of competitors'.

Historically, unprocessed agricultural products are sold as generic products unbranded. Agricultural product is frequently marketed as a commodity, where a particular grade is treated as a brand which can be compared against similar product from a different origin. Branding provides for a basis for non price competition by removing a product from a commodity category. A distinctive seller's brand and trademark make it possible to legally protect unique product features.

CASE STUDY

The Story of Building the Desi Brand of Dairy Products - Amul[1]

Creation of the AMUL brand and its famous mascot gave a rural revolution a durable competitive edge. The tubby little moppet in the familiar polka-dotted dress is not just the Amul Butter mascot.

The Amul Girl, who has entered urban lore with her regular appearance on billboards accompanied by clever catchphrases that comment on contemporary events, stands for the very fight its parent was born to counter.

The cooperative movement that began Gujarat back in 1946 was a movement against the atrocities of Polson Dairy, a locally-owned dairy in Anand, Gujarat, which allegedly procured milk from farmers at very low rates to sell to the Bombay (now Mumbai) government.

Amul's architect in almost every way was the late Dr. Verghese Kurien. Arriving at Anand in 1949 as a government employee to manage a dairy, he went from helping farmers repair their machinery to revolutionising the Indian dairy industry by scripting Operation Flood, a cooperative movement that turned India from a net importer of milk into one of the world's two

---

largest producers today. Not for nothing was Verghese Kurien called the Milkman of India, though his vision was a simple one of offering thousands of small dairy farmers's centralised marketing and quality control facilities, the missing links in the dairy economy at the time.

Thus, in 1973 the Gujarat Cooperative Milk Marketing Federation was established to market milk and milk products manufactured by six district cooperative unions of Gujarat. Branding also played a role, cleverly designed to add a tinge of nationalism to an essentially rural revolution.

When experts asked Kurien to choose brand names that would sound foreign, he wisely insisted on an Indian name. Thus, Amul (then short for Anand Milk Union Ltd) was born.

Amul was not just a milk and butter brand, it became an umbrella brand for all the products that GCMMF marketed. The first dairy, Kaira District Co-operative Milk Producers' Union, which created Amul in 1955, handed over the brand name to GCMMF in 1973.

By then, Amul had become a brand name in its own right. It is said that for Operation Flood, Kurien's idea of having farmers own the brand went a long way in creating a sense of ownership and, in turn, responsibility for the product's quality. This is the reason, state federations now have their own brands — Nandini in Karnataka, Verka in Punjab, Saras in Rajasthan and Mahananda in Maharashtra

Branding gave a farmers' cooperative a quasi-commercial strength, enabling it to adapt to competition. Brand AMUL is already present in over 50 countries. In India, it has 7,200 exclusive parlours.

Food branding has emerged as an effective strategy to make your food based products saleable, rather attract the 'hungry' customer to your product which has specifications than the competitor's product which may or may not have one. Branding creates an image of your company with the customer. If he is satisfied with your product, on his repurchase intention he is sure to look for your product which is clearly identified by the brand name creating a brand loyalty for your product.

It is important that you consider about branding in the early stages of your business, which can increase your chance of success. What you have to keep in mind is that your brand should have a personality which reflects the strengths and qualities of your product/business.

---

[1]Adapted from: Sohini Das, 40 Years Ago...and now: Utterly self-sufficient, Business Standard,Ahmedabad September 3, 2014

How to develop a brand?

Thinking of developing a brand for your product?. here are some tips that can be useful to you.

- Think about successful brands: Think about the successful brands in the market and reasons why they are successful. Internet and trade magazines can be rich sources of information for providing insights about leading brands.

- Conduct market research: It is important to find out what is the size of market, the type of customers you are targeting and who are your competitors. Work out on how will you present yourself before the customers so that they distinguish you from the competitors (or simply work out on your competitive advantage).

- Align your business to the brand: The brand is not just a logo; it should reflect the business name, customer service, quality, pricing and even promotion. You may know that the brand TATA is always remembered by the customer as 'trustable'.

- Be authentic, simple and consistent: As said earlier, brand can be attached to values that the business may deliver. Hence, it is possible that depending upon the business delivery, positive or negative value may be associated with the brand. For example, the food safety issue of Maggi noodles has created a negative brand image for Nestle's products.

- Seek professional advice: You may seek professional help to design and launch your own brand, if you feel that you do not have the expertise in doing so. A professional consultant may be able to deliver a workable brand image for your company, if you can offer to bear his hiring charges.

You can even brand 'Tomatoes'[2]

We have spoken about branding a value added product; how about branding the common tomato available in plenty with your grocers.

Looye is a family company that was set up in 1946 by J.M. Looije. The company started out growing a range of vegetables. Later in the 1970s, the company started to specialise in growing tomatoes. Presently, besides growing tomatoes, Looye is also involved in the other links in the supply chain, like sorting and packing.

In early days, the company sold tomatoes via auctions to begin with and later through cooperatives. In 2003, the company started selling product themselves.

According to the company

"We see selling our product ourselves as a core part of our business. It enables us to keep in close contact with those around us and it's the best way to get to know our consumer. And they are ultimately the people we are doing this for!"

In 2005, the company started to focus even more intensively on growing the tastiest tomatoes. LooijeSubliem is one of the branded tomatoes grown and marketed by the company.

Looije Subliem

Looije Subliem is a tomato that tastes the way tomatoes used to taste; one to really enjoy. These tomatoes are grown in Jas Loije's greenhouses with a great deal of care and attention. After picking tomatoes are sent straight to the packing hall, where our panels of tasters are ready to taste and check them. Not a single crop goes out of the door without having first been run past our panel of tasters. So, you can be confident that the tomatoes you buy are a very special product with a very special flavour.

²Adapted from http://looye.com/

Packaging

From the traditional function of physical protection of the product, packaging serves marketing the product too. The subject matter of this discussion shall focus on packaging as a marketing tool.

Product packaging has role in differentiating products, where there are large number of brands in the market.

Food package design uses creative graphics that may attract the attention of prospective customers. It should satisfy the informational content as insisted by the standards for food products. Moreover, the quality of packaging is often interpreted as indication of quality of contents. Hence, type of packaging can induce the customer to facilitate purchase decision based on the information available and suitability to his needs. Packing also serves as an aspect which differentiates one brand of a product from another brand.

Pricing

Pricing your product is one of the most difficult decisions that the entrepreneurs must make regarding the key to controlling costs and earn profit. Price conveys an image and affects demand as well as helps to target selected market segments.

A number of other factors can influence the entrepreneurs' ability to effectively price the product or service. These factors include.

- Number of competitors
- Seasonal or cyclical changes in supply and demand
- Production and distribution costs
- Customer services
- Markups.

There are various pricing methods that may be adopted by entrepreneurs. This differs according to whether it is retail, manufacturing or service oriented enterprise. However, some general rules of thumb may be adopted in case of pricing

Value of the product: Price should be fixed on the basis of value of the product or service to the customer. If the customer does not perceive the price to be reasonable, the entrepreneur should consider changing the price as well as altering the image of the product/service.

Rationale: Entrepreneurs should explain why their products differ from that of competitors; whether the price premium is applied because of higher quality materials used in production. A lower price in turn might be justified by efficiency in production process.

Positioning: One way to position a product is to charge a high price when competition and substitution are minimal. Another is to match the competition by pricing slightly under the competitor's rates to expand one's own market share. Third method is to substantially under price the market so as to exclude competitors altogether (Just like Reliance Jio has launched its 4G services).

Physical Distribution

Once you have priced the product appropriately, the next question would be how you are actually going to reach out to the ultimate consumer; i.e, what channels of distribution you are going to adopt?

There are many ways how you can reach the potential customer; however selection of the path will depend on the nature, type and scope of business. We will now examine different methods how you can reach the ultimate customer.

Direct sales: Products and services are sold directly to the end user. This is the most effective distribution channel and used when distance between producers and consumers are short. For example farmers may opt for direct sales to local consumers at farm gate to avoid transporting produce to distant markets.

Retailers: Retailers include a wide range of outlets such as merchants, equipment dealers, department stores, super markets and small grocers who deal with end user of product or service.

Wholesalers: wholesale traders are intermediate agencies who purchase fairly large quantities of goods and sell these items on to retailers in small quantities.

Sales agents and brokers: sales agents or brokers are different from other channel players that they do not take title to goods. The role of agents and brokers is to facilitate distribution by bringing sellers and buyers together.

Vertical marketing systems: In this system, producers, wholesalers and retailers act as a unified system. Usually one channel member owns the others or has contracts with them or has franchisees with others in the channel. For example we can see that franchisers operate by vertically linking several levels of the marketing system. 'Coca Cola' wholesales its syrup concentrate (product) to franchisees who carbonate and bottle and distribute the brand (processing, packaging and physical distribution) to consumers who have been targeted by Coca Cola's heavy advertising (promotion).

Agricultural equipment manufacturers supply machinery (product) to appointed distributors who are given exclusive rights to sell and maintain their whole goods and parts (physical distribution and service) within a specified geographical area. The manufacturer will provide sales support (promotion) to authorized dealers.

Horizontal marketing systems**:** When two or more organisations at the same channel level cooperate to pursue marketing opportunities, it can be termed as horizontal marketing systems. For example, a seed company and a grain merchant might set up a joint venture to offer farmers a complete package where he can buy certified seed from the merchant who guarantees to buy the grains harvested at agreed prices.

Online marketing or E-commerce**:** products and services are sold through website or through internet partner alliances. By the surge in internet users in India, and consequent of demonetization, usage of e-commerce is expected to boost up significantly. In addition to other products, agri based products are also sold through e commerce, which makes it convenient for the consumers in avoiding time spent for shopping and convenience. The advantage of internet based marketing is that you don't have to invest in physical store but a virtual address and application can allow you to kick start the business. However, you need to have considerable investment and co-ordination on processing the orders and arranging for logistics on the other side. However, the potential customers will be limited to those who depend on online shopping.

CASE STUDY

E-Commerce Prospects: Online marketing of Value added Jackfruit[3]

Messy jackfruit is no more a headache for the homemaker today. She can get dried (raw) jackfruit semi processed and packed in retort pouches, amenable for culinary recipes by simply opening the pouch. Also you need not waste your precious time in the supermarkets searching for the product; it is available at your fingertips in your online shop. Jackfruit 365, has made your favourite raw (processed) jackfruit online. You can shop through the shopping site of amazon in select cities in India.

M-commerce: Marketing through mobile phones, particularly specialized apps (applications) is becoming very popular. You may have noticed the *'Daily fish'* app that helps you to order fish from the application in the mobile phone. Another app, *Fresh to Home* helps you to get fresh fish from the landing centre.

Cashless payment options: Whatever be the distribution channel opted, as the economy is marching towards a digital payments system, you will have to ensure digital payment mechanisms in the form of Point of Sales terminals (POS), Aadhaar Enabled Payment System (AEPS), mobile banking, m-wallet payments etc. E-commerce allows payment gateways for making payments, while cash on delivery (COD) options also may be exercised through portable digital payment mechanisms. Most of the digital payment systems require smart phones. However, if you are a small scale operator dealing with common man having basic phones, you may depend on payment mechanisms like USSD (where basic GSM phone may be used to make payments using *99# Mobile number, MMID).

[1]http://www.jackfruit365.com/

Case Study

Mobile app based marketing

What Fresh to home app offers the customers?[1]

Over 190 varieties of fish & meat delivered to your doorstep

Available Fish

Fish   Meat

Special Kerala Sardine / Mathi
₹178.00/Kg   Buy

One of the cheapest yet the most tastier of our entire fish collection. Has too few available flavors, but what an amazing taste! Lots of Omega 3 to grow those budding brain cells

Buy Fresh Chemical-Free Fish, Antibiotic free Chicken, Mutton and Duck with the Fresh to home app. Halal-cut and delivered fresh to home in Bangalore, Delhi, Cochin and Trivandrum.

Fresh to home is India's first e-commerce venture for fresh, chemical-free seafood. We buy only from country boats and small trawlers that return within 24 hours from the sea to ensure you get the most fresh seafood possible. We give maximum value to the fishermen and consumer by cutting off the middle-men. We assure 100% chemical free fish - some of our premium fish is "hook-caught" or caught the traditional way using a line and a hook and thus will be much superior in taste than the mass produced trawler caught fish.

Fresh to home is a venture by seafood industry veterans who have been dealing with fish for decades. Our online system is powered by a dynamic website and app, with backend linked live to our processing plants. Products are delivered within hours of the catch landing in our plants. We only use natural ice for packing and delivery.

We deliver at customer's doorstep by our own specialized logistics team. Both cash on delivery and online payments are supported on Fresh to home app. We have thousands of happy customers in Bangalore Delhi, Cochin and Trivandrum.

Promotion

Marketing communications takes four forms; advertising, sales promotion, personal selling and publicity. This helps the consumer to be aware of products and services they need, who might supply the products and what benefits it might offer. Even the best product might not reach the consumers if it fails on the promotion front.

Marketing communication serves following five key objectives

- Provision of needed information to the consumer about the product
- Highlight product's value
- Differentiating the product/service
- Stimulation of demand
- Accelerating sales

Depending upon the geographical coverage, type of marketing communication may be selected. If your product is meant for mass consumption that is spread over geographically dispersed markets, advertising is the best suitable marketing communication tool.

Advertisements

Advertisement is a suitable means to send the same promotional message to large audiences. While determining what form of advertising is to be adopted, it is worthwhile to think about the reachability to your target market segment and the associated costs.

Advertising in media is often an entrepreneur's first choice. Such advertising can include television, radio, news paper ads and magazines. While large businesses can afford for prime time commercials, small businesses will have to depend on low budget local cable televisions. Whether to go for cable TV commercials depend upon the viewership of such channels in the concerned area. Similarly, business men can opt to advertise in newspapers for gaining wide geographical coverage; national daily may be preferred by big business, while regional or local newspapers may be opted by small entrepreneurs.

Creative advertisements done professionally can cost you high. So enterprises need to orient advertisement strategy on the basis of budget availability.

Digital marketing

With the expansion of e-commerce, digital strategy for keeping abreast with latest technologies is essential for any business to survive. Digital strategy is a roadmap for how you will use digital technologies to enhance business practices, increase productivity and enhance revenue. An important element of digital strategy is digital marketing, involves use of online resources and tools in digital marketing including websites, microsites, mobile apps and social media platforms. These may include advertisements in LinkedIn, blogs, guest blogs, newsletters, facebook, search engines, emails, mobile banner and e books.

Summary

Now you might be clear about what are the basics of marketing a food based value added product. This chapter establishes various tips for marketing food based products and recent developments in the marketing that may be encashed by an entrepreneur.

Preparation of marketing plan is a requisite for an entrepreneur and constitutes an element of business plan. Such a plan delineates what strategies will be adopted by the business in marketing the product. Each of the elements of marketing mix; product, price, physical distribution and promotion need to be adequately taken care of while designing the marketing strategies. A proper and well-conceived marketing strategy is the corner stone of any business venture. Hence, it is important for an entrepreneur to dedicate his time and resources to effectively market his product.

References

Crawford, I.M. 2016. Agricultural and Food Marketing Management, Food and Agriculture Organization of the United Nations, Rome, FAO 1997, Accessed From http://www.fao.org/ docrep/ 004/W3240eOn 22nd December 2016.

Das, S. 2014. 40 Years Ago...and Now: Utterly Self-Sufficient, *Business Standard*, Ahmedabad, September 3, 2014.

Kaplan, J. M. 2004. *Patterns of Entrepreneurship*, John Wiley & Sons Asia Pvt.Ltd.

http://looye.com/accessed on 24th December 2016.

https://www.freshtohome.comaccessed on 24th December 2016.

http://www.jackfruit365.com/accessed on 24th December 2016.

https://www.business.qld.gov.au accessed on 24th December 2016.

16

How to Prepare Business Plans

Sangeetha K Prathap and Sudheer K P

Once you have embarked upon the idea of doing (any) business, your concentration should be focused on doing things that should be relevant to maximizing returns and minimizing risks. Looking upon why businesses fail, one can find that it is usually because entrepreneurs are ill equipped to identify real business opportunities and take advantage of them, or they don't anticipate and adapt critical changes or they fail to recognize or minimize real threats to business. Successful entrepreneurs are found to intensively pursue the path of planning by preparing business plans, for having foresight of critical incidents that would affect business and possible ways of action.

What is a business plan?

Business plan is such a document which describes where a business is heading, gives an idea of its marketing and financial forecasts and how it hopes to achieve its goals and objectives. It is important to have a business plan, because it helps you to set realistic goals, secure funding from investors and delineate operational guidelines.

There are two major reasons for preparing a business plan. Firstly, business plan acts as a vehicle for communicating to others about what the business is all about. Attracting investors for your business will certainly depend on the strength of your business plan. Financing institutions are found to meticulously review the plans, gauging entrepreneur's ability to perceive successful running of the business. Secondly, the plan provides a basis for measuring actual performance against expected performance.

Preparation of business plan involves considerable time, effort and resources. Therefore it has to be useful to employees, management and potential investors. Now you may ask how long it would take to draft a business plan. It may take almost around 200 hours, but may have variations on either side depending on experience and knowledge of business. Collecting the background materials is the most time consuming and hilarious task while attempting for preparation of the plan.

Target the plan to selected groups

A business plan could be the best method to reach out to the target groups. Primarily businesses need to mobilize capital from investors. Apart from investors, other target audience for business plans may include business brokers, employees, investment bankers, suppliers etc. Let's look into each of the target groups and the purposes for which they may be approached.

- Bankers: to provide loans for expansion and equipment purchases
- Business brokers: to sell the business
- New and potential employees: to learn about the company
- Investors: to invest in the company
- Investment bankers: to prepare the prospectus for an IPO
- Suppliers: to establish credit for purchases

Using the Business Plan

Profile: Homemade Jack Delights

How Shyamala used the business plan to raise loan from the bank for her small business?

Shyamala, 49 years old, was a casual labourer in the construction sector. She lost her husband in her early age and has two children who attend the nearby secondary school. She also looks after her old mother who is bedridden. Shyamala is the lone bread winner for the family. She has 3 cents of land for which application has been lodged with authority for grant of title deed. Due to spine related problems she is unable to get involved with works that demand intense physical activity. She quit her job in the sector and was trying for an alternative source of income.

She came to know about a training conducted by the nearby Krishi Vigyan Kendra (KVK) on preparation of value added items from fruits. The programme was envisaged to impart handson training on preparation of value added products from commonly available fruits in the area, especially jackfruit. Shyamala was inspired by the programme and thought of encashing the plan of preparation of value added products. She pursued her idea of starting a small production unit of value added products of jackfruit which was available in plenty in her area. The enterprise development cell of the KVK offered all help in realizing the aspiration of the poor lady.

Though, the idea of starting a tiny unit was inspiring, the financial position was not sufficient to offer own funds for starting a unit. The KVK offered

---

help to pursue financial assistance from funding agencies. Shyamala was not very positive about getting a loan from the bank as she had no collateral to offer. Yet, she followed the advice of her mentor and approached the nearby bank. She came to know that under the new scheme of PradhaanManthri Mudra Yojana, there is provision for funding for starting new enterprises. All that is required is a project report (business plan) which describes the activities that may be taken up along with its technical and financial viability. Shyamala became happy and with the help of the mentor prepared a business plan and submitted to bank along with other procedures for availing the loan. The banker after carefully scruitinising the application along with the project report approved the loan of Rs 50,000/- as 'Sishu loan'. Shyamala started her business of preparing value added products and sells in the brand name 'Homemade Jack Delights' as suggested by her kids and mentor. Homemade Jack Delights is a hot cake in the nearby bakeries and confectionary stores and Shyamala is at present thinking of scaling up her business. The banker has offered her top up loans as she has completed the repayment of her installments promptly.

Preparation of business plan

Business Plan provides the blue print of activities proposed to be carried out by an enterprise. Preparation of the business plan requires knowledge of what are its content requirements. We will now examine what are the essentials of a business plan.

The executive summary

The executive summary can be viewed as a synopsis of the business plan or in other words it provides an overview of the main points in your business plan. It is very important part of the business plan and is positioned at the front of the document. It is the first section that the potential investor or lender will read and needless to say, further reading will be triggered only if it succeeds in arousing reader's interest.

Identifying your business opportunity (Overview of the company, industry, products and services)

This part of the plan gives an introduction of your business and its products and services. This section may give an overview of the business; minute details of which may be covered elsewhere in the document. The section begins with a general description of legal form of the organisation; whether the entrepreneur wants to do his business as a sole proprietor or register as a partnership or even start a company. This section also addresses questions such as; what is the business? What customers will it serve? Where is it located and where will it do business?

Further, it should also provide insight into what stage the company has reached. Is it a 'seed stage' company without a fully developed product line? Or is it a company with a developed product line but has not started its marketing or is it a company which has already launched product in the market and wants to scale up its operations? After describing about the organisation, section may concentrate upon product or service. This section may include a prototype, sample, or demonstration of how the products work. The section should include the following

- Physical description: Physical characteristics of a product usually include photographs, drawings or brochures or description of service

- Uniqueness of your product or service

- Description of the stage of development: prototype design, quality testing, implementation and so on

"Jackfruit365™¹ packs are handier than all season Jackfruit tree in your yard!"

Uniqueness of the Product

Background

Jackfruit is mainly used during its season, in various traditional Indian dishes. However, almost 80% of the Jackfruit in India is going waste as there is no organized market for Jackfruit based products. The professional chefs are reluctant to use Jackfruit for three reasons- Jackfruit is seasonal, messy to prepare and the aroma spreads quickly in the kitchen.

Freeze Dried Jackfruit

The whole process of taking out the pieces after cutting the jackfruit is cumbersome and very messy. Freeze drying is a process used widely to preserve fruits and vegetables during off season by removing water content by first freezing and then converting ice to vapour. This process reduces the weight of the fruit or vegetable and can be preserved in room temperature in an air tight packing, hence, making it easy to store. The rehydration process takes just about 15 minutes.

Most importantly, it is possible to increase the fiber in your meal, reduce glycemic load and calories without you even noticing it. All that is required is to add one third of powdered dried jackfruit to the most traditional carbohydrate meal we consume in the north and south India from Idli and Dosa to Roti and Aloo Paratha.

Jackfruit365™ Premium Quality and Hygiene Assured

Jackfruit365™ packs are freeze dried from the premium quality Jackfruits sourced at the peak season. Whole jackfruits are peeled, processed and packed at facilities fully certified for the highest safety standards for America, European Union, Japan and Australia to ensure we offer the best quality expected in your Kitchen.

Benefits of Jackfruit365™

Easy to cook from a reseal pack, easy storage for 365 days, quick rehydration in minutes, and the closest you can get to the real king of fruits and jack of fibre.

Market analysis

The section on market analysis indicates how business will react to market conditions. The investors keenly look into market analysis section, as this would reveal the actual situation when the product is launched in the market. No projections on financial front will be realized once the product has failed in the market. Hence, the marketing strategy and other miniscule elements in the market analysis will give a blueprint of the company's presence in the market which will guide the readers to the strength and weakness, thereby giving cues about chances of being success/failure.

Market research

Market research will give you the data you need to identify and reach your target market at a price customers are willing to pay. It helps you to understand the customer response to the proposed product, in terms of their needs and aspirations, expected utility, the results of pilot test of the product. Documentation of the market research can also form a part of the marketing plan.

Market opportunity

Identifying the marketing opportunity is very important when one think of starting his business. This is to find out whether the proposed product will have acceptance in the market. So, the first thing to do is to conduct a survey in the market about the product to be launched. There are specialized market research agencies whom you can approach for this purpose, however, that may turn out to be a costly preposition. You can do a short survey by going to the market by yourself, interacting with potential customers about the features of the new product and analyse their reactions. You will be able to spot out customers

---

[1]Adapted from www.jack365.com, the seller of jackfruit products online

aspirations of the new product and identify the shortcomings of the proposed product. No doubt, product acceptability as per the opinion emerging out in the market survey will boost up your confidence in launching your product.

Identifying the level of competition

The level of competition in business is one of the important factors that a businessman should look into while starting his business. He must definitely look forward to what market share he should be targeting at while starting his business. Many new entrepreneurs are ignorant about this kind of information and even if they want to get such information, they are not able to. Competitors' information may be gathered while conducting your market survey for assessing the potential product. This may be done by discussing with retailers as to what is the share of sales of different brands and pattern of demand during different seasons. Information may be gathered from magazines, newspapers and internet based information sources.

Is competitor analysis important in agri-business as well? Yes, of course, competitors' analysis cannot be avoided even for small business. Let's analyse a simple scenario in which we look upon competitors from the customer's angle in making choice between products. For example, a person who is thirsty has many choices before him. His choices can be bottled water or hot drinks (tea/coffee), cool drinks (juices) or even alcoholic beverages. If he chooses cool drinks, the next step will be to choose between packed or fresh juice. If he chooses for packed juice, choice has to be made between carbonated and non carbonated ones. These are the levels of competition that may arise in soft drink market. Finally, while picking up his juice pack, customer may prefer a particular brand while he picks his choice of non carbonated drinks (brand choice entailing competition between brands at the same level).

Entrepreneurs have to be alert of what competitors have to offer and they should always compare their product and strategies with their competitors to remain in the business.

Designing marketing strategies

This section defines how business will use its marketing tools. Based on the information from market research on market opportunities and level of competition, the overall marketing strategy of the firm is to be designed. A market strategy identifies target customers that a particular business can better serve than its competitors and tailors product, prices, distribution and promotion.

Market Analysis

Market Analysis of Coconut Neera[1]

Market demand

Coconut Neera already enjoys an excellent market potential in countries like SriLanka, Myanmar, Thailand, Philippines and other Pacific countries. Coconut neera, if introduced in India is bound to create a huge market potential as a health drink and as a base for manufacture for value added coconut products like concentrated syrup, sugar, honey *etc.* which has wide export potential in USA, Europe and African countries.

Market development

Coconut neera and its value added products can be positioned in the market for the targeted section like health conscious people, diabetic patients, kids and youngsters as a natural nutritive product to replace soft drinks, also sugar and honey after value addition.

There is a growing market trend for coconut neera and its value added products in beverages, food, bakery and confectionary industries.

Competitive advantage

Competitive advantages are essential for survival of any product in a competitive market. The term competitive advantage is the ability gained through attributes and resources to perform at a higher level than others in the same industry or market. Considering the glycemic index, coconut neera and its value added products have a higher competitive advantage over existing soft drinks/beverages. Not all carbohydrate foods are ranked equal, in fact they behave quite differently in our body due to individual metabolism. The glycemic index or GI describes this difference by ranking carbohydrates according to their effect on our blood glucose level. Low glycemic index foods produce only small fluctuations in our blood glucose and insulin level. So neera and its value added products can be safely used by diabetic patients.

Marketing and sales plan

A strong business plan will include a section on marketing and sales plan that describes specific activities that you will use to promote and sell your products or services. A strong sales and marketing section demonstrates that you have a clear idea of how you will get your product into the market. You can answer the following questions of the reader of your business plan.

---

[2]Adapted from DPR for setting up integrated unit for Coconut Neera and its Value Added Products

- Who are your customers?
- How are you going to reach your customers?
- How are you going to position your product or service?
- How are you going to price your product or service?
- What are the promotional strategies that you might use to attract your customers?

This analysis will guide you in establishing pricing, distribution and promotional strategies that will enable your enterprise to become profitable within a competitive environment.

Pricing

Pricing is an important element in the marketing strategy because it has a direct impact on business's success. When considering what price to be charged, one should understand that it is not always cost plus some profits that may be used to determine price of a product. The question that may prop up into your mind may be "How much to charge?" Various factors need to be analysed before fixing the price of the product.

- Consider comparable commercial products, price charged by others in your community for similar products, company's desired image
- Consult business people in the community
- Pricing your product to ensure a desired image in the customer's mind should be based on the firm's competitive advantage. This can be either through differential advantage (providing unique products and services) or through cost advantage (offering comparable products and services at cheaper prices)
- Prices should reflect the cost of production and provide adequate returns to the entrepreneur

Typically, there are different types of pricing that can be adopted by a firm. The business plan should specify what type of pricing strategy is followed by the firm; whether it is cost plus pricing, demand pricing, competitive pricing or markup pricing.

Distribution

The distribution strategy describes how your product will reach the ultimate consumer. The type of distribution channel will depend on the industry and the size of the market. Entrepreneurs should analyse competitor's distribution channels before deciding to use similar channels or alternatives. You may opt for direct or online selling methods. You may think about pushing your product to the market through one or more of the following channels.

- Direct sales methods: through the channel of retail institutions, wholesalers, sales agents and brokers.
- Vertical marketing systems
- Horizontal marketing systems
- E-commerce:
  - Web based
  - M-commerce

The reader of business plan should be convinced about the reach of the product to your targeted customer segment through the selected channels of distribution. Suitability of the channels of distribution depends on the product you are selling and the target customer segment. Detailed account of various channels of distribution is provided in Chapter 17.

Promotion

Promotion is the communication aspect of the marketing mix that includes whatever is done to tell the public or potential customers about your product or service.

Many small businesses rely on word of mouth as means of promoting their business. This is often effective, low cost method of promoting your business compared to advertising. However, there may be no control for the manufacturer on the message that is spread.

Advertising and other promotional strategies help you to inform the potential customers that you have a product that may match their needs. If you have a big budget for promotion, no doubt you will be able to hire professionals for doing the same. If the promotion kitty is not that large, you may have to adaptive strategies; you may think of guerrilla marketing strategies (low cost, creative marketing).

In order to popularize e-commerce, marketing through online media has gained prominence. Digital marketing has become the buzz word today and agriculture based products are increasingly being channelized through e-commerce channels. A detailed emphasis on e-commerce and online marketing has been included in Chapter 15.

Online Marketing Has Reduced Geographical Distance Among Markets

Jackfruit365™[3] is now available on amazon.in.

Order it online from the comforts of your home!

While home cooks didn't find jackfruit as a viable diet due to its seasonal nature; chefs were reluctant to use jackfruit for three reasons- jackfruits are seasonal, messy to cut and has a rich smell that could seep into their professional kitchen. Well not anymore!

Jackfruit365™ is available 365 days of the year in re-sealable packets that are convenient to use and store. Right before use, all that is required is to re-hydrate the pieces. In other words, soak it in plain warm water and they swell to their regular weight in about 10-15 minutes.

Operations

The operation section of your business plan will outline your daily operational and facility requirement for running the business. This section usually includes information like daily operational requirements, hours of operation, seasonality, suppliers and credit terms etc. Facility requirements includes things like size and location, information on lease agreements, supplier quotations etc. Further, details of Management Information Systems of the organisation including inventory control, management of accounts, quality control and customer tracking may be given.

Management team

It is important to know who are managing the business; the staff strength, their expertise, skills, experience etc. This section shall include organizational layout or chart of the business, biodata of managers, needed skills and job responsibilities of each person in position and other relevant information.

---

[3]Adapted from www.jack365.com, the seller of jackfruit products online

Financial forecast

Financing institutions appraise the project proposals on the basis of strength of the project in terms of concept, practicality, operational specifications, marketing strategies adopted and financial indicators. The financial forecast turns your plan into numbers. As part of any good business plan, you need to include financial projections for the business that provide a forecast for the next three to five years. The details included in a financial forecast are:

Cash flow statements: cash balance and first 12-18 monthly cash flow pattern including working capital, salaries and sales.

How to prepare a cash flow forecast?

A cash flow forecast shows the amount of cash coming in (receivables) and going out (payables) during a month. This shows the financial position of the business; how much liquidity is available with the business and whether cash in hand will be sufficient to make interest payments on a revolving line of credit or to cover shortfalls when payables exceed receivables. Cash flow forecasts can easily be made in an excel sheet. Cash outflow should be in terms of various expense categories like raw materials, other overhead expenses like wages and salaries, power, fuel etc. Cash inflow should be projected by realistically estimating the sales realized in each month. Cash inflow less outflow will provide you the net cash inflows with the firm; the amount that comes in for recouping your investment and once this is done, profits start accruing. An example of cash flow forecast is given in Chapter 17.

In addition to cash flow statements, you may provide statements including profit and loss forecast which gives the projected level of profit based on your projected sales, the costs of providing goods and services and your overhead costs. Projected balance sheet which indicates the assets and liabilities of the business also may be provided. Most important part of the financial forecast is working out of the indicators which reflect the financial strength of the proposed business.

How to calculate financial indicators?

Financial indicators points to financial position of the firm reflected from its income generation capacity in relation to its initial investment. The investment decision criteria (whether to deploy the capital available with you), commonly referred to as capital budgeting techniques can be worked out to find out whether an investment project (business investment) is worth investing. Sound appraisal criteria should be used to measure the economic worth of the project.

The commonly used capital budgeting techniques are

i. *Payback period (Non discounted cash flow method)*

Payback period is one of the simplest and popular methods of evaluating project proposals. Payback period is the number of years required to recover the original cash outlay invested in the project. A business proposal is said to be sound if it satisfies the criteria that the initial investment will be reaped back in less than half of the projected life period of the project.

ii. *Net Present Value (NPV) (Discounted[4] cash flow method)*

Net Present Value is a discounting method which measures investments profitability. NPV can be calculated by subtracting present value of cash inflows from the present value of cash outflows. The proposal may be accepted if NPV happens to be greater than or equal to zero.

iii. *Profitability Index (PI)*

Profitability Index is the ratio of present value of cash inflows, at the required rate of return, to initial investment cash flow. It is measured as the ratio of present value of cash inflows by the initial cash flow. A proposal of PI greater than one is acceptable.

iv. *Internal Rate of Return (IRR) (Discounted cash flow method)*

Internal Rate of Return is the rate that equates investment outlay with present value of cash inflow received after one period. This also implies that rate of return is the discount rate which makes NPV equal to zero.

An example of detailed financial forecast for a selected value added product is appended at the end of this text as Annexure

Establishing fund requirement

Some other things to consider include is the funding requirement of the business and how much funding is required for the business, where these funds will be spent and when it will be needed. To determine financing requirements, entrepreneurs must evaluate and estimate the funds needed for (but not limited to) research and development purchase of equipment and assets and working capital. The entrepreneurs should plan short term financing arrangements for funding working capital, whereas R&D should be financed out of long term

---

[4]A discounted method explicitly recognizes the time value of money. It postulates that cash flows arising at different time periods differ in value and are comparable only when their equivalents – present values are found out.

finances. The following questions may be answered by entrepreneurs so that lenders get a perspective of the firm's funding requirements, capacity to borrow and repayment capacity.

- How much capital do you need, if you are seeking external funding?
- What security can you offer to lenders?
- How do you plan to repay your debts?
- Forecasts covering a range of scenario.

Conclusion

This chapter establishes the value of business plans and the step by step procedure involved in its preparation. Entrepreneurs can use the business plan as a guide for establishing the direction of the company and the action steps needed in obtaining funding.

Preparation of business plan requires adequate ground information regarding specific sections that constitute the plan. Executive summary provides a brief account of the business that would summon the reader's interest in surfing through the detailed plan. Once the reader is attracted with the proposal, the business plan in detail gives an account of the marketing, financial, operational and organizational information regarding the project. It can be said that a well conceived and well written business plan can be helpful in launching the business as well as serve as an instrument in attracting funding for business.

References

Kaplan, J.M. 2004 . *Patterns of Entrepreneurship*, John Wiley & Sons Asia Pvt.Ltd.
Pandey, I. M. 2011. *Financial Management*, Vikas Publishing House Pvt Ltd, Noida.

17

Model Business Plan

Sangeetha K Prathap & Sudheer K P

Executive summary

The enclosed project report envisages the setting up of a modern rice mill complex with attached parboiling facilities having an average processing capacity of 16 MT of paddy per day. The initial investment of the project is estimated to be Rs 480.17 lakhs. Sales Realization is assumed as constant for all the 7 years; amounted Rs 1129.2 lakhs. Pay Back Period for the project is estimated as somewhat more than 3 years. Internal Rate of Return for the project is 31%. Net Present Value is Rs.367.92 lakhs which is a positive value. It indicates that present value of cash inflows is greater than present value of cash outflows. Projected working results of the project throws light on its viability.

Introduction

The proposed business unit produces parboiled rice with low glycemic index manufactured from the best rice variety. It is intended to market this product as a premium rice brand for the health conscious consumer in Kerala State. The unit will also focus on by-products of rice and also like to focus on nutritive value and compete with similar brand available in the market.

Vision

To be a leading brand in the cereal products range that concentrates on conserving nutritive values to suit health conscious consumer.

Mission

To cater the needs of the health conscious consumer by availing a technology that provides leeway to conserving the inherent qualities of rice.

Production process

Rice milling is the process that helps in removal of hulls and bran from paddy grains to produce polished rice. Paddy undergoes certain processing treatments prior to its conversion into edible form. It consists of various unit operations viz; pre-cleaning, parboiling, shelling, husk aspiration, paddy separation, polishing, grading, sorting etc (Fig.1). Various unit operations and equipment used in a modern rice mill and the product flow is explained in the process flow chart given below. Details are given under Chapter No.5.

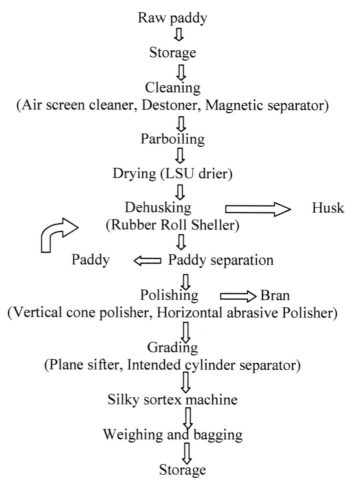

Raw paddy

⇩

Storage

⇩

Cleaning
(Air screen cleaner, Destoner, Magnetic separator)

⇩

Parboiling

⇩

Drying (LSU drier)

⇩

Dehusking   ⟹   Husk
(Rubber Roll Sheller)

⇩

Paddy   ⟸ Paddy separation

⇩

Polishing   ⟹ Bran
(Vertical cone polisher, Horizontal abrasive Polisher)

⇩

Grading
(Plane sifter, Intended cylinder separator)

⇩

Silky sortex machine

⇩

Weighing and bagging

⇩

Storage

Fig. 1: Flow chart of a typical modern rice mill

Operational units /machinery	Purpose
Cleaning/ air screen cleaner	For removing foreign matter from paddy
Parboiling	Hydrothermal treatment to increase head rice yield
Shelling/ rubber-roll sheller	For separating husk from paddy grain
Husk separation/ husk aspirator	Separation of husk from the product obtained from sheller
Paddy separation/ paddy separator	Separation of paddy from brown rice
Bran removal/ polisher/ whitener	Removal of bran layers from brown rice
Bran aspiration/bran aspirator	Removal of bran adhering to rice kernel
Grading /grader	Separation of broken rice from head rice
Silky sortex section	Separation of black and off colour materials
Handing equipment	Conveying of paddy- rice to various processing units

Quality control

The processing facility gives utmost importance to quality of raw and finished products which is monitored regularly for all shifts/lots. Graders and sortex machine used in this plant ensures good quality finished product. Finished product is checked for moisture content, head rice, colour, etc. ISO certification for processing centres will also be sought, for ensuring quality assurance at factory level.

Permission and licenses

The unit requires license from Grama Panchayat and SSI registration from Industrial department for the production. No other license is required.

By-products of rice

Paddy when milled usually yields apart from marketable rice (head rice, medium and big broken rice) about 17% husk and 6% bran. Economic utilization of by-products is essential to ensure viability of rice milling.

Husk

Husk, generated as by-product when paddy is processed, is not edible and cannot be fed even to animals. However, husk may be used as fuel, particularly in rice mill for parboiling and drying of paddy. Husk is used as fuel in boilers to produce steam and hot air required for the purpose. Husk has calorific value of about 3000 Kcal/kg (nearly one third of that of mineral oil and half that of coal). Husk is also used as loose particle boards and insulating materials in building and cold storage, in shipping and packing material etc. Fully burnt white ash of husk can be used for manufacturing sodium silicate, silica gel, insulating bricks and also used as diluents in manure etc.

Bran

Rice bran is the most valuable by-product of rice milling industry. It contains 12% to 15% protein, 14% to 20% oil and is rich in vitamin B. Bran is utilized in several ways. It is used as feed for animals. It is used as a more valuable source of vegetable oil and should be first solvent extracted to recover the oil. The extracted bran is used as animal feed.

Broken rice

The medium and large broken rice is mixed with the head rice depending on the marketing requirements. The small broken rice that is separated during grading of milled rice is sold as such or used for rice flour making.

Technical feasibility

Scientific utility

Use of modern machines like rubber roll sheller along with improved methods of drying, paddy separation, bran removal and graders can give higher out turn. The unit has an average processing capacity of 16 MT of paddy per day; with these capacities the unit can produce 400 MT of rice per month. Packing, sealing and labelling can be carried out in the unit itself.

Major raw material required is paddy, procured from paddy cultivators. Paddy is obtained at the point of production which is nearest to the location of the mill. Since paddy is seasonal, its proper storage facilities should be ensured in the unit itself, to ensure year round processing.

Uniqueness

Traditional huller type rice mills are found to be inefficient. Modern rice mills are highly capital intensive which prevents the prospective entrepreneur. Present model can help the potential entrepreneurs in bridging the information gap and opt for a semi capital intensive form which falls in the small scale industry category.

The uniqueness of this rice mill is that it contains machines for parboiling and sorting. Parboiling is a pre milling treatment given to paddy to achieve maximum recovery of head rice and to minimize breakage. Sortex machine helps to remove all kinds of black or off colour impurities from the finished rice. Parboiled rice is a better source of fibre, calcium, potassium and vitamin B_6 than regular white rice. Parboiled rice contains more bran compared to white rice which is highly nutritious.

The rice mill intends to tap the potentially stable demand for rice as staple food, trying to foresee the apparently increasing future market. We have also identified the lack of domestic players in the market.

Scope for commercialisation

Rice is the staple food for 65% of the population in India. It is the largest consumed calorie source among the food grains. With a per capita availability of 73.8 kg, rice meets 31% of the total calorie requirement of the population. Following liberalization, Indian rice has been identified as one of the major commodities for export. This provides us with ample opportunity for development of rice based value added products for earning more foreign exchange. Apart from rice milling, extracting of rice bran oil is also an important agro processing activity generating income. Many of the rice processing units are of the traditional huller type and are inefficient. Modern rice mills which have high capacity are generally capital intensive. Small modern rice mills with all plausible facilities are the solution to this problem.

Location

The processing unit should be located at a place where adequate availability of raw materials, access to cheap labour, power, water, transportation facility of input and output, storage facility, suitable waste management system are available. Low lying areas should be avoided, else proper land filling, compaction and consolidation should be done.

Infrastructure facilities

The area of proposed site for rice mill complex is about 2 acres, which will be sufficient for establishing a modern rice mill with a floor area of 5,000 square feet structure. This structure includes godowns for raw and finished products, office buildings, and rest rooms.

Sustainability of project

The project is located at a rice growing area which enhances the scope of the project in terms of low procurement cost of raw material.

Technical Parameters

Parameters	Remarks
Plant capacity	16 MT/day
Production per day	5.76 MT/day
Number of working days in a month	25 days
Production per month	200 MT
Number of labourers	23
Procurement source of paddy	Paddy cultivators
Cost of 1 MT paddy(purchased from farmers) (Rs.)	21000
Selling price per kg of rice(Rs.)	40
Selling quantity	1 kg, 2kg, 5 kg, 10kg, 25 kg
Sales per month (72% of paddy)	144000
Electricity	300 units per day
Unit cost for electricity (Rs.)	Rs. 8.5/unit
Monthly use of electricity	7500 units
Depreciation	10%

Marketing plan

The proposed business unit produces parboiled rice with low glycemic index manufactured from the best rice variety. It is intended to market this product as a premium rice brand for the health conscious consumer in Kerala State. The unit will also focus on by-products of rice and also like to focus on nutritive value and compete with similar brand available in the market.

Current market scenario

Rice is the staple food of Kerala. Generally, people use it twice in a day. Food habits of people are changing due to several reasons like health problems and adoption of fast food. Most of the people has reduced rice consumption to once in a day. The demand for rice in Kerala (Population of 3.18 crores as per 2011 census) is about 40.68 lakhs tons, at an average daily rice consumption rate of 350 g/person. Eighty per cent of the demand is outsourced from other states as the annual production is only 3.65 lakh tons.

Market potential of rice mill

At present, the paddy milling capacity available in the state is about 3000-3500 tons per day giving an output of only about 2000 tons of rice per day, leaving a huge gap which is being met by PDS and by the supply of rice from other States. As per figures from rice mills operating association, there are about 125 rice mills operating in the State, out of which about 50 per cent are modern units and others are partially modernized. Rice mills in Kerala are meeting only 20 per cent of the States requirements. So, the demand supply gap is huge and there is sufficient scope for more rice mills.

Target customers

Much of today's rice is consumed as parboiled rice. Milled rice is nutritionally superior to standard milled rice. Parboiled rice is better source of fibre, calcium, potassium and vitamin B_6 than white rice. The glycemic index is low, so it doesn't increase blood sugar level compared to white rice. It can be consumed by diabetic patientS too. So, the target customers include all people without any age limit.

Distribution channel

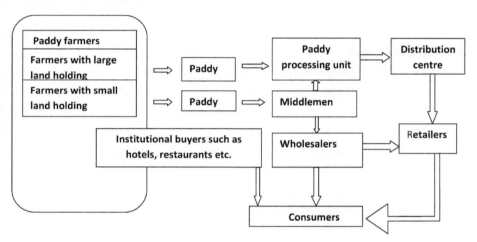

SWOT Analysis

Strength
* Support from the Govt. as agro processing is one of the important agribusiness.

* The by-products of rice have a huge market.

Opportunities
*The domestic and export consumers are sensitive and conscious about food quality and thus the use of nutritive premium quality of rice is growing at a higher rate.
*Since parboiled rice is rich in its nutritive content markets could be explored and exploited

Weakness
*Requirement of huge amount of working capital.
*Availability of raw material.

Threats
*Competition from local players and neighboring state may hinder the profit in the local market.
*Rice mills are categorized under polluting industries due to the dispersion of rice husk ash.

Competition

The threat of new entrants is very high and the bargaining power of the consumers is also very high. The major competitors of our products are Nirapara, Sabari, Palakkadan Matta, Pavizham, Ponni, Double Horse etc. Parboiled rice has less glycemic index which is a competitive advantage over white rice.

Marketing mix

Product	Price
Wide range of products like rice, rice bran, broken rice, rice husk	Sold @ Rs 40/ Kg. It is a traditional inflation indexed product, so more sensitive to price
Place	**Promotion**
All over Kerala	Sales promotion word-of-mouth

Future challenges

- Rice is a traditional inflation indexed product. Rice is a staple food for millions of people with high degree of sensitivity to price.

- Each district prefers a certain variety of rice for their consumption purposes.

- Convincing customer on the importance of glycemic index sell parboiled rice as a solution for this problem.

- Since it is a start-up project, the product promotion, nutritive information dispersion, publicity etc have to be at the lowest cost with the least possible returns.

Financial feasibility

Capital budgeting is concerned with the allocation of firm's scarce financial resources among the available market opportunities. The consideration of investment opportunities involves the comparison of the expected future streams of earnings from project with the immediate and subsequent streams of expenditures for it. This part of the report is organized under the following heads

- Techno-economic parameters
- Sources of finance
- Cash flow statement
- Capital budgeting techniques

Techno-economic parameters

Investment in business ventures requires commitments on fixed basis during the initial phase. Working capital is needed during the period of operation. The cash flow from operations is calculated on the basis of the expected expenditure and revenue out of the project.

Fixed capital requirement

In order to establish processing facility for rice milling, the following fixed capital requirements are necessary. It is assumed that the land is owned and no cost towards acquiring land is provided for. Building/modifying the facility/rooms required for the establishment of the unit is provided for. Purchase of two trucks worth Rs. 15 lakh each is provided for (Instead of owning the truck, it is possible to make alternate arrangements with carriage and freight agents which may be opted suitably). The expenses incurred for installation of the plant and machinery and other necessities for establishment of the processing facility can be accounted under capital expenditure. Details of plant and machinery along with price is explained in Chapter-5, hence, here we are giving only an approximate total value of plant of machinery in the business plan. The total fixed capital required for the proposed processing unit for rice mill is estimated as Rs. 480.17 lakh of which, detailed expenditure heads are furnished below.

Initial investment

Item	Quantity	Cost (Rs. Lakhs)
Land (owned)	30 cents	
Site preparation		2
Building cost		
(i) Office building	5000 sq ft.	50
(ii) Compound wall, open well & road		
(iii) Godown separately for raw paddy and finished product		
Office furniture		5
Mini trucks	2 No.	30
Plant and machinery including tax and installation related expenditure		150
Escalation and contingencies (10% of total above)		9.25
Preliminary and pre operative expenses		5
License and registration		0.1
Interest during project implementation period (construction and installation of machinery)		5.19
Working capital margin/month		223.63
Total (Initial Investment)		480.17

Estimation of working capital

The project is expected to operate at 50% capacity in the first year. Detailed estimate of raw materials and manpower along with the consolidated statement of working capital for the proposed rice mill is given below.

Raw material and utilities

Particulars	Quantity (daily)	Cost (Monthly) (Rs)
Paddy	8 MT at the rate of 21000/T (25days)	38,64,000
Electricity	8.5/ per unit 300 unit per day for 25 days	63,750
Packaging material		75,000
	Total	40,02,750

Man power

Position	Quantity	Salary/ month(Rs)	Monthly cost (Rs)
Technical			
General supervisor	2	25,000	50,000
Rice mill operators (skilled)	2	10,000	20,000
Mill helper (unskilled)	1	10,000	10,000
Boiling and par boiling unit			
Boiler operators (skilled)	2	10,000	20,000
Boiler helpers (unskilled)	2	10,000	20,000
Parboiling operator (skilled)	2	7,000	14,000
Parboiling helpers (unskilled)	1	6,000	6,000
Drier operator	2	10,000	20,000
Administrative staff			
Manager /Accountant	1	20,000	20,000
Assistant /Clerk	2	10,000	20,000
Labours	4	6,000	24,000
Drivers	2	13,000	26,000
Total			3,87,000

Working capital assessment

The working capital requirement of the proposed project on rice mill is summarized as follows.

Working capital computation	Amount (Rs)
Wages	3,87,000
Raw material	38,64,000
Packing material	75,000
Electricity	63,750
Insurance	4,000
Transportation	70,000
Selling and distribution	1,88,200
Repair and maintenance	15,00,000
office overheads – administration	40,000
Miscellaneous	20,000
Total working capital	62,11,950
Working capital financing for the year [70% of the cost]	5,21,80,380

Sources of finance

The proposed project comes under the classification of small enterprise. For the enterprises engaged in the manufacture or production, processing or preservation of goods, where investment in plant and machinery does not exceed Rs. 25 lakh is termed as micro enterprise. The capital requirement for the project may be financed out of owned funds and bank loans. At present, banks are providing requisite funding for the MSMEs at concessional terms. Schemes to promote industries are in vogue. The Ministry of MSME, Government of India and SIDBI set up the Credit Guarantee Fund Trust for Micro and Small Enterprises (CGTMSE) with a view to facilitate flow of credit to the MSE sector without the need for collaterals/third party guarantees. The main objective of the scheme is that the lender should give importance to project viability and secure the credit facility purely on the primary security of the assets financed.

Banks have, however, been advised to sanction limits after proper appraisal of the genuine working capital requirements of the borrowers keeping in mind their business cycle and short term credit requirement. As per Nayak Committee Report, working capital limits to SSI units is computed on the basis of minimum 20% of their estimated turnover up to credit limit of Rs.5 crore.

In terms of RBI circular RPCD.MSME & NFS.BC.No.5/06.02.31/2013-14 banks are mandated not to accept collateral security in the case of loans up to Rs 10 lakh extended to units in the MSE sector. Further, banks may, on the basis of good track record and financial position of MSE units, increase the limit of dispensation of collateral requirement for loans up to Rs.25 lakh with the approval of the appropriate authority.

Total capital requirement for the establishment of the unit is Rs.480.17 lakh. It is proposed to mobilize the required capital from the following sources.

- Bank loan

- Own contribution

It is proposed to mobilize Rs. 336.11 lakh as term loan from the commercial bank. Expected rate of interest is 12%. Repayment will be made in 7 year equal instalments.

Indicative proposal for term loan from commercial bank

Particulars	Amount (Rs. Lakhs)
Project cost for loan purpose	480.17
Margin money @ 30%	144.05
Loan amount	336.11
Interest rate	12%
Repayment period	7 years

The repayment schedule of the loan for a period of five years shall be as follows

Indicative loan repayment schedule

Year	Principal (Rs.)	Interest @ 12%(Rs.)	Total repayment/ annum(Rs.)	Balance(Rs.)
1.	4801702	4033430	8835132	28810212
2.	4801702	3457225	8258927	24008510
3.	4801702	2881021	7682723	19206808
4.	4801702	2304817	7106519	14405106
5.	4801702	1728613	6530315	9603404
6.	4801702	1152408	5954110	4801702
7.	4801702	576204	5377906	0

Working capital loan

Year	Principal (Rs.)	Interest @ 12%(Rs.)	Total repayment/ annum(Rs.)	Balance(Rs.)
1.	7454340	7827057	15281397	44726040
2.	7454340	6708906	14163246	37271700
3.	7454340	5590755	13045095	29817360
4.	7454340	4472604	11926944	22363020
5.	7454340	3354453	10808793	14908680
6.	7454340	2236302	9690642	7454340
7.	7454340	1118151	8572491	0

Cash flow statement

Cash flow includes cash inflows and out flows-cash receipts and cash payments-during a period. Movements of cash are of vital importance to the management. The short term liquidity and short term solvency positions of a firm are dependent on its cash flows.

For the present calculation, commercial taxes are not included; taxes apply as per taxation policy of the Government as and when applicable. At present, all food processing industries are exempted from taxes during initial years of establishment.

Capital budgeting techniques

The financial feasibility of the project is analyzed by the application of the following capital budgeting techniques:

- Payback period
- Net Present Value (NPV)
- Internal Rate of Return (IRR)
- Profitability Index/Benefit Cost Ratio (BCR)

Payback period

The payback period is one of the popular methods useful to ascertain the number of years required to recover the initial investment of the project. Thus, this method reveals the length of time required for the inflow of cash proceeds generated from the investment. Payback period of the current project is estimated to be more than half the life of the project, *ie* 3 years 0.1 months (3 days). Thus, the present proposal may be accepted.

Net present value

NPV may be defined as the excess of present value of project cash inflows over that of cash outflows. The discount rate is 12%.

(Rs. Lakhs)

Sl.No	Particulars	0th year	1st year	2nd year	3rd year	4th year	5th year	6th Year	7th Year
1.	Initial investment	480.17							
2.	Cash inflow	-	1129.2	1129.2	1129.2	1129.2	1129.2	1129.2	1129.2
3.	Cash outflow	-	986.60	969.66	952.71	935.77	918.83	901.88	884.94
4.	Net cash inflows	-	142.60	159.54	176.49	193.43	210.37	227.32	244.26

Year	Discounting factor @12%	Cash inflows in the project (Rs. Lakhs)	Cash outflows (Rs. Lakhs)
1.	0.893	142.60	127.32
2.	0.797	159.54	127.19
3	0.712	176.49	125.62
4.	0.636	193.43	122.93
5.	0.567	210.37	119.37
6.	0.507	227.32	115.17
7.	0.452	244.26	110.49
a.Total present value of cash flows (Rs. Lakh)			848.09
b. Initial investment (Rs. Lakh)			(480.17)
c. Net present Value [a-b] (Rs. Lakh)			367.92

The present value of cash inflow is greater than zero. Hence, the project satisfies the criteria of having positive value for NPV and hence viable.

Internal rate of return

Internal rate of return (IRR) is the interest rate at which the net present value of all the cash flows from a project or investment equal zero. Internal rate of return is used to evaluate the attractiveness of a project or investment. If the IRR of a new project exceeds a company's required rate of return, that project is desirable. If IRR falls below the required rate of return, the project should be rejected and the investment with the highest IRR is usually preferred.

Particulars	Amount (Rs. Lakhs)
Initial Investment	480.17
Cash Inflow for year 1	142.60
Cash Inflow for year 2	159.54
Cash Inflow for year 3	176.49
Cash Inflow for year 4	193.43
Cash Inflow for year 5	210.37
Cash Inflow for year 6	227.32
Cash Inflow for year 7	244.26
Internal Rate of Return 31%	

IRR of the present project is high, hence the project is feasible

Profitability index (PI) or benefit cost ratio (BCR)

Profitability index is the ratio of the present value of cash inflows, at the required rate of return, to the initial cash outflow of the investment. It is the ratio of payoff to investment of a proposed project. It is a useful tool for ranking projects because it helps to quantify the amount of value created per unit of investment. A profitability index of 1 indicates breakeven. Value of lower than one indicates that the project's present value of cash inflows is less than the initial investment. As the value of the profitability index increases, the financial attractiveness of the proposed project also tends to increase.

Sl. No	Particulars	Amount (Rs. Lakhs)
1.	Present value of cash inflows	793.36
2.	Initial investment	480.17
	Profitability Index [a/b]	1.77

Profitability index obtained is 1.77 which is greater than the optimum criteria of 1. So, the project can be accepted.

Conclusion

The financial, technical and economical analysis of the proposed rice milling unit is viable and has good market potential. The tools used for the financial analysis including Payback Period, Net Present Value, Internal Rate of Return and Profitability Index show the worthiness and bankability. Investment in the proposed project will be worthwhile. There are lot of schemes offered by the Government to support agri-business entrepreneurship, which takes the form of tax concessions and subsidies and other forms of support. Model business plan depicted herein has not taken into account subsidy component in initial investment as available under various schemes. Needless to say, an earlier payback period, higher NPV, IRR and PI may be expected, with subsidisied capital expenditure for agri-business projects.

Appendix

Appendix -1 (Chapter 1) - Government Schemes

Prime Minister's Employment Generation Programme (PMEGP)

Prime Minister's Employment Generation Programme (PMEGP) is a credit-linked subsidy programme launched by Ministry of MSME in 2008-09 for creation of employment in both rural and urban area of the country.

The PMEGP assists entrepreneurs who are willing to start enterprises in the fields of agro based food processing industry, forest based industry, handmade paper and fibre industry, mineral based industry, polymer and chemical based enterprise, rural engineering and biotech industry and service and textile industry. A negative list also is available as some activities are being continued even after strict enforcement which is available in the website.

If you are an entrepreneur interested in agro based food processing sector, the activities eligible for assistance under the scheme include manufacturing of bakery products, bedana/raisin industry/seeds processing, cashew/chironji processing (dry fruits), cattle feed, charolie making, coconut and arecanut products, daliya making, fruits and vegetable processing, ghani oil industry, groundnut decordicator (seeds/oil purpose), Indian sweets making, khava&chakka unit, manufacture of mahendi, manufacture of cane-gur and khandsari/jaggery making, manufacturing of chips from banana(raw)/potato, manufacturing of ice/ice candy, masala udyog, milk products making units, mini rice shelling unit/rice mill, noodles making, paddy unit (PCPI), palm gur making and other palm products industry, papad making, pepsi unit/ cold/soft drinks, poha making unit/popcorn, power atta chakki/flour mill, processing of maize and ragi, production of ice box, raswanti- sugarcane juice catering unit, soda mfg. products, supari making unit, threshing and vermicelli (shyarige) machine.

Eligibility conditions of beneficiaries

- Any individual, above 18 years of age
- There will be no income ceiling for assistance for setting up projects under PMEGP

- For setting up of project costing above Rs.10 lakh in the manufacturing sector and above Rs. 5 lakh in the business /service sector, the beneficiaries should possess at least VIII standard pass educational qualification

- Assistance under the scheme is available only for new projects sanctioned specifically under the PMEGP

- Self Help Groups (including those belonging to BPL provided that they have not availed benefits under any other Scheme) are also eligible for assistance under PMEGP

- Institutions registered under Societies Registration Act,1860

- Production co-operative societies, and charitable trusts

- Existing Units (under PMRY, REGP or any other scheme of Government of India or State Government) and the units that have already availed Government subsidy under any other scheme of Government of India or State Government are not eligible.

Norms of assistance

Margin money constitutes of 10% of the project cost in case of general category entrepreneur, whereas, it is only 5 % for special categories including SC/ST/OBC minorities etc.

Table 1: Subsidy and owner's contribution under PMEGP scheme

Categories of beneficiaries under PMEGP	Beneficiary's own contribution (of project cost)	Rate of Subsidy	
		Urban	Rural
General Category	10%	15%	25%
Special (including SC/ST/OBC /Minorities/ Women, Ex-Servicemen, Physically handicapped, NER, Hill and Border areas etc)	05%	25%	35%

Activities not permitted: There are some activities which are excluded from the financing by PMEGP. The negative list is given in Table. 2.

Table 2: Negative List of Activities under PMEGP

1.	Any industry/business connected with Meat(slaughtered),i.e. processing, canning and/or serving items made of it as food, production/manufacturing or sale of intoxicant items like Beedi/Pan/ Cigar/Cigarette etc., any Hotel or Dhaba or sales outlet serving liquor, preparation/producing tobacco as raw materials, tapping of toddy for sale.
2.	Any industry/business connected with cultivation of crops/ plantation like Tea, Coffee, Rubber etc. sericulture (Cocoon rearing), Horticulture, Floriculture, Animal Husbandry like Pisciculture, Piggery, Poultry, Harvester machines etc.
3.	Manufacturing of Polythene carry bags of less than 20 microns thickness and manufacture of carry bags or containers made of recycled plastic for storing, carrying, dispensing or packaging of food stuff and any other item which causes environmental problems.
4.	Industries such as processing of Pashmina Wool and such other products like hand spinning and hand weaving, taking advantage of Khadi Programme under the purview of Certification Rules and availing sales rebate.
(5)	Rural Transport (Except Auto Rickshaw in Andaman & Nicobar Islands, House Boat, Shikara & Tourist Boats in J&K and Cycle Rickshaw).

Implementing agencies

The Scheme is implemented by Khadi and Village Industries Commission (KVIC) at the national level and through State Directorates of KVIC, State Khadi and Village Industries Boards (KVIBs) and District Industries Centres in rural areas. In urban areas, the Scheme is implemented by the State District Industries Centres (DICs). KVIC and DICs will also involve NSIC, Udyami Mitras empanelled under Rajiv Gandhi Udyami Mitra Yojana (RGUMY), Panchayati Raj Institutions and other NGOs of repute in identification of beneficiaries under PMEGP.

Financial institutions & bank finance

The bank will sanction 90% of the project cost in case of General Category of beneficiary/institution and 95% in case of special category of the beneficiary/ institution, and disburse full amount suitably for setting up of the project. Following are the financing institutions involved in financing of projects under PMEGP

(i)   27 public sector banks.

(ii)   All regional rural banks.

(iii)   Co-operative banks approved by State Level Task Force Committee headed by Principal Secretary (Industries)/Commissioner (Industries).

(iv)   Private sector scheduled commercial banks approved by State Level Task Force Committee headed by Principal Secretary (Industries)/ Commissioner (Industries).

(v)   Small Industries Development Bank of India (SIDBI)

Contact information

State Office

Khadi and Village Industries Commission

Gramodaya, M.G. Road P.B. No. 198

City/Vill:Thiruvanantapuram

District:Thiruvananthapuram

Pin: 695001, State: Kerala

Ph.:0471-23316251

e-mail:kvictvm@gmail.com

Khadi and Village Industries Board

Vanchiyoor

City/Vill: Thiruvananthapuram

District: Thiruvananthapuram

Pin: 695035

State: Kerala

Ph.:0471-4723791

District Industries Centre

VIKASBHAVAN THIRUVANANTHAPURAM

City/Vill: THIRUVANANTHAPURAM

District: Thiruvananthapuram

Pin: 695033

State: Kerala

Ph.:047123027741

Website

http://www.kviconline.gov.in/pmegp

Pradhan Mantri Mudra Yojana (PMMY)

MUDRA which stands for Micro Units Development and Refinance Agency Ltd. is a refinance agency and not a direct lending institution. MUDRA provides refinance support to its intermediaries viz. banks/micro finance institutions/non banking finance companies, who are in the business of lending for income generating activities in the non farm sector in manufacturing, trading and services sector and who in turn will finance the beneficiaries.

Under the aegis of Pradhan Mantri MUDRA Yojana, three loan products have been launched and are known as 'Shishu', 'Kishor' and 'Tarun' to signify the stages of growth/development and funding needs of the beneficiary of micro unit/entrepreneur and also provide a collateral free loan for sure. The three variants are

i).   Shishu : covering loans upto 50,000/-

ii). Kishor : covering loans above 50,000/- and upto 5 lakh

iii). Tarun : covering loans above 5 lakh to 10 lakh

MUDRA Yojana operates as a refinancing institution through State/Regional level intermediaries. MUDRA's delivery channel is conceived to be through the route of refinance primarily to NBFCs/ MFIs, besides other intermediaries including banks, primary lending institutions etc.

A large number of 'Last Mile Financiers' in the form of companies, trusts, societies, associations and other networks are already providing informal finance to small businesses.

Eligible borrowers

Non-Corporate Small Business Segment (NCSBS) comprising of millions of proprietorship/ partnership firms running as small manufacturing units, service sector units, shopkeepers, fruits/ vegetable vendors, truck operators, food-service units, repair shops, machine operators, small industries, artisans, food processors and others, in rural and urban areas.

Any Indian Citizen having business plan for a non-farm sector income generating activity such as manufacturing, processing, trading or service sector and whose credit need is less than 10 lakh can approach a bank, MFI, or NBFC for availing of MUDRA loans under Pradhan Mantri Mudra Yojana (PMMY). The usual terms and conditions of the lending agency may have to be followed for availing of loans under PMMY. The lending rates are as per the RBI guidelines issued in this regard from time to time. Generally, loans upto 10 lakh issued by banks under Micro Small Enterprises is given without collaterals.

Pradhan Mantri Mudra Yojana (PMMY) loans are offered by all public sector banks, private sector banks, foreign banks, select regional rural banks and cooperative banks, micro finance institution, and non-banking finance companies. All loans up to a loan size of 10 lakh provided for non-farm sector income generating activities since April 08, 2015 is treated as PMMY.

There is no subsidy for the loan given under PMMY. However, if the loan proposal is linked to some existing Government scheme, wherein the Government is providing capital subsidy, it will be eligible under PMMY also.

Source: Mudra FAQ has been taken from its webpage and we acknowledge the support of Department of Financial Services.

Small Farmers Agribusiness Consortium (SFAC)

Small Farmers Agribusiness Consortium (SFAC) is an autonomous society promoted by Ministry of Agriculture, Cooperation and Farmers' Welfare, Government of India. It was registered under Societies Registration Act XXI of 1860 on 18th January, 1994 and as Non-Banking Financial Institution by Reserve Bank of India. SFAC is implementing the central schemes of Government of India namely VCA, EGCGS for economic inclusion of small and marginal farmers in agribusiness activities.

Society is pioneer in organising small and marginal farmers as Farmers Interest Groups, Farmers Producers Organisation and Farmers Producers Company for endowing them with bargaining power and economies of scale. It provides a platform for increased accessibility and cheaper availability of agricultural inputs to small and marginal farmers and in establishing forward and backward linkages in supply chain management. This initiative has triggered mobilization of farmers for aggregation across the country with ultimate aim of sustainable business model and augmented incomes.

Following are the schemes under offer by SFAC.

i). Equity Grant Scheme

Scheme extends support to the equity base of Farmer Producer Companies (FPCs) by providing matching equity grants subject to maximum of Rs. 10 lakh per FPC in two tranche and to address nascent and emerging FPCs which have paid up capital not exceeding Rs. 30 lakh

ii). Credit Guarantee Fund Scheme

Under this scheme, Credit Guarantee Cover to Eligible Lending Institutions (Scheduled Commercial Banks included in 2nd schedule of RBI Act), RRBs, NCDC, NABARD and its subsidiaries) is provided to enable them to provide collateral free credit to FPCs by minimising their lending risks in respect of loans with the primary objectives:

- To provide protection ELIs by extending credit guarantee and covering their lending risks upto Rs.1 Crore.

- To enable FPC to get collateral free loan by providing credit guarantee to ELIs

iii). Venture Capital Assistance Scheme

Venture Capital Assistance is financial support in the form of an interest free loan provided by SFAC to qualifying projects to meet shortfall in the capital requirement for implementation of the project. The main objectives of the Scheme are:

- To facilitate setting up of agribusiness ventures in close association with all Notified Financial Institutions notified by the Reserve Bank of India where the ownership of the Central/State Government is more than 50% such as Nationalized banks, SBI & its subsidiaries, IDBI, SIDBI, NABARD, NCDC, NEDFi, Exim Bank, RRBs & State Financial Corporations.

- To catalyze private investment in setting up of agribusiness projects and thereby providing assured market to producers for increasing rural income & employment.

- To strengthen backward linkages of agribusiness projects with producers.

- To assist farmers, producer groups, and agriculture graduates to enhance their participation in value chain through Project Development Facility.

- To arrange training and visits, etc. of agripreneurs in setting up identified agribusiness projects.

- To augment and strengthen existing set up of State and Central SFAC.

Eligibility conditions

Assistance under the Scheme will be available to Individuals; Farmers; Producer Groups; Partnership/Proprietary Firms; Self Help Groups; Companies; Agripreneurs; units in agri export zones, and Agriculture graduates Individually or in groups for setting up agribusiness projects. For professional management and accountability the groups have to preferably form into companies or producer companies under the relevant Act.

Salient features of the scheme

SFAC would provide Venture Capital to qualifying projects on the recommendations of the Notified Financial Institution financing the project. This venture capital will be repayable back to SFAC after the repayment of term loan of lending Notified Financial Institution as per original repayment schedule or earlier.

SFAC would provide venture capital to agribusiness projects by way of soft loan to supplement the financial gap worked out by the sanctioning authority of term loan under Means of Finance with respect to cost of project subject to the fulfilment of the following conditions:

(i) Qualifying projects under Venture Capital:

- Project should be in agriculture or allied sector or related to agricultural services. Poultry and dairy projects will also be covered under the Scheme.
- Project should provide assured market to farmers/producer groups.
- Project should encourage farmers to diversify into high value crops, to increase farm incomes.
- Project should be accepted by Notified Financial Institution for grant of term loan.

(ii) The quantum of SFAC Venture Capital Assistance will depend on the project cost and will be the lowest of the following:

26% of the promoter's equity or Rs. 50.00 lakhs. Provided that for projects located in North-Eastern Region, Hilly States (Uttarakhand, Himachal Pradesh, Jammu & Kashmir) and in all cases in any part of the country where the project is promoted by a registered Farmer Producers Organisation, the quantum of venture capital will be the lowest of the following:

40% of the promoter's equity or Rs. 50.00 lakhs. The cost of proposed agribusiness project would have to be Rs. 15 lakh & above, subject to a maximum of Rs. 500 lakh. However, projects valuing Rs. 10 lakh and above, proposed to be located in backward districts as notified by Planning Commission, hilly and North-Eastern States could also be considered for PDF and VCA.

Contact Details

Small Farmers' Agri-Business Consortium

Head Office, NCUI Auditorium Building, 5th Floor, 3, Siri Institutional Area

August Kranti Marg, Hauz Khas, New Delhi - 110016.

(T) +91-11- 26966017, 26966037, (F) +91-11- 26862367, sfac@nic.in
www.sfacindia.com

Directorate of Micro Small and Medium Enterprises (MSME)

DC(MSME) facilitates credit support through schemes operated via financial institutions including SIDBI, NSIC, banks and others. A glimpse of the schemes operated is given below

Credit Guarantee Fund Scheme for Micro and Small Enterprises

To deal with the impending problem of credit availability, particularly caused by non-availability of collateral to offer for a loan, Credit Guarantee Scheme for MSEs was launched by GoI in the year 2000.

The guarantee cover available under the scheme is to the extent of maximum 85% of the sanctioned amount of the credit facility. The guarantee cover provided is up to 75% of the credit facility up to Rs.50 lakh (85% for loans up to Rs. 5 lakh provided to micro enterprises, 80% for MSEs owned/ operated by women and all loans to NER including Sikkim) with a uniform guarantee at 50% for the entire amount if the credit exposure is above Rs.50 lakh and up to Rs.100 lakh. In case of default, Trust settles the claim up to 75% (or 85% / 80% / 50% wherever applicable) of the amount in default of the credit facility extended by the lending institution. For this purpose the amount in default is reckoned as the principal amount outstanding in the account of the borrower, in respect of term loan, and amount of outstanding working capital facilities, including interest, as on the date of the account turning Non-Performing Asset (NPA).

The credit facilities which are eligible to be covered under the scheme are both term loans and/or working capital facility up to Rs.100 lakh per borrowing unit, extended without any collateral security and / or third party guarantee, to a new or existing micro and small enterprise. For those units covered under the guarantee scheme, which may become sick owing to factors beyond the control of management, rehabilitation assistance extended by the lender could also be covered under the guarantee scheme. Any credit facility in respect of which risks are additionally covered under a scheme, operated by Government or other agencies, will not be eligible for coverage under the scheme.

Micro Finance Programme

To help MFIs/NGOs to avail loan, scheme of microfinance has been launched wherein contribution is done (GoI contributes a 'Portfolio Risk Fund') towards maintaining security deposits on the basis of which loans may be availed. The scheme is operationalised by SIDBI. At present, share of security deposits to be furnished by MFIs is 2.5% of the loan amount while balance 7.5% of the funds required will be contributed from the fund provided by GoI.

Initiatives for Women Entrepreneurs

The Micro, Small & Medium Enterprises Development Organisation (MSME-DO), the various State Small Industries Development Corporations (SSIDCs), the nationalised banks and even NGOs are conducting various programmes including Entrepreneurship Development Programmes (EDPs).

- To cater to the needs of potential women entrepreneurs, who may not have adequate educational background and skills, MSME-DO has introduced process/product oriented EDPs in areas like TV repairing, printed circuit boards, leather goods, screen printing etc.

- A special prize to "Outstanding Women Entrepreneur" of the year is being given to recognise achievements made by and to provide incentives to women entrepreneurs.
- The Office of DC (MSME) has also opened a Women Cell to provide coordination and assistance to women entrepreneurs facing specific problems
- The Small Industries Development Bank of India (SIDBI) has been implementing two special schemes for women namely Mahila Udyam Nidhi which is an exclusive scheme for providing equity to women entrepreneurs and the Mahila Vikas Nidhi which offers developmental assistance for pursuit of income generating activities to women.
  - The SIDBI has also taken initiative to set up an informal channel for credit needs on soft terms giving special emphasis to women.
  - Over and above this, SIDBI also provides training for credit utilisation as also credit delivery skills for the executives of voluntary organisations working for women.
- Grant for setting up a production unit is also available under Socio-Economic Programme of Central Social Welfare Board.

Trade Related Entrepreneurship Assistance and Development Assistance to Women Entrepreneurs [TREAD]

TREAD scheme, originally introduced in the 9th plan period for women has been reintroduced by GoI, which envisages economic empowerment of women through trade related training, information and counselling extension activities related to trades, products, services etc. It has been envisaged that the credit will be made available to women applicants through NGOs who would be capable of handling funds in an appropriate manner. These NGOs will not only handle the disbursement of such loans needed by women but would also provide them adequate counselling, training and assistance in developing markets. Following are the components provided under the TREAD scheme.

Credit: Under the credit component, Government Grant up to 30% of the total project cost as appraised by lending institutions which would finance the remaining 70% as loan Assistance to applicant women, who have no easy access to credit from banks due to their cumbersome procedures and the inability of poor & usually illiterate/semi-literate women to provide adequate security demanded by banks in the form of collaterals. GOI Grant and the loan portion from the lending agencies to assist such women shall be routed through eligible NGOs engaged in assisting poor women through any kind of income generating activities in non-farm sector.

Training & Counseling: Training organizations viz. Micro, Small and Medium Enterprises (MSMEs), Entrepreneurship Development Institutes (EDIs), NISIET and the NGOs conducting training programmes for empowerment of women beneficiaries identified under the scheme would be provided a grant upto maximum limit of Rs. 1.00 lakh per programme provided such institutions also bring their share to the extent of minimum 25% (10% in case of NER) of the Government grant. The batch size for such a training activity will be at least 20 participants. Duration of the training programme will be minimum one month.

Eliciting Information on Related Needs: Institutions such as Entrepreneurship Development Institutes (EDIs), NIMSME, NIESBUD, IIE, MSME-DIs EDIs sponsored by State Govt. and any other suitable institution of repute will be provided need based Government grant primarily for undertaking activities aiming at empowerment of women such as field surveys, research studies, evaluation studies, designing of training modules, etc. covered under the scheme. The grant shall be limited up to Rs. 5 lakhs per project.

NABARD

NABARD has set up a special fund for food processing units in designated food parks for providing direct term loans at affordable rates of interest to the designated food parks (DFPs) and food processing units in the DFPs.

Eligible borrowers: State Governments, entities promoted by State Governments or Government of India, Joint Ventures, SPVs, Co-operatives, Federation of co-operatives, Farmers' Producer Organisations, Corporates, Companies, Entrepreneurs etc are eligible for term loan assistance from NABARD.

Designated food parks include

- Food Parks promoted by MOFPI
- Mega Food Parks promoted by MOFPI
- Food Parks/exclusive Food Processing Industrial Estates promoted by State Governments
- Food Processing/agro processing/multi product Special Economic Zones including de-notified areas of these SEZs designated by MoFPI or
- Any other area having developed enabling infrastructure and designated Food Park by MoFPI

Term loan from the fund will be eligible for

- Development /establishment of all infrastructure required in the DFPs
- Augmentation/modernisation/creation of additional infrastructure in the DFPs

- Setting up of individual food processing units or any other unit that is established for supporting the operations of the food processing units within the DFPs and

- Modernisation of the existing processing units in the DFPs resulting in process technology upgradation, automation, increased efficiency, improvement in product quality, reduction in cost etc

Scope of processing activities eligible for term loan

The products of processing/manufacturing undertaken by the units may include

- Fruits vegetables, mushrooms, plantation crops and other horticulture crops
- Milk and milk products
- Poultry and meat
- Fish and other aquatic and marine products
- Cereals, pulses, oilseeds and oil crops
- Herbs, medicinal and aromatic plants, forest produce etc
- Consumer food products, such as bakery items, confectionery, snack etc
- Any other ready to eat food/convenience foods etc
- Beverages, non alchoholic drinks, energy drinks, carbonated drinks, packaged drinking water, soft drinks etc
- Food flavours, food colours, spices, condiments, ingredients, preservatives and any other item which may be required in food processing
- Nutraceuticals, health foods, health drinks etc

The eligible activities may cover range of post harvest processes resulting in value addition and or enhanced storage life, such as cleaning, grading, waxing, controlled ripening, labelling, packing and packaging, warehousing, canning, freezing, freeze drying, various levels of product processing (primary/secondary sector) etc.

The term loan can be extended upto 75% of the eligible project cost assessed by NABARD, while it can go upto 95% in case of State Govt promoted entities.

Contact Address for information

NABARD

Head Office, Mumbai

Department of Storage and Marketing

Plot No. C-24, G block, BandraKurla Complex, Bandra E, Mumbai- 400 051

Email.dsm.fpf@nabard.org, nabardfpf@gmail.com

National Institute for Entrepreneurship and Small Business Development (NIESBUD)

The National Institute for Entrepreneurship and Small Business Development is a premier organisation of the Ministry of Skill Development and Entrepreneurship, engaged in training, consultancy, research, etc. in order to promote entrepreneurship. The major activities of the Institute are Training of Trainers, Management Development Programmes, Entrepreneurship-cum-Skill Development Programmes, Entrepreneurship Development Programmes and Cluster Intervention. The Institute has conducted a total of 36,752 different training programmes, till March, 2016, covering 9,37,438 participants which includes 206 International programmes with 3,993 participants from more than 135 countries. Activities of NIESBUD include evolving effective training strategies and methodology, standardizing model syllabi for training various target groups, formulating scientific selection procedures, developing training aids, manual, tools and facilitating and supporting central/state/other agencies in organizing entrepreneurship developing programmes. A detailed overview of the activities is given below.

Training

- Assessing training needs as well as gaps therein and accordingly facilitating organizing of training programmes, orienting and motivating youth towards entrepreneurship.

- Evolving, designing and helping the use of various media for promoting the culture of entrepreneurship among different strata of society in the country.

- Playing a supportive and catalytic role by helping organizations which are directly or indirectly engaged in developing and promoting entrepreneurship and self employment in the country.

Consultancy

- Offering consultancy services in the area of entrepreneurship especially for MSMEs. Offering advice and consultancy to other Institutions engaged in entrepreneurial training either in the Government or in the private sector.

- Conceptualizing, designing and standardizing course curriculum for entrepreneurship and skill development programmes.

- Playing a supportive and catalytic role by helping organizations which are directly or indirectly engaged in developing and promoting entrepreneurship and self employment in the country.

Research & Development

- Promoting research and development activities in the area of entrepreneurship, particularly in MSME sector. Undertaking documentation and disseminating information related to entrepreneurship/ enterprise development;

- Preparing and publishing literature and information material related to entrepreneurship/enterprise development/ MSMEs;

- Providing a forum for interaction and exchange of views/experiences for different groups mainly through seminars, workshops, conferences etc;

- Studying problems and conducting research/ review studies etc. for generating knowledge, accelerating the process of entrepreneurship development culminating in establishment of new economic ventures. The Institute's training activities are focused on areas of stimulation, support and sustenance of entrepreneurship development. The programmes initiated/sponsored by NIESBUD are constantly evaluated and revised to enable it to adapt to the changing needs of entrepreneurship and small business development. The Institute is engaged in creating an environment conducive to the development of entrepreneurship and in creating a favourable attitude amongst the general public in support of those who opt for an entrepreneurial career by removing the prevalent myth and misconceptions that entrepreneurs are born.

Small Industries Development Organization (SIDO)

Small Industries Development Organisation functions under the Department of SSI and ARI, Government of India. The main functions of SIDO include framing of policy, co-ordination, monitoring, industrial development and extension. It is a nodal agency for development of small scale industries and its above mentioned functions are performed through a large number of field outfits and subordinate formations providing a wide range of techno economic consultancy, services and support. All small scale industries, except those falling under the specialised boards and agencies like KVIC, Coir Boards, Central Silk Board fall under the purview of SIDO.

Entrepreneurship Development Institute of India (EDII)

Entrepreneurship Development Institute of India (EDII), an autonomous and not-for-profit institute, set up in 1983, is sponsored by apex financial institutions - the IDBI Bank Ltd., IFCI Ltd., ICICI Bank Ltd. and the State Bank of India (SBI).

The thrust areas of EDI include Entrepreneurship education and research, SMEs & Business Development Services, Social Entrepreneurship and Corporate Social Responsibility, Women Entrepreneurship and gender studies, Cluster competitiveness, growth and technology, micro entreprises, microfinance and sustainable livelihood. Various centres have been established to carry out the vision and mission envisaged by EDI.

i). Centre for Entrepreneurship Education & Research

The centre is established for running educational and research programmes in the field of entrepreneurship

- Doctoral level Fellow Programme in Management (FPM) with focus on entrepreneurship.
- Post-Graduate Diploma Programmes in (i)Business Entrepreneurship (ii) Development Studies.
- Distance Education Bureau, UGC approved, diploma in Entrepreneurship and Business Management.
- Capacity building of educational institutions to promote entrepreneurship.
- National Summer Camps on inculcating entrepreneurial spirit among children & youth.
- Support for research to spear head entrepreneurship.
- Start-up support for the budding entrepreneurs.
- Biennial conference on entrepreneurship research.

ii). Centre for SMEs & Business Development Services

- Institution building and capacity enhancement programmes at National/ International levels.
- Identifying & profiling business opportunities.
- Training-cum-counselling for existing entrepreneurs.
- Human Resource Development in SME sector.
- Facilitating Science & Technology based innovations.

iii). Centre for Social Entrepreneurship & Corporate Social Responsibility

- Online programme on CSR in partnership with Indian Institute of Corporate Affairs.
- Knowledge partner for setting up techno social business incubator.

- Spear heading social entrepreneurship agenda with Ministry of Skill Development and Entrepreneurship, Govt. of India.
- Research in social entrepreneurship.
- Distance Education Programme in social entrepreneurship.

iv). Centre for Women Entrepreneurship & Gender Studies

- Women Entrepreneurship Development Programmes (WEDPs).
- Education/awareness programmes in life skills, gender equality & empowerment.
- Seminars/workshops for women entrepreneurs on possible business opportunities, sustaining and growing their businesses.
- Case studies of selected successful women entrepreneurs.
- Management development and performance improvement programmes.

v). International Programmes

- Industrial, infrastructure and sustainable project preparation & appraisal.
- Empowering women through entrepreneurship development.
- Capital markets & investment banking.
- Entrepreneurship & small business promotion.
- Informal sector enterprise, entrepreneurship & local economic development.
- Business research methodology & data analysis.
- Entrepreneurial management.
- Promoting innovations & entrepreneurship through incubation.
- Cluster Development Executives (CDEs) programme.
- SME banking & financial services.
- Sustainable livelihoods & mainstreaming with market.
- Agri-entrepreneurship & supply chain management.
- Entrepreneurship education to strengthen emerging economies.

Source: Website of EDI[1], Ahmedabad

The National Science and Technology Entrepreneurship Development Board (NSTEDB)

The National Science & Technology Entrepreneurship Development Board (NSTEDB), established by Government of India in 1982 is an institutional mechanism, with a broad objective of promoting gainful self-employment amongst the Science and Technology (S&T) manpower in the country and to setup knowledge based and innovation driven enterprises.

NSTEDB functions under the aegis of Department of Science & Technology. It has representation from socio-economic and scientific departments/Ministries, premier entrepreneurship development institutions and all India Financial Institutions.

The major objectives of NSTEDB are:

- To promote knowledge based and innovation driven enterprises.
- To facilitate generation of entrepreneurship and self-employment opportunities for S & T persons.
- To facilitate the information dissemination.
- To network with various Central & State Government agencies for S&T based entrepreneurship development.
- To act as a policy advisory body to the Government agencies for S&T based entrepreneurship development.
- To generate employment through technical skill development using S & T infrastructure.

The programmes conducted by NSTEDB have created awareness among S&T persons to take up entrepreneurship as a career. The academicians and researchers have started taking a keen interest in such socially relevant roles and have engaged themselves in several programmes initiated by NSTEDB. About 100 organisations, most of which are academic institutions and voluntary agencies, were drafted in the task of entrepreneurship development and employment generation.

For further information, please contact :

The Member Secretary, National S&T Entrepreneurship Development Board (NSTEDB), Department of Science & Technology, Technology Bhawan, New Mehrauli Road, New Delhi 110016, Ph: (011) 26517186, Fax : (011) 26517186, E-mail : hk.mittal@nic.in

Ministry of Food Processing Industries (MoFPI)

Ministry of Food Processing Industries is concerned with the formulation and implementation of the policies and plans for the food processing industries within the overall national priorities and objectives. Of late the Ministry has announced the Central Sector Scheme: SAMPADA (Scheme for Agro Marine Processing Sectors and Development of Agro Processing Clusters) for the period of 2016-20 co-terminus with 14th Finance Commission Cycle.

The objective of SAMPADA is to supplement agriculture, modernise processing and decrease agri- waste. SAMPADA is an umbrella scheme incorporating ongoing schemes of the Ministry. It is a comprehensive package to give a renewed thrust to the food processing sector in the country. It includes new schemes of the Ministry of Mega Food Parks, Integrated cold chain and value addition infrastructure and also new schemes like infrastructure for agro processing clusters, creation of backward and forward linkages, creation/ expansion of food processing & preservation capacities. The scheme aims at development of modern infrastructure to encourage entrepreneurs to set up food processing units based on cluster approach, provide effective and seamless backward and forward integration for processed food industry by plugging gaps in supply chain and creation of processing and preservation capacities and modernisation and expansion of existing food processing units.

Some of the schemes put forward by the Ministry are listed below.

i). Mega Food Parks

Mega Food Park scheme aims at providing a mechanism to link agricultural production to the market by bringing together farmers, processors and retailers so as to ensure maximizing value addition, minimizing wastage, increasing farmers' income and creating employment opportunities particularly in rural sector. The Mega Food Park scheme is based on "Cluster" approach and envisages a well-defined agri/horticultural-processing zone containing state-of-the art processing facilities with support infrastructure and well-established supply chain.

The scheme envisages a onetime capital grant of 50% of the project cost (excluding land cost) subject to a maximum of Rs. 50 crore in general areas and 75% of the project cost (excluding land cost) subject to a ceiling of Rs. 50 crore in difficult and hilly areas i.e. North East Region including Sikkim, J&K, Himachal Pradesh, Uttarakhand and ITDP notified areas of the States.

Contact Details

Sh. Sanjay Kumar Singh, Under Secretary, Ministry of Food Processing Industries, Panchsheel Bhawan, August Kranti Marg, New Delhi-110049, Tel:011-26406531, Email:sk[dot]singh[at]nic[dot]in

ii). Cold Chain

The objective of the scheme of cold chain, value addition and preservation infrastructure is to provide integrated cold chain and preservation infrastructure facilities without any break from the farm gate to the consumer.

It covers pre-cooling facilities at production sites, reefer vans, mobile cooling units as well as value addition centres which include infrastructural facilities like processing/multi-line processing/ collection centres, etc. for horticulture, organic produce, marine, dairy, meat and poultry etc. Individual, groups of entrepreneurs, cooperative societies, Self Help Groups (SHGs), Farmers Producer Organisations (FPOs), NGOs, Central/State PSUs etc. with business interest in Cold Chain solutions are eligible to set up integrated cold chain and preservation infrastructure and avail grant under the Scheme.

Financial assistance (grant-in-aid) of 50% the total cost of plant and machinery and technical civil works in General areas and 75% for NE region including Sikkim and difficult areas (J&K, Himachal Pradesh and Uttarakhand) subject to a maximum of Rs.10 crore.

Endnotes

Source: EDI@ a glance. www.ediindia.org

For Product Safety Concerns and Information please contact our EU
representative GPSR@taylorandfrancis.com
Taylor & Francis Verlag GmbH, Kaufingerstraße 24, 80331 München, Germany

www.ingramcontent.com/pod-product-compliance
Ingram Content Group UK Ltd.
Pitfield, Milton Keynes, MK11 3LW, UK
UKHW021123180425
457613UK00006B/204